Biotechnology in Agriculture and Forestry

Springer

Berlin
Heidelberg
New York
Barcelona
Hong Kong
London
Milan
Paris
Tokyo

Biotechnology in Agriculture and Forestry 50

Cryopreservation of Plant Germplasm II

Edited by L.E. Towill and Y.P.S. Bajaj

With 102 Figures and 100 Tables

Springer

Series Editor
Prof. Dr. Y.P.S. Bajaj†
New Delhi, India

Volume Editors

Dr. Leigh E. Towill
USDA-ARS National Center for
Genetic Resources Preservation
1111 S. Mason St.
Fort Collins, CO 80521
USA

Prof. Dr. Y.P.S. Bajaj†
New Delhi, India

ISSN 0934-943-X
ISBN 3-540-41676-5 Springer-Verlag Berlin Heidelberg New York

Library of Congress Cataloging-in-Publication Data. . . .

Cryopreservation of plant germplasm II / edited by L.E. Towill.
    p.   cm. – (Biotechnology in agriculture and forestry; 50)
    Includes bibliographical references.
    ISBN 3540416765 (alk.paper)
    1. Crops – Germplasm resources – Cryopreservation.   2. Germplasm resources,
Plant – Cryopreservation.   I. Towill, Leigh Edward.   II. Series.

Springer-Verlag Berlin Heidelberg New York
a company of BertelsmannSpringer Science + Business Media GmbH

http://www.springer.de

© Springer-Verlag Berlin Heidelberg 2002
Printed in Germany

The use of general descriptive names, registered names, trademarks, etc. in this publication does not imply, even in the absence of a specific statement, that such names are exempt from the relevant protective laws and regulations and therefore free for general use.

Cover design: *Design & Production* GmbH, Heidelberg

Typesetting: SNP Best-set Typesetter Ltd., Hong Kong

SPIN: 10691007   31/3130/50 5 4 3 2 1 0-Printed on acid-free paper

Preface

Ex situ preservation of germplasm for higher plant species has been accomplished using either seeds or clones, but storage of these under typical conditions does not provide the extreme longevities that are needed to minimize risk of loss. Costs of maintenance and regeneration of stocks are also high. Systems that provide virtually indefinite storage should supplement existing methods and it is within this context that cryopreservation is presented.

The use of low temperature preservation was initially more a concern of medicine and animal breeding, and was expanded to plants in the 1970s. Survival after cryogenic exposure has now been demonstrated for diverse plant groups including algae, bryophytes, fungi and higher plants. If survival is commonplace, then the eventual application is a cryopreservation system, whereby cells, tissues and organs are held indefinitely for use, often in the unforeseen future. The increasing interest and capabilities for application could not have occurred at a more opportune time since expanding human populations have placed unprecedented pressures on plant diversity. This book emphasizes cryopreservation of higher plants and was initially driven by the concern for loss of diversity in crops and the recognized need that this diversity would be essential for continued improvement of the many plants used by society for food, health and shelter. The interest in cryopreservation has been expanded by conservationists and their concerns for retaining, as much as possible, the diversity of natural populations. The need for cryopreservation, thus, is well established.

I was first introduced to cryobiology as a postdoctoral fellow with Dr. Peter Mazur, whose pioneering studies on the biophysical processes that occur during a cell's excursion to and from low temperature, and on the components of injury, led to many approaches for practical cryopreservation. During this time, I also became familiar and extremely impressed with the works of Dr. Akira Sakai and Dr. Peter Steponkus. The former has elucidated detailed information on cold acclimation, particularly of woody species, and has utilized that knowledge to provide a system for cryopreservation. The latter has provided a mechanistic understanding of cold acclimation and a detailed analysis of cellular injury. Although this understates the accomplishments of all three cryobiologists, their works are fundamental to the understanding of cryobiology in plant systems, and ultimately to the development of cryopreservation.

This book advances the goals that the late Dr. Y.P.S. Bajaj presented in Springer's Agriculture and Biotechnology Series, Volume 32, Cryopreservation of Plant Germplasm I. The aim is to highlight achievements in cryopreservation, chronicle method development, and describe relevant literature. Since the area is methodology rich, the presentation of detailed information should help practitioners develop and improve methods for desired species. Cryopreservation has expanded beyond being an interesting research area. The generality of certain procedures and the understanding of the important processes that is emerging are very encouraging and mean that cryopreservation can be now be applied for conservation purposes.

Dr. Y.P.S. Bajaj had invited researchers throughout the world to contribute to this volume. The book is divided into four parts:

I. Cryopreservation of Germplasm
II. Herbaceous Species (Barley, Chicory, Celery, Ginseng, Garlic, Chamomile, Mint, Hops, Horseradish, Taro, Wasabi)
III. Woody Species (*Eucalyptus*, Poplar, Oak, Guazuma, Horsechestnut, Neem, *Prunus*, Olive, *Ribes*, Rose, Coffee)
IV. Australian Species

Fort Collins, Colorado, December 2001   LEIGH E. TOWILL
                                                                 Volume Editor

Contents

Section III Woody Species

III.1 Cryopreservation of Somatic Embryos
   from *Aesculus hippocastanum* L. (Horse chestnut)
Z. Jekkel, J. Kiss, G. Gyulai, E. Kiss, and L.E. Heszky
(With 3 Figures)

III.2 Cryopreservation of *Azadirachta indica* A. Juss. (Neem) Seeds
D. Dumet and P. Berjak

III.3 Cryopreservation of *Coffee* (Coffee)
S. Dussert, N. Chabrillange, F. Engelmann, F. Anthony,
N. Vasquez, and S. Hamon (With 4 Figures)

III.4 Cryopreservation of *Eucalyptus* spp. Shoot Tips
   by the Encapsulation-Dehydration Procedure
M. Pâques, V. Monod, M. Poissonnier, and J. Dereuddre
(With 3 Figures)

Section IV Australian Species

IV.1 Cryostorage of Somatic Tissue of Endangered Australian Species
D.H. TOUCHELL, S.R. TURNER, E. BUNN, and K.W. DIXON
(With 3 Figures)

IV.2 Cryopreservation of Australian Species –
     The Role of Plant Growth Regulators
D.H. TOUCHELL, S.R. TURNER, T. SENARATNA, E. BUNN,
and K.W. DIXON (With 7 Figures)

List of Contributors

ANTHONY, F., CATIE, Ap. 59, 7170 Turrialba, Costa Rica

ARROYO GARCÍA, R., Dep. de Genética Molecular de Plantas, Centro Nacional de Biotecnología, CSIC, Campus de la Universidad Autónoma de Madrid, Cantoblanco, 28049 Madrid, Spain

DE BOUCAUD, M.-T., Laboratoire de Physiologie Cellulaire Végétale, Université Bordeau des Facultés, 33405 Talence Cedex France

BERJAK, P., School of Life and Environmental Sciences, University of Natal, Durban 4041, South Africa

BRISON, M., Laboratoire de Physiologie Cellulaire Végétale, Université Bordeaux 1, Avenue des Facultés, 33405 Talence Cedex France

BUNN, E., Kings Park and Botanic Garden, West Perth, WA 6005, Australia

CELLÁROVÁ, E., Department of Experimental Botany and Genetics, Faculty of Science, P.J. šafárik University, Mánesova 23, 041 67 Košice, Slovakia

CHABRILLANGE, N., IRD (previously ORSTOM), 911 Av. Agropolis, BP 5045, 34032, Montpellier, France

DEMEULEMEESTER, M., Laboratory of Plant Culture, Faculty of Agricultural and Applied Biological Science, Katholieke Universiteit Leuven, Willem de Croylaan 42, 3001 Heverlee, Belgium

DE PROFT, M., Laboratory of Plant Culture, Faculty of Agricultural and Applied Biological Science, Katholieke Universiteit Leuven, Willem de Croylaan 42, 3001 Heverlee, Belgium

DEREUDDRE, J., Université P. et M. Curie, Laboratoire de Cryobiologie Végétale, 12 rue Cuvier, 75230 Paris Cedex 05, France

DIXON, K.W., Kings Park and Botanic Garden, West Perth, WA 6005, Australia

DUMET, D., School of Life and Environmental Sciences, University of Natal, Durban 4041, South Africa

Dussert, S., IRD (previously ORSTOM), 911 Av. Agropolis, BP 5045, 34032, Montpellier, France

Engelmann, F., IPGRI, Via delle Sette Chiese 142, 00145 Rome, Italy

González-Benito, M.E. Departamento de Biología Vegetal, Escuela Universitaria de Ingeniería Técnica Agrícola, 28040 Madrid, Spain

Gyulai, G., Department of Genetics and Plant Breeding, St. István University, 2103 Gödöllö, Hungary

Hamon, S., IRD (previously ORSTOM), 911 Av. Agropolis, BP 5045, 34032, Montpellier, France

Helliot, B., Laboratoire de Physiologie Cellulaire Végétale, Université Bordeaux 1, Avenue des Facultés, 33405 Talence Cedex, France

Hervé-Paulus, V., Laboratoire de Physiologie Cellulaire Végétale, Université Bordeaux 1, Avenue des Facultés, 33405 Talence Cedex, France

Heszky, L.E., Department of Genetics and Plant Breeding, St. István University, 2103 Gödöllö, Hungary

Hirata, K., Environmental Bioengineering Laboratory, Graduate School of Pharmaceutical Sciences, Osaka University, 1-6 Yamada-oka, Suita, Osaka 565-0871, Japan

Huang, C.-N., College of Life Science, Zhejiang University, 232 Wensan Road, Hangzhou 310012, China

Hummer, K.E., USDA-ARS National Clonal Germplasm Repository, 33447 Peoria Rd., Corvallis, Oregon 97333-2521, USA

Iriondo, J.M., Departamento de Biología Vegetal, Escuela Universitaria de Ingeniería Técnica Agrícola, Universidad Politécnica de Madrid, Ciudad Universitaria, 28040 Madrid, Spain

Jekkel, Z., Department of Genetics and Plant Breeding, St. István University, 2103 Gödöllö, Hungary

Keller, E.R.J., Genebank Department, Institute of Plant Genetics and Crop Plant Research, IPK, Corrensstr. 3, 06466 Gatersleben, Germany

Kimáková, K., Department of Experimental Botany and Genetics, Faculty of Science, P.J. šafárik University, Mánesova 23, 041 67 Košice, Slovakia

Kiss, E., Department of Genetics and Plant Breeding, St. István University, 2103 Gödöllö, Hungary

Kiss, J., Department of Genetics and Plant Breeding, St. István University, 2103 Gödöllö, Hungary

Lambardi, M., National Research Council (CNR), Istituto sulla Propagazione delle Specie Legnose, via Ponte di Formicola 76, 50018 Scandicci (Firenze), Italy

Lynch, P.T., Division of Biological Sciences, School of Environmental and Applied Sciences, University of Derby, Kedleston Road, Derby DE22 1GB, UK

Martín, C., Departamento de Biología Vegetal, Escuela Técnica Superior de Ingenieros Agrónomos, Universidad Politécnica de Madrid, Ciudad Universitaria, 28040 Madrid, Spain

Martínez, D., Departamento Biología de Organismos y Sistemas, Facultad de Biología, C/C. Rodrigo Uría s/n, Universidad de Oviedo, 33071 Oviedo, Spain

Martínez-Zapater, J.M., Dep. de Genética Molecular de Plantas, Centro Nacional de Biotecnología, CSIC, Campus de la Universidad Autónoma de Madrid, Cantoblanco, 28049 Madrid, Spain

Maruyama, E., Bio-Resources Technology Division, Forestry and Forest Products Research Institute, Box 16, Tsukuba Norinkenkyudanchi-nai, Ibaraki 305-8687, Japan

Matsumoto, T., Shimane Agricultural Experiment Station, Ashiwata 2440, Izumo, Shimane, 693-0035, Japan

Miyamoto, K., Environmental Bioengineering Laboratory, Graduate School of Pharmaceutical Sciences, Osaka University, 1-6 Yamada-oka, Suita, Osaka 565-0871, Japan

Monod, V., AFOCEL, Station de Biotechnologies, Domaine de l'Etançon, 77370 Nangis, France

Monthana, P., Environmental Bioengineering Laboratory, Graduate School of Pharmaceutical Sciences, Osaka University, 1-6 Yamada-oka, Suita, Osaka 565-0871, Japan

Pâques, M., AFOCEL, Station de Biotechnologies, Domaine de l'Etançon, 77370 Nangis, France

Poissonnier, M., AFOCEL, Station de Biotechnologies, Domaine de l'Etançon, 77370 Nangis, France

Reed, B.M., USDA-ARS National Clonal Germplasm Repository, 33447 Peoria Rd., Corvallis, Oregon 97333-2521, USA

REVILLA, M.A., Departamento Biología de Organismos y Sistemas, Facultad de Biología, C/C. Rodrigo Uría s/n, Universidad de Oviedo, 33071 Oviedo, Spain

SAKAI, A., 1-5-23 Asabu-cho, Kita-ku, Sapporo 001–0045, Japan

SENARATNA, T., Kings Park and Botanic Garden, West Perth, WA 6005, Australia

SHIMOMURA, K., Tsukuba Medicinal Plant Research Station, National Institute of Health Sciences, 1 Hachimandai, Tsukuba, Ibaraki, 305-0843 Japan

TAKAGI, H., Japan International Research Center for Agricultural Sciences, 1–2 Ohwashi, Tsukuba, Ibaraki 305, Japan

THINH, N.T., Department of Biotechnology and Nuclear Techniques, Nuclear Research Institute, 1 Nguyen Tu Luc St., Dalat City, Vietnam

TOUCHELL, D., Kings Park and Botanic Garden, West Perth, WA 6005, Australia

TOWILL, L.E., USDA-ARS National Center for Genetic Resources Preservation, 1111 S. Mason St., Fort Collins, Colorado 80521, USA

TURNER, S.R., Kings Park and Botanic Garden, West Perth, WA 6005, Australia

URBANOVÁ, M., Department of Experimental Botany and Genetics, Faculty of Science, P.J. Šafárik University, Mánesova 23, 041 67 Košice, Slovakia

VANDENBUSSCHE, B., Laboratory of Plant Culture, Faculty of Agricultural and Applied Biological Science, Katholieke Universiteit Leuven, Willem de Croylaan 42, 3001 Heverlee, Belgium

VASQUEZ, N., CATIE, Ap. 59, 7170 Turrialba, Costa Rica

WANG, J.-H., College of Life Science, Zhejiang University, 232 Wensan Road, Hangzhou 310012, China

YOSHIMATSU, K., Tsukuba Medicinal Plant Research Station, National Institute of Health Sciences, 1 Hachimandai, Tsukuba, Ibaraki, 305-0843 Japan

Section I
Cryopreservation of Plant Germplasm

I.1 Cryopreservation of Plant Germplasm: Introduction and Some Observations

Leigh E. Towill

1 Germplasm Conservation

It has been amply stated that preservation of germplasm is important not only for plant improvement and utilization for food, fiber, medicinal and forest crops, but also for conservation of rare and endangered species (Knutson and Stoner 1989; Falk and Holsinger 1991; Reaka-Kudla et al. 1997). As documented in contributed chapters in this volume and in comprehensive compendia, reasons for species or landrace loss are diverse but usually related to pressures from increasing human populations and the attendant habitat loss, land use changes and desire for productive crops. Whatever the root causes, the final effect is certainly the loss of genotypes. In efforts to stave off further loss, preservation systems have been devised to retain as much genetic diversity for the species as possible. Broadly classed, both in situ and ex situ preservation systems have been proposed and have been implemented in varying degrees for different species. In situ preservation allows for evolutionary forces to continue and can be argued to be important, for example, for disease-resistance development. However, in situ preservation still requires management and is not suitable for more domesticated lines. Preservation ex situ (within genebanks) is, therefore, used as a system around the world for many species, both commercially important and endangered.

For effective ex situ conservation, diversity acquisition, maintenance, distribution and evaluation are needed. Seeds are the predominant storage propagules for many species, but a considerable number are maintained as collections of clones and are vegetatively propagated as needed. As described in this volume, storage of the clone may be by field or greenhouse plantings, or as in vitro plants. Often combinations of these are used. However, maintaining individuals in the growing condition subjects them to risk of loss due either to biotic or abiotic stress and is usually costly due to labor and land uses.

Cryogenic storage, as used here meaning storage below temperatures such as $-130\,°C$, has been argued to be an effective and safe mode for preservation. Such storage should enhance the efficacy of preservation for both seed and clonally maintained germplasm and provide for considerable, if not indefinite,

USDA-ARS National Seed Storage Laboratory, 1111 S. Mason St., Fort Collins, Colorado 80521, USA

Biotechnology in Agriculture and Forestry, Vol. 50
L.E. Towill and Y.P.S. Bajaj (Eds.) Cryopreservation of Plant Germplasm II
© Springer-Verlag Berlin Heidelberg 2002

longevity. In the case of seeds, longevity at temperatures such as −20 °C may be considerable and the use of cryogenic storage may not be cost effective. For preservation of clones, the risk of loss under growth conditions may be great and the development of cryogenic storage is desirable as a backup system. Longevity of shoot tips or scions held at temperatures between about −4 and −80 °C is short, so cryogenic storage is the only option for long-term preservation. Because materials are continually requested from a genebank, this storage is best viewed as a backup should materials be lost. It would not be efficient to routinely distribute from cryogenic storage since replenishment would be periodically needed.

2 Framework

In developing a cryopreservation system to deal with genetic diversity, an initial question is what propagule to cryopreserve. Seed, somatic embryos, pollen and shoot tips or buds are currently amenable to cryopreservation and can be useful for conservation. Protoplasts, cells and some differentiated tissues (hairy roots, hypocotyls and leaf disks) may also survive cryogenic exposure, but are not routinely used for conservation purposes. These might be useful once plant regeneration in all genotypes is possible and offtypes minimized.

There have been several treatises on cryopreservation of germplasm and related topics (Kartha 1985; Bajaj 1995; Engelmann and Takagi 2000; Razdan and Cocking 2000) and numerous articles, particularly in the journals *Cryobiology* and *Cryo-Letters*. This introductory chapter will emphasize practical aspects for cryopreservation of vegetatively propagated materials where shoot tips or buds are the excised propagule. The intent is to provide an introduction and overview of the process with selected references. There has been no attempt to list all references in plant cryobiology, nor to compile the many species studied. However, cryopreservation is equally applicable, and usually simpler, for seed and pollen samples and will be briefly discussed. What one uses depends on the characteristics of the plant, and the seed, pollen and vegetative propagule it produces, and on whether one is interested in preserving genes or genotypes. It should also be emphasized that cryopreservation is only one component of germplasm conservation and that a preservation or storage system may utilize more than one propagule. Thus, a coordinated approach is desirable. Realistically, the time and money available for developing and operating the system may dictate how extensive it is and what route is taken.

Successful cryopreservation involves consideration of three phases that comprise the overall process (Fig. 1), and it is helpful to consider each in devising an efficient, effective procedure. Overall success usually requires some optimization, or at least adjustment, of the parameters in each phase. Plant material often must be adapted to, or exist in, a desirable physiological state before the propagule is harvested and this is especially so for the preservation

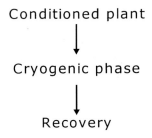

Fig. 1. Three phases in developing a cryopreservation protocol

of clonal lines. For example, orthodox seed should be mature and recalcitrant seed in the correct developmental state at collection. For quality production, plants must receive the appropriate photoperiod for flower induction and not be stressed. Species which are cold hardy are best used when acclimated.

Once the propagule has been isolated from the stock plant, the cryobiology phase consists of processes needed to expose and retrieve the sample from low temperatures. Most studies in cryopreservation have emphasized steps in this phase. Lastly, cryoprotectant exposure, desiccation and temperature excursions are potentially damaging events which may require careful recovery of the treated propagule. Appropriate procedures for utilization require some studies in this area, whether it is testing pollen for fertility or regrowth of treated shoot tips for viability. There is considerable evidence to suggest that overall success requires coordination among the phases. Whereas cryoreservability may be enhanced with suitable plant adaptation, use of a suboptimal cryogenic procedure may not take advantage of this ability. Likewise, inadequate recovery systems may hinder development and understanding of the former two phases.

3 Cryopreservation of Seed

The desiccation characteristics of seed differ among species, being broadly grouped as desiccation-tolerant (orthodox), intermediate and desiccation-sensitive (recalcitrant) behaviors.

It is well appreciated that seed longevity of the first group can be increased by carefully lowering and controlling seed moisture and storage temperatures (Vertucci and Roos 1990, 1993; Vertucci et al. 1994). After desiccation to a fairly low moisture level, seeds within an appropriate container can be directly placed into liquid nitrogen (LN) or the vapor phase over LN. Generally, because free water does not exist in seeds at the moisture levels used, the cooling rate is not important, although slower rates may be beneficial for seeds with hard seed coats or with high oil levels (Stanwood 1985, 1987). Warming rates are, likewise, not usually critical, and containers can be placed directly to room temperature. Samples can be recycled back to liquid nitrogen, but care is needed to avoid condensation during warming since this may increase seed moisture, rendering them susceptible to subsequent cooling. Viability levels after LN exposure and storage are often very similar to the initial viability level of the sample. There are a number of reports showing that seeds from many species survive low temperature exposure, but longevity data are sparse for the obvious reason that cryogenic storage for plant materials is relatively recent. Systematic application of cryogenic storage in genebanks around the world is somewhat difficult to assess since data are not published in research literature. One example of practical seed cryopreservation is at the USDA/ARS National Seed Storage Laboratory, the long-term storage facility for crops important to US agriculture. Currently, about 33,000 lines from a range of species are stored in the vapor phase over LN.

There are numerous factors to consider when deciding whether seeds should be maintained in cryogenic storage or in more traditional storage, such as at about −20°C, where longevities are still often great. Costs of seed regeneration, longevity at existing storage temperatures, size of seed, and survival after LN-exposure tests are common criteria. For example, with optimal moisture content, barley seeds may be stored for many decades but lettuce for only a few years at −20°C. The storage system of choice becomes an issue of what longevities are desired and the cost, and a thorough analysis should be carried out to determine what storage system should be used to assure safety yet operate within cost constraints.

Desiccation-sensitive seeds require considerably more care (Pammenter and Berjak 1999; Farnsworth 2000). Seeds need to be collected at an appropriate maturity, which may represent a rather narrow time window in their development, and thus make studies difficult. If careful desiccation, often rapid, is carried out to a moisture range whereby free water is diminished, but desiccation injury is minimized, then rapid cooling often gives some level of viability (Touchell and Walters 2000). Discrepancies between seed and embryonic axis moisture contents may exist for larger seed during dehydration and the larger size reduces cooling rate during rapid or plunge cooling methods. To counter these problems, embryonic axes are excised and processed. Unfortunately, plants are often not recovered from samples initially deemed viable; this may relate to inadequate recovery systems and the inability to retain viability in all cells of the embryonic axis. Obviously, a suitable culture medium for recovery is needed. Application of vitrification solutions has been successful in specific instances (Thammasiri 1999), but has not been extensively reported.

Seeds with an intermediate desiccation or storage behavior often tolerate somewhat more desiccation than sensitive seed but are processed similarly to sensitive seed. Germination of coffee seeds after cryopreservation was enhanced by controlled rehydration, suggesting a recovery from sublethal injury (Dussert et al. 2000). For both sensitive and intermediate behavior seed, use of cryoprotectants may be beneficial, but studies are few. Permeation and toxicity are obvious concerns and the system must be rehydrated or imbibed before application of the cryoprotectant.

4 Cryopreservation of Pollen

Pollen is handled very similarly to seed, noting that the moisture content adjusts much more rapidly (Towill 1985; Hoekstra 1995; Hanna and Towill 1995). Pollen quality is important. Pollen collected before or well past anthesis does not store well. Heat or drought-stressed plants often produce pollen with low fertility. Pollen storage requires little space but mainly preserves nuclear genes. Maternal factors would be lost if this were the sole means of preservation.

The desiccation classes of pollen are also very similar to seed, but species with desiccation-sensitive seed may have desiccation-tolerant pollen, and vice versa. Tolerant pollens can be desiccated to low moisture levels (ca. 5–10% fresh weight basis), usually using either saturated salt solutions or by air desiccation on a surface. Samples within vials are directly placed into cryogenic storage and are warmed to room temperature before use, avoiding moisture condensation on the pollen. Cooling and warming rates are not critical and samples can be recycled back to LN. Like seed, pollen can also be preserved at temperatures such as −20, −4 and +4 °C, but longevities become progressively shorter. Although cryogenic storage can provide great longevities, and in spite of its simplicity, routine use for germplasm efforts is sparse, except for some forest conifer programs (Mercier 1995).

Desiccation-sensitive pollen is more problematic, and moisture content must be critically controlled. The pollen is desiccated to a moisture window where rapid cooling gives survival, but before desiccation injury occurs. Although cryoprotectants may in concept be added, their removal and possible interference with fertilization mitigate the process.

Again, it should be noted that good quality pollen needs to be obtained for storage; heat and drought stress of the plant may reduce pollen production and viability. From a functional viewpoint, fertility is the ultimate test of successful preservation, but in vitro germination and stains provide an estimate of viability. Desiccation-sensitive pollen usually has very short lifetimes at temperatures such as −20 °C and warmer. Longevities at temperatures between −20 and ca. −160 °C have not been well described for either class of pollen.

5 Cryopreservation of Clones – What to Preserve?

Shoot tips or buds have been used in order to faithfully regenerate the clone, but the form and physiological status of the stock plant used differs considerably among species. One framework for decision is shown in Fig. 2. It is well appreciated that most species are held as ex vitro plants, and those that possess considerable cold hardiness are easier to cryopreserve with existing methodology. Isolated buds, nodal sections or whole scions from winter-hardy materials can be cooled. Where xylem ray parenchyma are not injured by supercooling, the twig may be rooted, if this ability exists, after cryostorage, for example with poplars and willows. If xylem ray parenchyma exhibit extensive supercooling, exposure to LN is lethal. However, the vegetative bud and cambial regions usually exhibit freeze dehydration during cooling and, if viable, can be recovered by grafting or in vitro culture. Budding would give a more useful plant in a shorter duration. If grafting is not feasible, recovery in vitro is an alternative route, but major considerations are minimizing contamination and whether culture recovery and micropropagation systems are available and are applicable across genotypes.

If the species does not possess the ability to cold harden, or is obtained in a non-hardened state, the choices are to produce an in vitro plant and utilize shoot tips or buds from this for cryopreservation or to utilize buds, if enough are available, from the ex vitro plant directly for cryopreservation and then recover in vitro. Micrografting is an option, but methods are not well developed for non-woody materials and recoveries from even untreated samples are often low. It is obvious that in either case the development of in vitro systems for plant conservation is crucial for cryopreservation (Benson 1999).

Somatic embryos can be used for vegetatively propagated lines and could provide the basis for germplasm preservation if they could be induced in each genotype with minimal offtype production. Where examined, once suitable maturation has occurred somatic embryos exhibit the desiccation tolerance of the zygotic embryos. Thus, the advantage for use is storage characteristics similar to seed, but preserving the clonal identity of the line. Somatic embryos in various developmental stages have been used and papers have described

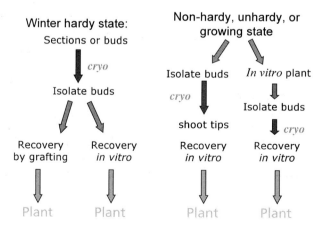

Fig. 2. Options in using vegetative buds or shoot tips from clonal lines for cryopreservation

cryopreservation (see below) by rapid cooling (Dumet et al. 1993; Mycock et al. 1995), two-step cooling (Jekkel et al. 1998), and encapsulation-dehydration (Shibli and Al-Juboory 2000). Drying prior to application of cryoprotectants has given increased survival in cassava somatic embryos (Stewart et al. 2001).

5.1 Physiological Adaptation

Many researchers have shown that some treatment of the stock plant is beneficial for cryopreservation success. This is most obvious with cold-hardy species, where, if used in the growing condition, buds without cryoprotectant application are killed at about −5 °C, whereas after suitable cold acclimation buds survive very low temperatures, depending on species and genotype (Sakai and Larcher 1987). This observation has been extended to in vitro plants that also possess the ability to cold acclimate (Chang and Reed 1999), but the manner and extent of cold acclimation necessary for in vitro plants has not been well defined (Chang and Reed 1999, 2000). Addition of abscisic acid to in vitro stock plants has also improved cryopreservability for cold-hardy species (Ryynanen 1998).

Tropical plants cannot undergo cold acclimation and many may suffer chilling injury at temperatures between +5 and +15 °C. Adaptation in this group has not been examined extensively. Culture of the stock plant on elevated sucrose levels has been beneficial for representatives in this group (Takagi 2000).

Overall plant quality and uniformity of the explant also influence survival. Several studies have noted that shoot tips from vigorous plants give better results, and a system whereby nodal sections are cultured for a few days in advance of harvesting the shoot tip has been described for several species (Bouafia et al. 1996). This culture allows for development of the axillary bud into a growing apical shoot tip, but reasons for the improved response have not been determined. In some species there appears to be a differential response of axillary and apical shoot tips to some stages of cryopreservation. Light treatment of the stock plant also affects survival (Benson et al. 1989; Pennycooke and Towill 2000), but it is uncertain if this is a general phenomenon across species.

Thus, there are a myriad of possible stock plant treatments that might improve survival after cryogenic exposure. Growth conditions, heat, drought, salt stress and growth regulators have all been observed to alter plant growth habit, physiology and biochemistry. It would not be surprising if some application would improve subsequent survival after low temperature exposure, but most studies have been quite limited in approach and treatment application. The lack of understanding of what makes a shoot tip more cryopreservable is apparent.

5.2 Methods of Cryopreservation

5.2.1 Preculture

Many researchers have observed that culturing the excised propagule for a period of time (hours to days) in defined solutions enhances subsequent survival after low temperature treatment. This exposure is termed 'preculture'. As with many steps in plant cryopreservation, descriptions of the best treatments have come from empirical tests and there is no consensus procedure. There is also no demonstrated mechanism for enhancing survival, although it is supposed that culture in sugar lowers water content and increases endogenous sugar concentrations (Matsumoto et al. 1998a; Hitmi et al. 1999). Sugars protect membranes from desiccation events that are inherent in any preservation protocol and are also known to enhance glass formation during cooling. Some have found sugar alcohols to be better than sugars, but many others have described the opposite. Concentrations ranges often differ; sometimes 0.09 M sucrose is best, whereas other times 0.3 M is desirable. Although no consensus exists for specifics, some preculture regime is usually beneficial.

5.2.2 Cooling

Many methods for processing relatively hydrated samples to and from low temperatures have been described in the literature for a whole range of species including microbes; animal embryos, semen and cultures; and plant protoplast, cell suspensions, callus, somatic embryos, twigs and shoot tips. With examination, it is surprising how similar some of these methods are, with the main variables being substances applied, kinetics of exposure, and cooling and warming rates. For plant systems a sequence of treatments is used (Fig. 3). The details of these methods vary considerably. In most cases, the exception being use of winter scions or buds, a cryoprotectant, defined here as an added compound that improves survival after low temperature exposure, is added before cooling and removed after warming.

5.2.2.1 Two-Step Cooling

This procedure is sometimes referred to as slow cooling, optimum rate cooling, or conventional cooling. The biophysical basis for the appearance of an optimum cooling rate has been well described by Mazur (1966, 1984) whereby, when using low or moderate cryoprotectant concentrations, injury at suboptimal rates is due to "solution effects" and at supraoptimal rates by intracellular ice formation. Solution effect injury has been ascribed to pH alterations, intracellular solute concentration, dehydration, membrane alterations, and protein denaturation, but is not well defined and obviously has a kinetic component. In elegant studies by Steponkus and coworkers, both light and freeze fracture microscopy has been used to show alterations that occur in mem-in-vitro or ex-vitro plant

     ↓   temperature, growth regulator, light treatment

   conditioned plant

        ↓   excision

 isolated shoot tips/buds

      ↓   preculture {type, conc., time, temp.}

      ↓   application of cryoprotectants {conc., rate, time, temp.}

      ↓   cooling {nucleation, rates}

      ↓   storage {temp., time}

      ↓   warming {rates}

      ↓   removal of cryoprotectant {rates, time, temp.}

 isolated shoot tips/buds

      ↓   post-thaw treatment (medium; time; culture conditions)

      ↓   viability assessment, growth treatment

   plant

Fig. 3. Generalized flowchart for cryopreservation of shoot tips

branes in plant cells cooled slowly and that these depend on cold acclimation state and cryoprotectant (Uemura and Steponkus 1999).

    Procedurally, after suitable pretreatment and/or preculture, shoot tips are exposed to a cryoprotectant solution for a defined length of time. Dimethyl sulfoxide (DMSO), ethylene glycol (EG), or sucrose are common cryoprotectants for plant systems and often are applied together or in some combination with polyethylene glycol, proline or glycinebetaine. Exposure may be at 0 or 25 °C in one step, or may be gradual to minimize osmotic shock. DMSO and EG are quite permeable to plant cells, but sucrose, glycerol and polymers are either not permeable or only slowly permeable in the time frames used. Proline and glycinebetaine are permeable, but apparently less so than DMSO and EG. There are few studies in plant cells on the kinetics of permeation of these compounds and none describing permeation kinetics across multicellular tissues.

    After cryoprotectant application, samples are cooled to a temperature just below the freezing point of the cryoprotectant solution and nucleated with an ice crystal to avoid supercooling (Keefe and Henshaw 1984; Fahy 1995). The temperature is then decreased at a defined rate to an intermediate temperature, typically between −30 and −40 °C. During this gradual temperature reduction, cells progressively dehydrate due to water loss to the extracellular ice mass. It should be noted, however, that there is little evidence that the cell in a suspension, or cells within a shoot tip, are in water potential equilibrium with the external ice mass during cooling. Although modeling cell osmotic behavior has been possible in some animal cells and has aided defining cooling conditions (McGrath 1997; Bischof 2000), such predictions have not been performed for plant systems. The hydraulic conductivity of the plasmalemma and its temperature coefficient are not known at lower temperatures, there is incomplete knowledge of cryoprotectant permeation, and cell volume measurements on plant cells during cooling are very imprecise. Although protoplasts from several species have been used for some mechanistic studies (Steponkus et al. 1992), it is uncertain whether values developed from these are more broadly applicable to cells within shoot tips or to cells from different species.

Survival usually decreases considerably with continued slow cooling below ca. −30 to −40°C, and thus samples are transferred to LN or the vapor phase above it in this temperature range, during which it is supposed that the remaining cellular contents either vitrify or form few, small non-disruptive ice crystals. Maximum survival usually occurs with somewhat rapid warming (brief exposure to water at 25–40°C).

For application of the method a suitable cooling rate to the intermediate temperature must be used. For cells and shoot tips this is often less than 0.5°C/min (range about 0.1–1°C/min), but has not been critically determined for many species. Overall, this method has been useful for species that are cold acclimated, but has not given consistently high levels of survival with many others, notably tropical species. Lack of success in specific studies may be due to suboptimal cryoprotectant addenda, and inadequate cooling rates. An interpretation is that the dehydration injury occurs before the cellular contents have reached an adequate composition that can vitrify with second step cooling to LN. Whether altering additional pretreatments and/or cryoprotectants and the use of more rapid rates to LN can circumvent or avoid this injury is uncertain.

A devise that cools at defined rates is useful for these studies, but, in practice, a desired rate can often be obtained by placing insulated containers directly into chest freezers set at the intermediate temperature. A linear rate of cooling is often implied as being desirable, but again this is due to expediency and allows for a reproducible treatment. Since water permeability declines as temperature is lowered, it can be argued that slower rates may be more desirable at lower temperatures in order to keep cells in approximate water vapor pressure equilibrium with the external ice.

5.2.2.2 Rapid Cooling

Rapid cooling has been used to cryopreserve shoot tips from some species, more recently asparagus shoot tips (Mix-Wagner et al. 2000), and has been successfully applied at a genebank for potato (Schafer-Menuhr et al. 1997). The method is sometimes called droplet freezing and employs a rather low concentration of cryoprotectant (i.e., 10% DMSO). Samples using droplets containing the shoot tips on small foil strips are directly immersed into LN from ice or room temperature. Rapid warming is necessary. Mechanistically, it is uncertain what the incidence and extent of intracellular ice formation are with such a method. Conceptually, this may be measured with differential scanning calorimetry (DSC), but such cooling rates are not attainable in the apparatus and the small extent of ice formed within the shoot tip may be below the limit of detection.

5.2.2.3 Vitrification

An alternative route for cryopreservation has been adopted because of the low viability encountered in application of two-step and rapid cooling methods across a wide range of taxa. In the late 1980s vitrification as a method for cryopreservation was reexamined in animal systems, particularly expounded by Fahy and associates (Fahy et al. 1984, 1987). Vitrification refers to the solidification of a liquid as a glass during cooling, and the physical aspects of ice nucleation, crystallization and solution vitrification have been well described (Mehl 1996). Inherent in this concept is the lack of ice crystal occurrence, but it should also be noted that this is not an all-or-none event. Small ice crystals may nucleate, but not grow, as the viscosity greatly increases with decreasing temperature. If the cell contents can be adjusted with supplied or naturally synthesized compounds, the contents may vitrify upon exposure to low temperatures. Two different routes have been described for vitrification. The first approach uses exposure to a series of solutions to create the condition which could vitrify upon low temperature exposure. The second method uses encapsulation in alginate, sucrose exposure and dehydration to attain the vitrifiable state.

Solution-Based Vitrification: The general procedure comprises two steps. After preculture, a dilute solution of a permeating cryoprotectant is applied (loading phase), followed by a vitrification solution (dehydration phase). A discussion of vitrification temperature and the relationship to concentration of solutes is presented in Fahy et al. (1984, 1987).

Several points are worth mentioning at this juncture. The first is that loading is not always used or advantageous, and may be circumvented by application of other solutions. For example, Sakai and coworkers (Matsumoto et al. 1998a) have reported that a short duration (20–60 min) application of 2 M glycerol + 0.4 M sucrose before addition of the vitrification solution is often beneficial. Neither glycerol nor sucrose is readily permeable in such an exposure time frame, and where examined cells within the shoot tip are plasmolysed (Matsumoto et al. 1998a). Thus, such a solution is not really a loading solution, implied by the term to mean allowing permeation of desired solutes, but nevertheless the application is effective.

A second issue is the constitution of the vitrification solution. Since the osmotic strength of the vitrification solution is very great (>8 osmol) and the duration of application fairly short, the main function is to dehydrate the sample, concentrating the permeable components and other cytoplasmic contents within the cell. Sakai et al. (1991) devised a solution termed PVS2 [30% glycerol, 15% EG, 15% DMSO (w/v %) and 0.4 M sucrose], which has been adopted by many researchers. Other formulations have also been presented including those by Helliot and de Boucaud (1997) and Nishizawa et al. (1993), Towill (1990), and Steponkus et al. (1992). These solutions can be shown to vitrify using DSC, but the crucial issue is whether cellular contents treated with these solutions can vitrify during cooling and avoid devitrification during warming. If the main function is dehydration, several other concentrated solutions also should be effective, but the kinetics of dehydration may be altered. Dehydration and survival kinetics with these other vitrification solutions have not been examined sufficiently to allow a firm conclusion as to whether their sole function is dehydration. The issue has also been raised that some components of vitrification solutions are toxic, implying chemical toxicity, but this, too, has not been proven to be the reason for loss in viability in plant cells during extended exposure.

To vitrify samples with less than ideal solute concentration requires rapid cooling and several researchers have noted that survival percentages are increased if cooling to LN is faster. This observation comes from data where cooling on copper grids or foil strips has given better survival than cooling within straws or plastic cryovials. This observation suggests that the cellular contents are not ideal and both cooling and warming rates are critical. Studies have been insufficient to determine if additional alterations of preculture or loading may allow relatively slower rates to be used. One should note that the size of the specimen dictates the rate of heat transfer and that rates that are needed for vitrification of cellular contents may not be attainable.

Warming rates for vitrified samples, likewise, should be rapid to avoid devitrification and are usually accomplished by placing the vessel in water (ca. 25–40 °C). Devitrification refers to the formation of crystallized ice as the solution is warmed above the glass formation temperature (Macfarlane 1986). Since this is a kinetic and probabilistic event, if warming is sufficiently rapid, both homogeneous nucleation and crystal regrowth are minimized.

Removal of the vitrification solution is performed by placing the cells or tissues into an elevated osmotic solution, for example, 1.2 M sucrose. This is done to avoid rapid water uptake into the cells while still allowing some of the permeable components to diffuse out. As with warming, there are few studies that define critical kinetic conditions for these events.

Encapsulation–Dehydration. The development of an encapsulation technique by Dereuddre and associates (Dereuddre et al. 1990) was a conceptual breakthrough for handling small samples from cultures and has been applied to protoplasts, cells, somatic embryos, and shoot tips. The encapsulation technique is simple and allows easy handling of the propagule, gradual exchange of solutes and a slower and more controlled rate of dehydration. As usually used, shoot tips are encapsulated in an alginate solution and then dropped into a CaCl$_2$ solution to form a gel. The encapsulated shoot tip is then exposed, usually gradually (2–5 days), to a final sucrose concentration of ca. 0.75–1.25 M and desiccated in air or over desiccants for varying periods of time. Preculture of shoot tips prior to the encapsulation process is sometimes beneficial.

During subsequent cooling to LN, samples at higher moisture levels ostensibly are damaged due to intracellular ice formation and, at lower moisture levels, due to extensive desiccation. At a moisture content of about 20–24%, samples survive exposure to LN by direct immersion, usually within cryovials.

However, there are notable exceptions, some species show better viability if slow cooling is employed (Plessis et al. 1993), whereas others show better survival if more rapid cooling is used (beads directly immersed in LN, not within a vial) (Pennycooke and Towill, unpubl.).

Samples are retrieved by plating either the bead or the shoot tip dissected from the bead on a suitable growth medium.

Other Procedures and Comparisons. Many modifications of the two basic vitrification procedures described above have been reported. For example, encapsulated shoot tips have been exposed to vitrification solutions prior to LN immersion (Tannoury et al. 1991; Matsumoto et al. 1995). Encapsulated shoot tips have also been exposed to glycerol/sucrose solutions prior to desiccation and cryogenic exposure (Sakai et al. 2000).

When comparisons have been made among procedures, no general trend of effectiveness can be found. For example, with *Prunus domestica* L., vitrification with PVS2 or by encapsulation-vitrification gave higher levels of viability after cryoexposure than encapsulation-dehydration (De Carlo et al. 2000). Tobacco suspensions survived better with vitrification than two-step cooling (Reinhoud et al. 1995). Shoot tips of in vitro silver birch showed better regrowth after two-step cooling than vitrification (Ryynanen 1996). Survival after cryoexposure with *Limonium* was similar among three methods (Matsumoto et al. 1998b). For any comparison to be meaningful, conditions should be optimized for each of the methods used, but this requires a large number of experiments. Since conditions are often not optimized, the results merely state that, under the conditions described, one method works better than another, and not that one method is inherently better than the other.

5.3 Recovery

There is a concatenation of potentially injuring events that might occur during excision, dehydration, osmolyte and cryoprotectant exposure, and temperature excursions in the overall cryopreservation process. It is not surprising then that materials processed are injured, and that subsequent treatment of the propagule will affect recovery. This has mainly been examined in cell suspensions (Watanabe 2000); however, in several cases, recovery has been demonstrated with shoot tips. Growth of treated shoot tips is best under low light conditions. Temporal omission of ammonium has improved rice culture viability (Kuriyama et al. 1989; Watanabe 2000) and has been used for shoot tips (Niino et al. 1992; Ryynanen and Haggman 2001). Culture on elevated sucrose levels for a short duration has improved survival. Application of desferrioxamine has improved cryosurvival, probably by minimizing the continuation of damage due to oxidative stress (Benson et al. 1995). Treated shoot tips often require different growth regulators to minimize callus growth and maximize shoot growth (Chang and Reed 1999). The nature of the sublethal injury induced in cryopreservation is not defined and is worthy of further analysis.

Certainly, several culturing options for enhancing survival and regrowth exist (Anthony et al. 2000). Some examination of recovery should be included in any comprehensive cryopreservation study.

6 Conceptual Questions

The preceding discussion has been general and applied, without extensive citations; many excellent references are cited within the chapters in this book. The empirical nature of the process of cryopreservation has been highlighted, especially for hydrated propagules such as shoot tips. Many interesting questions can be posed, which, for the most part, are still without answers. Although there is some information that addresses aspects of these questions, for no system is there a complete description or understanding of the physiological and physical events that occur in the overall process we term cryopreservation. A few of these queries are described below, and hopefully will provide thought for study.

Physiological adaptation is important, but what does this mean mechanistically? Certainly the buildup of endogenous sugar levels seems important, but what levels are beneficial, why, and can this be accomplished by other routes or treatments? How important are lipid and membrane alternations during this adaptation phase. From a physiological view, is a growing shoot tip more amenable to cryopreservation or one that has been correlatively suppressed, and why?

The cryobiology phase is not completely understood. What are the key components associated with preculture? Is it merely sugar buildup, and what sugars? And what are the kinetics of this, the critical levels needed and the differences that exist among genotypes and species? Cryoprotectant type and kinetics of exposure are also complex. How does a cryoprotectant function, why do some combinations seem better than others, and what are the kinetics of uptake of the components into single cells and within a complex tissue? Is gene activation important for cellular conversions during the phases? Since both two-step cooling and vitrification involve desiccation, albeit by different routes, what are tolerance levels and how do they differ amongst species and genotypes. Enhanced intracellular sugar levels, such as sucrose in higher plants and trehalose in microbes, are associated with increased desiccation tolerance and cryopreservability, but how can this be quantified with regard to cryopreservability? Cooling and warming are appreciated as being important for success, but what are critical rates and how do they reflect solute and water permeation characteristics, both for dehydration and rehydration?

Viability measurements are critical, but, for obvious reasons, most studies rely on regrowth as the sole criterion for comparisons in developing a practical procedure. Does this reliance obscure some meaningful information? During the cryopreservation, what and how is the cell perturbed in each phase? Injury to the plasma membrane is a key event for loss of viability, but how extensive is this to be lethal and what extent and manner of repair occurs for sublethal injury? Can these processes be further manipulated to enhance viability and recovery of shoot forming capacity? All compartments within the cell must be protected for normal growth, but permeability characteristics differ among the plasma membrane, tonoplast and organellar membranes. How are endogenous solutes or permeating cryoprotecants distributed in these compartments and what are these relationships to function and viability decline.

The above are only examples of questions whose examination would benefit a theoretical understanding of injury and how it can be alleviated during cryopreservation.

7 Practical Issues

There are several issues that need to be addressed before a method can be applied. It is reasonable to expect that no single method will give high levels of survival for every genotype of a species after cryogenic exposure. Since a method cannot be devised for each genotype, once a useful procedure is identified the next step would be to apply it to different genotypes to determine overall utility. The selection of genotypes should be judicious. Obviously, a method developed for cold-hardy species would not be expected to be successful with semitropical representatives. Therefore, the test should examine genotypes that are not too closely related but represent some extremes with regard to origin and physiological traits. From a practical viewpoint, the method can be applied to a select group and those for which it does not work would form a subgroup for further method development.

From data within the literature, it is often difficult to determine how useful and generally applicable a method is, even if it has been tested with several genotypes. It is often important to describe some negative data – that is, what did not work (assuming the experiment was conducted appropriately) – this type of information would be very helpful to other researchers using similar crops. Unfortunately, this information is often not reported in journal articles.

Another consideration for usefulness is ease of application of the method. Two-step cooling systems require a suitable apparatus. Solution-based vitrification is very time exacting, and processing large numbers of shoot tips may be difficult. But this is not insurmountable since a smaller number could be processed each day, or staggered in exposure. The encapsulation procedure allows more propagules to be handled at one time, and timing is not overly critical; however, desiccation usually requires several hours to attain the desired moisture level. If air-drying is used, the duration desired will differ depending on the relative humidity of the atmosphere.

Cryogenic storage is often referred to as a safe system, but this is dependent on reliable procedures and subsequent handling. For many seeds, where uncontrolled warming is generally not detrimental, cryogenic storage may be safe since the inadvertent loss or temporary unavailability of cryogen does not affect viability. However, for propagules from clones or desiccation-sensitive seed, uncontrolled temperature fluctuations, especially above perhaps −120 °C, may drastically affect viability. Hence, if cryogenic storage tanks are not carefully monitored for temperature or liquid nitrogen level, safety may be more apparent than real. Proper handling is, likewise, crucial for safe storage. Samples cannot be allowed to warm while searching for a desired line. Storage in the LN vapor phase may subject the sample to greater temperature fluctuations and require better monitoring of the storage container's LN content. But, storage in the liquid phase may allow seepage into sample vessels, causing a potential explosion possibility upon warming, as well as facilitating microbial or viral cross contamination amongst samples. The size of the LN storage unit also affects protocols; samples can be easily sorted in large tanks, but only with difficulty in smaller units (e.g., 50-l LN refrigerators). Each laboratory may develop somewhat different guidelines, but strict adherence to proper handling and storage protocols and training of personnel should assure safe, long-term storage.

One other issue will not be addressed in detail here but will be mentioned because of its importance. This is the concern for the genotype stability of materials processed through all the steps of cryopreservation. Several reports cite cell lines, after cryopreservation, retaining some critical function or plants appearing morphologically true-to-type, yet it is apparent that this is a rather coarse evaluation of stability. Molecular techniques now allow a much more detailed examination. Although reports are few, studies with potato have shown plants regenerated from cryopreserved shoot tips have similar microsatellite profiles as controls (Harding and Benson 2001). Although specific techniques have limitations for comparisons, such analyses are needed to assure that genebank methods preserve genetic diversity.

8 Summary

The numerous research reports on cryopreservation over the past few years have been very encouraging and provide the basis for application to plant systems. For desiccation-tolerant seeds and pollen this is straightforward and more extensive use is envisioned. For more hydrated propagules, methods have been described for a very wide range of taxa, but repeatability and effectiveness across diversity needs to be further assessed before adoption. Although the physical basis for cryoprotection and events that occur during cooling and warming are elusive, an understanding of the overall process is emerging such that some control over viability is possible. Manipulations of the stock plant physiological status, cryobiology phase and recovery steps are providing information that suggest some generalities in approaching the overall problem of cryopreservation of germplasm.

References

Anthony P, Davey MR, Azhakanandam K, Power JB, Lowe KC (2000) Cryopreservation of plant germplasms: new approaches for enhanced postthaw recovery. In: Razdan MK, Cocking EC (eds) Conservation of plant genetic resources in vitro, vol 2. Applications and limitations. Science Publishers, Enfield, New Hampshire

Bajaj YPS (ed) (1995) Cryopreservation of plant germplasm I. Biotechnology in agriculture and forestry, vol 32. Springer, Berlin Heidelberg New York

Benson EE, Harding K, Smith H (1989) Variation in recovery of cryopreserved shoot-tips of *Solanum tuberosum* exposed to different pre- and post-freezing light regimes. Cryo-Letters 10:323–344

Benson EE, Lynch PT, Jones J (1995) The use of iron chelating desferrioxamine in rice cell cryopreservation: a novel approach for improving recovery. Plant Sci 110:249–258

Benson EE (ed) (1999) Plant Conservation Biotechnology. Taylor and Francis, London

Bischof JC (2000) Quantitative measurement and prediction of biophysical response during freezing in tissues. Annu Rev Biomed Eng 2:257–288

Bouafia S, Jelti N, Lairy G, Blanc A, Bonnel E, Dereuddre J (1996) Cryopreservation of potato shoot tips by encapsulation-dehydration. Potato Res 39:69–78

Chang Y, Reed BM (1999) Extended cold acclimation and recovery medium alteration improve regrowth of *Rubus* shoot tips following cryopreservation. Cryo-Letters 20:371–376

Chang Y, Reed BM (2000) Extended alternating-temperature cold acclimation and culture duration improve pear shoot cryopreservation. Cryobiology 40:311–322

De Carlo A, Benelli C, Lambardi M (2000) Development of a shoot-tip vitrification protocol and comparison with encapsulation-based procedures for plum (*Prunus domestica* L.) cryopreservation. Cryo-Letters 21:215–222

Dereuddre J, Scottez C, Arnaud Y, Duron M (1990) Resistance of alginate-coated axillary shoot tips of pear tree (*Pyrus communis* L. cv. Beurre Hardy) in vitro plantlets to dehydration and subsequent freezing in liquid nitrogen: effects of previous cold hardening. CR Acad Sci Paris 310:317–323

Dumet D, Engelmann F, Chabrillange N, Richaud F, Beule T, Durand-Gassellin T, Duval Y (1993) Development of cryopreservation for oil palm somatic embryos using an improved process. Oleagineux 48:273–278

Dussert S, Chabrillange N, Vasquez N, Englemann F, Anthony F, Guyot A, Hamon S (2000) Beneficial effect of post-thawing osmoconditioning on the recovery of cryopreserved coffee (*Coffea arabica* L.) seeds. Cryo-Letters 21:47–52

Engelmann F, Takagi H (eds) (2000) Cryopreservation of tropical plant germplasm: current research progress and application. Japan International Research Center for Agricultural Sciences, Tsukuba, Japan, 496 pp

Fahy G (1995) The role of nucleation in cryopreservation. In: Lee RE, Warren GJ, Gusta LJ (eds) Biological ice nucleation and its applications. American Phytopathology Press, St. Paul, MN, pp 315–336

Fahy GM, MacFarlane DR, Angell CA, Meryman HT (1984) Vitrification as an approach to cryopreservation. Cryobiology 21:407–426

Fahy GM, Levy DI, Ali SE (1987) Some emerging principles underlying the physical properties, biological actions and utility of vitrification solutions. Cryobiology 24:196–213

Falk DA, Holsinger KE (eds) (1991) Genetics and conservation of rare plants. Oxford Univ Press, New York

Farnsworth E (2000) The ecology and physiology of viviparous and recalcitrant seeds. Annu Rev Ecol Syst 31:107–138

Hanna WW, Towill LE (1995) Long-term pollen storage. Plant Breeding Rev 13:179–207

Harding K, Benson EE (2001) The use of microsatellite analysis in *Solanum tuberosum* L. in vitro plantlets derived from cryopreserved germplasm. Cryo-Letters 22:199–208

Helliot B, de Boucaud MT (1997) Effect of various parameters on the survival of cryopreserved *Prunus* Ferlenain in vitro plantlets shoot tips. Cryo-Letters 18:133–142

Hitmi A, Coudret A, Barthomeuf C, Sallanon H (1999) The role of sucrose in freezing tolerance in *Chrysanthemum cinerariaefolium* L. cell cultures. Cryo-Letters 20:45–54

Hoekstra FA (1995) Collecting pollen for genetic resources conservation. In: Guarino L, Rao VR, Reid R (eds) Collecting plant genetic diversity. Technical guidelines. CAB International, Wallingford, Oxon, UK, pp 527–550

Jekkel Z, Gyulai G, Kiss J, Kiss E, Heszky LE (1998) Cryopreservation of horse-chestnut (*Aesculus hippocastanum* L.) somatic embryos using three different freezing methods. Plant Cell Tissue Organ Cult 52:193–197

Kartha KK (ed) (1985) Cryopreservation of plant cells and organs. CRC Press, Boca Raton

Keefe PD, Henshaw GG (1984) A note of the multiple role of artificial nucleation of the suspending medium during two-step cryopreservation procedures. Cryo-Letters 5:71–78

Knutson L, Stoner AK (eds) (1989) Biotic diversity and germplasm preservation, global imperatives, Kluwer, Dordrecht

Kuriyama A, Watanabe K, Ueno S, Mitsuda H (1989) Inhibitory effect of ammonium ion on recovery of cryopreserved rice cells. Plant Sci 64:231–235.

Macfarlane DR (1986) Devitrification in glass-forming aqueous solutions. Cryobiology 23:230–244

Matsumoto T, Sakai A, Nako Y (1998a) A novel preculturing for enhancing the survival of in vitro-grown meristems of wasabi (*Wasabia japonica*) cooled to –196 °C by vitrification. Cryo-Letters 19:27–36

Matsumoto T, Takahashi C, Sakai A, Nako Y (1998b) Cryopreservation of in vitro-grown apical meristems of hybrid statice by three different procedures. Sci Hort 76:105–114

Matsumoto T, Sakai A, Takahashi C, Yamada K (1995) Cryopreservation of in vitro-grown meristems of wasabi (*Wasabia japonica*) by encapsulation-vitrification method. Cryo-Letters 16:189–196

Mazur P (1984) Freezing of living cells: mechanisms and implications. Am J Physiol Cell Physiol 247:C125–142

Mazur P (1966) Physical and chemical basis of injury in single-celled micro-organisms subjected to freezing and thawing. In: Meryman HT (ed) Cryobiology. Academic Press, New York, pp 213–315

McGrath JJ (1997) Quantitative measurement of cell membrane transport: technology and applications. Cryobiology 34:315–334

Mehl P (1996) Crystallization and vitrification in aqueous glass-forming solutions. In: Steponkus PL (ed) Advances in low-temperature biology. JAI Press, Greenwich, CT, pp 185–255

Mercier S (1995) The role of a pollen bank in the tree genetic improvement program in Quebec (Canada). Grana 34:367–370

Mix-Wagner G, Conner AJ, Cross RJ (2000) Survival and recovery of asparagus shoot tips after cryopreservation using the "droplet" method. N Z J Crop Hort Sci 28:283–287

Mycock DJ, Wesley-Smith J, Berjak P (1995) Cryopreservation of somatic embryos of four species with and without cryoprotectant pretreatment. Ann Bot 75:331–336

Niino T, Sakai A, Yakuwa H, Nojiri K (1992) Cryopreservation of in vitro grown shoot tips of apple and pear by vitrification. Plant Cell Tissue Organ Cult 28:261–266

Nishizawa S, Sakai A, Amano Y, Matsuzawa T (1993) Cryopreservation of asparagus (*Asparagus officinalis* L.) embryogenic suspension cells and subsequent plant regeneration by vitrification. Plant Sci 91:67–73

Pammenter NW, Berjak P (1999) A review of recalcitrant seed physiology in relation to desiccation-tolerance mechanisms. Seed Sci Res 9:13–37

Pennycooke JC, Towill LE (2000) Cryopreservation of shoot tips from in vitro plants of sweet potato (*Ipomoea batatas* (L.) Lam.) by vitrification. Plant Cell Rep 19:733–737

Plessis P, Leddet C, Collas A, Dereuddre J (1993) Cryopreservation of *Vitis vinifera* L. cv. Chardonnay shoot tips by encapsulation-dehydration: effects of pretreatment, cooling and postculture conditions. Cryo-Letters 14:309–320

Razdan MK, Cocking EC (eds) (2000) Conservation of plant genetic resources in vitro, vol 2. Applications and limitations. Science Publisher, Enfield, NH

Reaka-Kudla ML, Wilson DE, Wilson EO (eds) (1997) Biodiversity II. Joseph Henry Press, Washington, DC

Reinhoud PJ, Schrijnemakers WM, van Iren F, Kijne JW (1995) Vitrification and a heat-shock improve cryopreservation of tobacco cell suspensions compared to two-step freezing. Plant Cell Tissue Organ Cult 42:262–267

Ryynanen L (1996) Cold hardening and slow cooling: tools for successful cryopreservation and recovery of in vitro shoot tips of silver birch. Can J For Res 26:2015–2022

Ryynanen L (1998) Effect of abscisic acid, cold hardening, and photoperiod on recovery of cryopreserved in vitro shoot tips of silver birch. Cryobiology 36:32–39

Ryynanen LA, Haggman HM (2001) Recovery of cryopreserved silver birch shoot tips is affected by the pre-freezing age of the cultures and ammonium substitution. Plant Cell Rep 20:354–360

Sakai A, Matsumoto T, Hirai D, Niino T (2000) Newly developed encapsulation-dehydration protocol for plant cryopreservation. Cryo-Letters 21:53–62

Sakai A, Larcher W (1987) Frost survival of plants. Responses and adaptation to freezing stress. Springer, Berlin Heidelberg New York, 321 pp

Sakai A, Kobayashi S, Oiyama I (1991) Survival by vitrification of nucellus cells of navel orange (*Citrus sinensis* var. *brasiliensis* Tanaka) by vitrification. Plant Cell Rep 9:30–33

Schafer-Menuhr A, Schumacher H-M, Mix-Wagner G (1997) Long-term storage of old potato varieties by cryopreservation of shoot tips in liquid nitrogen. Plant Genet Resour Newslett 111:19–24

Shibli RA, Al-Juboory KH (2000) Cryopreservation of 'nabali' olive (*Olea europea* L.) somatic embryos by encapsulation-dehydration and encapsulation-vitrification. Cryo-Letters 21: 357–366

Stanwood PC (1985) Cryopreservation of seed germplasm for genetic conservation. In: Kartha KK (ed) Cryopreservation of plant cells and organs. CRC Press, Boca Raton, pp 199–226

Stanwood PC (1987) Survival of sesame seeds at the temperature (−196 °C) of liquid nitrogen. Crop Sci 27:327–331

Steponkus PL, Langis R, Fujikawa S (1992) Cryopreservation of plant tissues by vitrification. In: Steponkus PL (ed) Advances in low-temperature biology. JAI Press, London, pp 1–61

Stewart P, Taylor M, Mycock D (2001) The sequence of the preparative procedures affects the success of cryostorage of cassava somatic embryos. Cryo-Letters 22:35–42

Takagi H (2000) Recent developments in cryopreservation of shoot apices of tropical species. In: Engelmann F, Takagi H (eds) Cryopreservation of tropical plant germplasm: current research progress and application. Japan International Center for Agricultural Sciences, Tsukuba, Japan, pp 178–193

Tannoury M, Ralambosoa J, Kaminski M, Dereuddre J (1991) Cryoconservation by vitrification of alginate-coated carnation (*Dianthus caryophyllus* L.) shoot tips of in vitro plants. CR Acad Sci Paris 313:633–638

Thammasiri K (1999) Cryopreservation of embryonic axes of jackfruit. Cryo-Letters 20:21–28

Touchell D, Walters C (2000) Recovery of embryos of *Zizania palustris* following exposure to liquid nitrogen. Cryo-Letters 21:261–270

Towill LE (1985) Low temperature and freeze-/vacuum-drying preservation of pollen. In: Kartha KK (ed) Cryopreservation of plant cells and organs. CRC Press, Boca Raton, pp 171–198

Towill LE (1990) Cryopreservation of isolated mint shoot tips by vitrification. Plant Cell Rep 9:178–180

Uemura M, Steponkus PL (1999) Cold acclimation in plants: relationship between the lipid composition and the cryostability of the plasma membrane. J Plant Res 112:245–254

Vertucci CW, Roos EE (1990) Theoretical basis of protocols for seed storage. Plant Physiol 94: 1019–1023

Vertucci CW, Roos EE (1993) Theoretical basis of protocols for seed storage II. The influence of temperature on optimal moisture levels. Seed Sci Res 3:201–213

Vertucci CW, Roos EE, Crane J (1994) Theoretical basis of protocols for seed storage III. Optimum moisture contents for pea seeds stored at different temperatures. Ann Bot 74: 531–540

Watanabe K (2000) Effect of postthaw treatments on viability of cryopreserved plant cells. In: Razdan MK, Cocking EC (eds) Conservation of plant genetic resources in vitro, vol. 2. Applications and limitations. Science Publishers Inc, Enfield, New Hampshire, pp 3–19

I.2 Implementing Cryopreservation for Long-Term Germplasm Preservation in Vegetatively Propagated Species

Barbara M. Reed

1 Introduction

Vegetatively propagated fruit, vegetables, or forestry related crops are important in the agriculture of every country. Maintaining the genetic diversity of these crops is more demanding than for most seed-producing plants, because the specific genotype must be maintained. Some plants are maintained as clones because: they do not form viable seeds; seeds are short lived or do not tolerate drying; seeds are large and require too much space to store; long juvenile periods limit usefulness of seedlings; clones are heterozygous and do not produce true to type seed; select clones are more productive than seed-derived lines (Towill 1988). Vegetatively propagated crops are normally stored as clonal material; sometimes, however, seed-derived materials may be used as well. Primary collections of clonal crops are in a field or screened enclosure; however, backups for these materials are needed to provide security in case of a disease or environmental disaster. In vitro gene banks provide alternative storage for a number of crops. Active in vitro gene banks exist for temperate and tropical crops in several countries (IPGRI/CIAT 1994; Ashmore 1997; Engelmann 1999). These in vitro gene banks, however, are not ideal for a base collection, as the plantlets require repropagation at 6-month to 4-year intervals and may be lost due to contamination or technical difficulties. Cryopreservation is the preferred option for the long-term storage of clonal germplasm (Engelmann 2000). Cryopreservation may also be the best option for long-term germplasm storage of some seed-propagated species (e.g., coconut, coffee, papaya). Cryopreservation is an excellent storage method for genetic variants with special medicinal or industrial value, recalcitrant seed, rare germplasm, disease-free plants, pollen, and embryogenic cultures (Bajaj 1995; Razdan and Cocking 1997).

Cryopreservation techniques have been developed over the last 25 years and can be implemented for routine storage of germplasm (Reed et al. 2000a,b). Techniques of controlled freezing, vitrification, encapsulation-dehydration, dormant bud preservation, and combinations of these are now directly applicable with plant genotypes representing hundreds of species.

USDA-ARS National Clonal Germplasm Repository, 33447 Peoria Rd., Corvallis, Oregon 97333-2521, USA

Biotechnology in Agriculture and Forestry, Vol. 50
L.E. Towill and Y.P.S. Bajaj (Eds.) Cryopreservation of Plant Germplasm II
© Springer-Verlag Berlin Heidelberg 2002

Unfortunately, few laboratories are actually instituting the techniques. Initial implementation of cryopreservation procedures can be daunting where financial and human resources are lacking. The experimental protocols tested on one or a few genotypes in a genus will provide a starting place for storage, but may require modification before application to a wide range of germplasm such as that available in national or breeder collections. Initial steps must be taken to adapt these protocols and set up procedures for testing, screening, and ultimately storing the range of genotypes in each collection (Reed et al. 1998a).

The cooperation of government and foundation administrators is important in providing the necessary infrastructure, labor, and international collaboration necessary for the security of a base clonal collection. Interest in germplasm security at the highest levels of government is often needed to secure stable funding and provide adequate resources to realize secure long-term germplasm storage (Benson 1999b).

2 Initial Planning

Early decisions in planning for plant germplasm cryopreservation include the choice of accessions to be stored, number of each accession per storage unit, number of replicates, protocol development, location of storage, viability testing, records, and proper control groups (Fig. 1). Emphasis should be placed on selecting a secure storage site and compiling complete records needed for the recovery of plant material. Secure remote storage, duplicate locations, and secure, accurate records are all important in insuring the safety and usefulness of base collections. Evaluation of cryostored collections should be initiated to determine the longevity of plants and stability of storage conditions.

2.1 Cryopreservation and Storage Records

Storage records must be designed to link cryopreserved propagules to the original plant accession and to all information related to that original plant (passport information). Important cryogenic information must also be linked because propagules are to be retrieved 50, 100, or 500 years in the future. Each cryopreserved accession must have information on preparation, pretreatment, cryopreservation method, thawing method, and the recovery medium. Thawing methods and recovery media are especially critical to recovering the germplasm. These two items should be readily accessible in the accession database records for easy access by future scientists wishing to recover plants. Complete protocol information that is not critical to recovering plants but may be of scientific interest could be stored in a secondary database.

DECISION FLOW CHART FOR CRYOPRESERVED
STORAGE OF CLONAL GERMPLASM

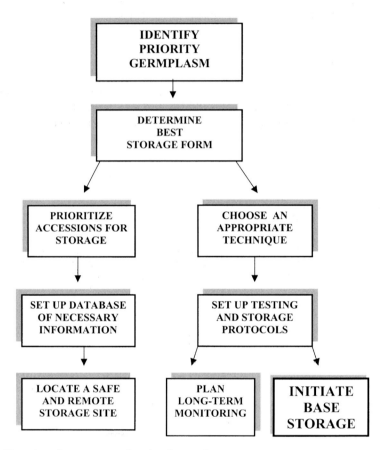

Fig. 1. Flow chart for cryopreserving clonal germplasm

2.2 Choice of Accessions

The choice of material to store will vary with the genus involved. Material
presently stored in cryopreserved collections varies from randomly chosen
selections to carefully chosen clones that represent morphological and genetic
variation as well as specific geographic criteria, sometimes referred to as
"core" collections (IPGRI/CIAT 1994; Reed et al. 1998a,b; Golmirzaie and
Panta 2000). Priority accessions will vary with the crop, country, and facilities
available. For any collection early attention should be paid to "at risk" accessions. These "at risk" accessions may be those susceptible to diseases, climatic
conditions, insect damage, or other environmental conditions that increase the

chance of loss from a collection. Plant material from areas where changes in farming practices or clearing of wild land may cause loss of local cultivars might also be placed in this category. Plants listed as "threatened" or "endangered" in their wild or native range should be conserved.

2.3 Storage Form

Seed, pollen, shoot apices, dormant buds, excised embryonic axes, zygotic or somatic embryos may be the form of choice depending on the species involved (Table 1). Pollen provides a way to store part of the genetic diversity in a form

Table 1. Storage forms for cryopreserved germplasm

Best Storage Form	Plant groups	Advantages	Disadvantages
Seed	Small seeded, desiccation tolerant, cold tolerant, not clonally propagated	Easy to use for appropriate seed types (orthodox)	Not useful for large, cold sensitive or desiccation intolerant seeds (recalcitrant) or clonally propagated plants
Pollen	Many	Easy for many plant genera, useful for plant breeding	Preserves only half of the genome
Dormant buds	Temperate woody plants	Dormant buds are readily available from field genebanks	Degree of cold hardiness varies with season and genotype, requires more storage space than other techniques, requires grafting and budding expertise for recovery
In vitro shoot tips	Many	Available at any time of year, easy to manipulate physically and physiologically	Techniques are not developed for all plants, requires a laboratory and skilled workers
Embryogenic cultures	Various plant groups	Callus is generally easy to cryopreserve	Not all plants produce somatic embryos, techniques are not broadly applicable across species/accessions
Embryonic axes	Some marginally recalcitrant seed types	Easy to remove and process	In vitro systems are needed to recover whole plants
Zygotic embryos	Some marginally recalcitrant seed types	Removal of the seed coat may improve recovery	Time consuming and often technically difficult

easily accessible and useable by plant breeders and may be a quick first step for starting a cryopreserved gene bank (Towill and Walters 2000). Dormant buds of cold hardy plants that can be grafted or budded may be a good option for temperate trees and shrubs (Forsline et al. 1998; Niino et al. 2000). Shoot apices are successfully recovered in vitro for many plant types, require little room for storage, and preserve the exact genotype of both tropical and temperate plants (Reed and Chang 1997; Benson 1999a). For some species the storage of excised embryonic axes provides the only long-term storage option, while for others the production of somatic embryos provides large amounts of material for preservation (Engelmann 2000).

2.4 Numbers, Numbers, Numbers

The number of propagules needed for storage will vary with the plant type and the recovery potential of the accession. Ideally, enough propagules should be stored in each container to produce several living plants and enough containers to allow for several recoveries over time. If periodic testing is anticipated, then additional vials should be included for that purpose so that the vials in the base collection remain untouched. Storage of somatic embryos could include large numbers of vials with large numbers of embryos due to their normally prolific production. Dormant bud wood is easy to collect; however, it may require more storage space so the number of replicates may be limited. In vitro shoot tips are relatively easy to produce but require excision before freezing, thus limiting the number available for storage at one time. At the extreme, zygotic embryonic axes require removal from the seeds, which may be difficult, and the seed may also be limited in availability, so few may be available for storage at one time.

2.5 Protocol Testing

The amount of initial protocol testing varies greatly from facility to facility. If methods development involved several species and cultivars, then less testing may be needed at the storage phase. There are two schools of thought on the amount of testing needed for each accession prior to cryopreserved storage. The first uses several storage trials and their controls as the testing phase. In this scenario a laboratory might store five vials of 10 propagules and use one as the control. If the control recovers with a high percentage, then four vials (40 propagules) remain in storage. If the control shows low recovery, then another group of vials should be stored to increase the number of viable propagules in storage (Schafer-Menuhr 1996). The second method stores 20–50 propagules in vials and thaws them after short storage. If the regrowth percentage is greater than a set percentage (i.e., 40%–50%), then base-storage shoot tips are processed and stored. If the percentage is lower, improvements are made to the culture or cryopreservation protocols to improve performance before storage (Reed et al. 1998a; Golmirzaie and Panta 2000).

2.6 Viability Testing and Controls

Both internal and external controls are needed for any storage format. Internal controls include regrowth of plants at critical stages of the protocol. For shoot tip cryopreservation this might occur after dissection from the mother plant, after cryoprotection, and after liquid nitrogen (LN) exposure. If cryopreserved vials are transferred from one storage container to another or from one location to another, a control vial should be regrown to monitor the effect of those transfers. No studies are available on the effect of insertion and removal of samples from the dewar on viability of stored propagules; however, frequent removal and replacement of storage racks may impact the storage life of cryopreserved plants. Proper techniques for inserting and removing vials from storage containers should be developed before storage begins. Improper handling at the storage or removal point can easily kill propagules.

2.7 Storage Location

The location of the storage dewar is very important for long-term storage. As a container for a base collection of important germplasm, the dewar should be located at a site remote from the field gene bank. The storage site needs to be secure and under the control of dependable management to insure that LN is added to the dewar as needed. Alarms should be installed to monitor the LN level in the dewars to insure constant temperature in the storage container. A remote storage location is especially important when field collections are at risk due to environmental or political problems (natural disasters, severe weather, civil unrest).

2.8 Genetic Stability

Limited data is available on the long-term storage of clonally propagated plants. Plant tissues were first stored in the 1970s and initial tests with short-term storage of pea (2 years), strawberry (10 years) and potato (1 year) shoot tips indicated no loss of viability (Haskens and Kartha 1980; Kartha et al. 1980; Towill 1981). With strawberry, fluctuations in viability were noted after various LN storage periods, but the changes were attributed to differences in the physiological status of the shoot tips used in different experiments, not to LN storage. Three generations of field-grown strawberries produced runners, exhibited vigorous growth, and produced normal fruit (Kartha 1985). Normal potato plants were recovered from cryopreserved shoot tips (Towill 1981) and strawberry plants of 10 cultivars evaluated in the field following LN storage all produced normal leaves, flowers and fruit (Reed and Hummer 1995; Reed and Chang 1997). An 8-year study of dormant buds from mulberry cv. Kenmochi found no change in shoot formation with increased length of $-135\,°C$ storage (Niino et al. 2000). Nearly 74% of 376 mulberry cultivars preserved for 5 years had 50% or more bud regrowth when grafted onto 1-year-old seedling rootstocks. Only 6% had regrowth of less than 30% of buds (Niino et al. 2000).

Genetic change of plants grown in tissue culture is termed somaclonal variation (Larkin and Scowcroft 1981). These variations are most common in plants regenerated from single cells or callus. Variation occurs, but is not common, in micropropagated or shoot tip propagated plants derived from existing shoot tips. If care is taken in the clonal multiplication of plant materials, the frequency of variation in propagules should be similar to that of field-propagated plants. Genotypes that often produce mutations/sports will also produce them in vitro. Familiarity of germplasm scientists with the plants in question can assist in identifying variable genotypes so that they can be monitored carefully prior to the cryopreservation process. The establishment of descriptor characteristics for each genotype allows for rogueing of off types upon regrowth. Any genotype with known variability should be flagged in the database and propagules carefully selected for storage. Field and genetic analyses are needed to determine whether instability is a problem; however, several studies have shown little need for concern (Towill 1990; Harding 1991, 1994; Harding and Benson 1993, 1994; Kumar et al. 1999).

2.9 Physical and Biochemical Stability

Another factor relating to viability of propagules stored for long periods involves the physical and biochemical stability of the system. Vitrified solutions are known to crack from physical shocks or in response to certain warming procedures. Thermal stress-induced fractures of biological materials may cause serious damage to stored samples. Slow cooling rates minimize thermal stress from non-uniform temperature distribution, and cryoprotectants contribute to reduce stress by changing the microstructure of the ice formed (Gao et al. 1995). Fractures can also occur as random events during cryopreservation (Najimi and Rubinsky 1997). Fractures typically occur in large organs such as whole seeds and are less common in cell suspensions and shoot tips. Most reports of physical cracking are with animal organs (liver slices, veins, and arteries) rather than plant specimens.

The stability of vitrified biological materials (liposomes and red blood cells), held at temperatures above the glass transition, was predicted to decrease by a factor of 10 for every 15 °C rise in temperature (Sun 1998). As a glassy (vitrified) system cools toward its glass transition temperature, processes become increasingly slow or nearly arrested because of high viscosity, which also slows molecular movement. Additional studies of the relationship between the glass transition temperature and the stability of cryopreserved organisms are needed to adequately predict storage life of biological collections (Sun 1998).

Cryopreserved samples held in dewars are exposed to warming and cooling cycles as other samples are added or removed. There are no studies that quantify the effects of these temperature variations on storage life. The stability principles discussed above could be applied to this phenomenon and with the resulting hypothesis that these fluctuations in temperature would have an impact on the storage life of a cryopreserved sample. It is not known if stability varies among the various procedures used to cryopreserve a propagule.

Biochemical stability of cryopreserved plant cells is well documented through the analysis of cell cultures that produce secondary products (Chen et al. 1984a,b; Yoshimatsu et al. 2000). Studies of cell and shoot tip cultures also suggest that the stability of these systems is maintained by cryopreservation (Harding 1991, 1994; Harding and Benson 1993, 1994).

3 Implementation

Curators, with teams of expert advisors such as crop germplasm committees, will need to determine the amounts and types of germplasm to be stored based on plant characteristics. A rough measure of the number of propagules needed should be based on the expected percent recovery following cryopreservation, the ease of regrowth of the plant, and the number of times samples will be removed from storage. For most clonally propagated crops recovery of five or more plants from a vial would provide adequate material for micropropagation. If the control recovery were 50% or more, then 10 shoot tips per vial would be adequate for long-term storage. If more individuals are desired or the recovery percentage is lower, 25 shoot tips per vial might be warranted. Most accessions could be stored as four or five vials, thus allowing four or five uses over 100 years. Any accessions used for viability testing over time should include more vials. One test vial for each 5–10 years of proposed storage might be a good guideline (i.e., 10–20 vials for testing a clone over 100 years). Designated accessions should be used for viability tests, so only a few genotypes out of hundreds would be stored in larger quantities. These test accessions should have high recovery percentages so any losses of viability over time would be readily apparent.

3.1 Existing Cryopreserved Storage

Cryopreserved storage of important crop plants has begun in many countries throughout the world. The storage form varies with the crop, facilities, and expertise available (Table 2). Additional genera (mint, strawberry, sweet potato, and taro) are currently being tested prior to LN storage, and while not yet officially stored, they will be in the near future. Cryopreserved storage comes in many forms and involves many techniques with varied advantages (Table 1).

Cryopreservation of in vitro shoot tips is frequently recommended because they are easy to multiply, available any time of year, easy to manipulate physically and physiologically, and can be recovered in culture. While in

Table 2. Cryopreserved collections of clonally propagated plant germplasm stored as dormant buds, *in vitro* shoot tips, or excised embryonic axes

Taxon	Country/Institute	Technique	Number Accessions/Replicates
Dormant Buds			
Apple	USA (NSSL)	E-D + CF	2100 accessions
Elm	France (AFOCEL)		101 accessions
Mulberry	Japan (NIAR)	CF	45 accessions
In Vitro Shoot tips			
Apple	China (CI)	CF/E-D	20 accessions/50 shoot tips each
Blackberry (species and cultivars)	USA (NCGR)	CF	17 accessions/100 shoot tips
Cassava	Columbia (CIAT)	E-D	95 accessions/30 shoot tips
Grass	USA (NCGR)	CF/E-D	10 selections/100 shoot tips
Hops	USA (NCGR)	CF	2 accessions/100 shoot tips
Pear	USA (NCGR)	CF	106 accessions/100 shoot tips
Potato	Germany (DSM/FAL)	Droplet	219 accessions/40–350 shoot tips
	Peru (CIP)	Vit	197 accessions/250 shoot tips
Currant/	USA (NCGR)	Vit	5 accessions/100 shoot tips each
Gooseberry	Scotland (UAD)	E-D	31 accessions/25–30 shoot tips
Embryonic Axes			
Almond	India (NBPGR)	D-FF	29 accessions/20 axes each
Citrus	India (NBPGR)	D-FF	12 accessions of 6 species/50–100
Hazelnut	USA (NCGR)	D-FF.	5 species/100–300 axes each
Jackfruit	India (NBPGR)	D-FF/Vit	3 accessions/25 axes each
Litchi	India (NBPGR)	D-FF	2 accessions/30 axes each
Tea	India (NBPGR)	D-FF	85 accessions/25 axes each
Trifoliate Orange	India (NBPGR)	D-FF	1 accession/30 axes each

Facilities: AFOCEL-Association Forêt-Cellulose; CI – Changli Institute of Pomology; CIAT – International Center for Tropical Agriculture; CIP – International Center for the Potato; DSM/FAL – Deutsche Sammlung von Mikroorganismen und Zellkulturen/Institute fur Pflanzenbau, Bundesforschungsanstalt fur Landwirtschaft; NBPGR – National Bureau of Plant Genetic Resources. NCGR – National Clonal Germplasm Repository-Corvallis; NIAR – National Institute of Agrobiological Resources; NSSL – National Seed Storage Laboratory; UAD – University of Abertay-Dundee.

Techniques: CF – Controlled Freezing; D-FF – Dehydration-Fast Freezing; Droplet – Droplet Freezing; E-D – Encapsulation-Dehydration; Vit – Vitrification.

vitro systems require some additional input before storage, the ease of recovering and propagating the shoot tips has many advantages. Storage space required is very small so many accessions can be stored in a small dewar. Shoot tips taken directly from source plants are less desirable as they may carry contaminants, are more difficult to handle, and are less uniform in their recovery from storage.

Embryonic axes are stored for some vegetatively propagated species and for some recalcitrant seeds. Removal and drying of the axes is time consuming but not usually difficult, and the resulting propagules can be stored in a small space and recovered in vitro. Embryonic axes are best used for preserving the genetic diversity of a species when seeds cannot be stored.

Somatic embryos are a good form for cryopreserved storage of selected genetic lines. Many forest-tree production systems depend on embryogenic callus cultures to produce plants from somatic embryos for field-testing. Over 5000 genotypes of 14 conifer species are cryostored in one facility alone. Storage of these cultures allows for use of an embryogenic culture line after an extended period of testing in the field (Cyr 2000). Continued subculturing of embryogenic cultures can lead to somaclonal variation or loss of embryogenic potential, so cryopreservation of important lines from freshly initiated callus is a high priority.

Pollen storage in liquid nitrogen is an important tool for plant breeders. Pollen storage requires drying pollen to a low moisture content. Many plant pollens are easily stored and can be available at any time of year for crosses that are not possible with fresh pollen due to timing constraints (Inagaki 2000; Towill and Walters 2000).

4 Conclusions

The first cryopreserved collections of clonally propagated germplasm are now established for at least 20 economically important crop genera. Cryopreservation is an ideal backup for collections in field gene banks, but is not intended to be the only form of a clonal plant accession. Bull semen was first cryopreserved in the 1940s and the production of healthy calves from semen stored for 60 years shows the reliability of this storage method. Long-term cryostored-plant viability studies monitoring the recovery of designated plant accessions will provide additional information on the stability and viability of this storage form.

Curators planning cryostorage for their crops should first determine the most practical technique for their facility and the crops involved. Off-site storage should be arranged well in advance with a trusted facility. Information management should be an important consideration because the recovery of plants from cryostorage will depend on knowing the proper techniques for thawing and regrowth of each accession.

When considering cryopreservation for long-term storage of germplasm collections, curators should determine the best storage form for the crop in question, prioritize accessions to be stored, determine the best technique to apply to these accessions, set up a database for needed information, make arrangements for offsite storage, plan long-term monitoring, and finally initiate storage.

References

Ashmore SE (1997) Status report on the development and application of in vitro techniques for the conservation and use of plant genetic resources. International Plant Genetic Resources Institute, Rome, Italy

Bajaj YPS (1995) Cryopreservation of plant cell, tissue and organ culture for the conservation of germplasm and biodiversity. In: Bajaj YPS (ed) Biotechnology in agriculture and forestry. Cryopreservation of plant germplasm I, vol 32. Springer, Berlin Heidelberg New York

Benson EE (1999a) Cryopreservation. In: Benson EE (ed) Plant conservation biotechnology. Taylor and Francis, London

Benson EE (ed) (1999b) Plant conservation biotechnology. Taylor and Francis, London

Chen THH, Kartha KK, Constabel F, Gusta LV (1984a) Freezing characteristics of cultured *Catharanthus roseus* (L). G. Don cells treated with dimethylsulfoxide and sorbitol in relation to cryopreservation. Plant Physiol 75:720–725

Chen THH, Kartha KK, Leung NL, Kurz WGW, Chatson KB, Constabel F (1984b) Cryopreservation of alkaloid-producing cell cultures of periwinkle (*Catharanthus roseus*). Plant Physiol 75:726–731

Cyr DR (2000) Cryopreservation: roles in clonal propagation and germplasm conservation of conifers. In: Engelmann F, Takagi H (eds) Cryopreservation of tropical plant germplasm. Current research progress and application. Japan International Research Center for Agricultural Sciences/International Plant Genetic Resources Institute, Rome

Engelmann F (ed) (1999) Management of field and in vitro germplasm collections. Proceedings of a consultation meeting, 15–20 Jan 1996, CIAT, Cali, Columbia. International Plant Genetic Resources Institute, Rome

Engelmann F (2000) Importance of cryopreservation for the conservation of plant genetic resources. In: Engelmann F, Takagi H (eds) Cryopreservation of tropical germplasm. Current research progress and application. Japan International Research Center for Agricultural Sciences/International Plant Genetic Resources Institute, Rome, Italy

Forsline PL, Towill LE, Waddell J, Stushnoff C, Lamboy W, McFerson JR (1998) Recovery and longevity of cryopreserved dormant apple buds. J Am Soc Hortic Sci 123:365–370

Gao DY, Lin S, Watson PF, Critser JK (1995) Fracture phenomena in an isotonic salt solution during freezing and their elimination using glycerol. Cryo Lett 32:270–284

Golmirzaie AM, Panta A (2000) Advances in potato cryopreservation at CIP. In: Engelmann F, Takagi H (eds) Cryopreservation of tropical plant germplasm. Current research progress and application. Japan International Research Center for Agricultural Sciences/International Plant Genetic Resources Institute, Rome

Harding K (1991) Molecular stability of the ribosomal RNA genes in *Solanum tuberosum* plants recovered from slow growth and cryopreservation. Euphytica 55:141–146

Harding K (1994) The methylation status of DNA derived from potato plants recovered from slow growth. Plant Cell Tissue Organ Cult 37:31–38

Harding K, Benson E (1993) Biochemical and molecular methods for assessing damage, recovery and stability in cryopreserved plant germplasm. In: Grout BWW (ed) Genetic preservation in vitro. Springer, Berlin Heidelberg New York

Harding K, Benson EE (1994) A study of growth, flowering, and tuberisation in plants derived from cryopreserved potato shoot-tips: implications for in vitro germplasm collections. Cryo Lett 15:59–66

Haskens RH, Kartha KK (1980) Freeze preservation of pea meristems: cell survival. Can J Bot 58:833–840

Inagaki M (2000) Use of stored pollen for wide crosses in wheat haploid production. In: Engelmann F, Takagi H (eds) Cryopreservation of tropical plant germplasm. Current research progress and application. Japan International Research Center for Agricultural Sciences/International Plant Genetic Resources Institute, Rome

IPGRI/CIAT (1994) Establishment and operation of a pilot in vitro active genebank. Report of a CIAT-IBPGR Collaborative Project using cassava (*Manihot esculenta* Crantz) as a model. International Plant Genetic Resources Institute and International Center for Tropical Agriculture, Rome

Kartha KK (1985) Meristem culture and germplasm preservation. In: Kartha KK (ed) Cryopreservation of plant cells and organs. CRC Press, Boca Raton

Kartha KK, Leung NL, Pahl K (1980) Cryopreservation of strawberry meristems and mass propagation of plantlets. J Am Soc Hortic Sci 105:481–484

Kumar MB, Barker RE, Reed BM (1999) Morphological and molecular analysis of genetic stability in micropropagated *Fragaria ◊ ananassa* cv. Pocahontas. In Vitro Cell Dev Biol Plant 35:254–258

Larkin PJ, Scowcroft WR (1981) Somaclonal variation – a novel source of variability from cell cultures for plant improvement. Theor Appl Genet 60:197–214

Najimi S, Rubinsky B (1997) Non-invasive detection of thermal stress fractures in frozen biological materials. Cryo Lett 18:209–216

Niino T, Seguel I, Murayama T (2000) Cryopreservation of vegetatively propagated species (mainly mulberry). In: Engelmann F, Takagi H (eds) Cryopreservation of tropical plant germplasm. Current research progress and application. Japan International Research Center for Agricultural Sciences/International Plant Genetic Resources Institute, Rome

Razdan MK, Cocking EC (1997) Biotechnology in conservation of genetic resources. In: Razdan MK, Cocking EC (eds) Conservation of plant genetic resources in vitro. Science Publishers, Enfield

Reed BM, Chang Y (1997) Medium- and long-term storage of in vitro cultures of temperate fruit and nut crops. In: Razdan MK, Cocking EC (eds) Conservation of plant genetic resources in vitro, vol 1. Science Publishers, Enfield

Reed BM, Hummer K (1995) Conservation of germplasm of strawberry (*Fragaria* species). In: Bajaj YPS (ed) Biotechnology in agriculture and forestry, cryopreservation of plant germplasm I, vol 32. Springer, Berlin Heidelberg New York

Reed BM, DeNoma J, Luo J, Chang Y, Towill L (1998a) Cryopreservation and long-term storage of pear germplasm. In Vitro Cell Dev Biol Plant 34:256–260

Reed BM, Paynter CL, DeNoma J, Chang Y (1998b) Techniques for medium-and long-term storage of (*Pyrus* L.) genetic resources. Plant Gen Res Newslett 115:1–4

Reed BM, Brennan RM, Benson EE (2000a) Cryopreservation: an in vitro method for conserving *Ribes* germplasm in international gene banks. In: Engelmann F, Takagi H (eds) Cryopreservation of tropical germplasm. Current research progress and application. Japan International Research Center for Agricultural Sciences/International Plant Genetic Resources Institute, Rome, Italy

Reed BM, DeNoma J, Chang Y (2000b) Application of cryopreservation protocols at a clonal genebank. In: Engelmann F, Takagi H (eds) Cryopreservation of tropical germplasm. Current research progress and application. Japan International Research Center for Agricultural Sciences/International Plant Genetic Resources Institute, Rome, Italy

Schafer-Menuhr A (1996) Refinement of cryopreservation techniques for potato. Final report for the period Sept 1991–1993 Aug 1996. International Plant Genetic Resources Institute, Rome

Sun WQ (1998) Stability of frozen and dehydrated cells and membranes in the amorphous carbohydrate matrices: the Williams-Landel-Ferry kinetics. Cryo Lett 19:105–114

Towill LE (1981) *Solanum etuberosum*: a model for studying the cryobiology of shoot-tips in the tuber-bearing *Solanum* species. Plant Sci Lett 20:315–324

Towill LE (1988) Genetic considerations for germplasm preservation of clonal materials. HortScience 23:91–95

Towill LE (1990) Cryopreservation of isolated mint shoot tips by vitrification. Plant Cell Rep 9:178–180 (Vitrification of shoot tips)

Towill LE, Walters C (2000) Cryopreservation of pollen. In: Engelmann F, Takagi H (eds) Cryopreservation of tropical plant germplasm. Current research progress and application. Japan International Research Center for Agricultural Sciences/International Plant Genetic Resources Institute, Rome

Yoshimatsu K, Touno K, Shimomura K (2000) Cryopreservation of medicinal plant resources: retention of biosynthetic capabilities in transformed cultures. In: Engelmann F, Takagi H (eds) Cryopreservation of tropical plant germplasm. Current research progress and application. Japan International Research Center for Agricultural Sciences/International Plant Genetic Resources Institute, Rome

Section II
Herbaceous Species

II.1 Cryopreservation of *Allium sativum* L. (Garlic)

E.R.J. KELLER

1 Introduction

The genus *Allium* comprises about 700 species. Several of them are important vegetables, spices and medicinal plants. The most important crop species are onion and shallot, *Allium cepa* L., garlic, *Allium sativum* L., leek, *Allium ampeloprasum* L. s.l., bunching onion, *A. fistulosum* L., chives, *A. schoenoprasum* L., Chinese chives, *A. tuberosum* Rottl. ex Spr., and rakkyo, *A. chinense* G. Don. Many other species are collected from the wild as spices or vegetables, or they are planted as ornamentals. Vegetative propagation is predominant in some of these species either because no seeds are set (garlic, great-headed garlic, rakkyo) or because of tradition, as in shallots. Seed-sterile hybrids are used as vegetables (top onions, gray shallots) or in the breeding of new ornamentals (as in subgenus *Melanocrommyum*). In all such cases, vegetative maintenance of genotypes is necessary. Due to continual vegetative propagation, virus occurrence is common in garlic. Meristem culture allows production of virus-free lines and, hence, is able to circumvent this problem as long as the material is kept permanently in vitro. In vitro storage is, therefore, preferable for this vegetatively propagated material.

Although today's garlic is a clonal crop, it underwent an intensive phase of diversification in the past before it lost its fertility and after that, perhaps, by mutation. Thus, there are some distinctly diverse groups within this species. Discriminating morphological characters are such as the structure of the bulb (arrangement, number and size of the cloves) and the ability to form inflorescences (the height of the inflorescence stalk, position, number and size of inflorescence bulbils and number of flower buds). The most original group has often been considered a separate species, i.e., *A. longicuspis* Rgl., but there are no morphological characters that separate these forms from true garlic. The present results of DNA analyses support these observations. Therefore, the previous opinion can no longer be followed (Maass and Klaas 1995; Maass 1996; Al-Zahim et al. 1997). This recently so-called *longicuspis*-group possesses high stalks with well-developed inflorescences, in which a high number of small bulbils are formed together with many flower buds. This group is located in

Genebank Department, Institute of Plant Genetics and Crop Plant Research, IPK, Corrensstr. 3, 06466 Gatersleben, Germany

Biotechnology in Agriculture and Forestry, Vol. 50
L.E. Towill and Y.P.S. Bajaj (Eds.) Cryopreservation of Plant Germplasm II
© Springer-Verlag Berlin Heidelberg 2002

Table 1. Some morphological characters of the garlic groups observed in the Gatersleben collection, being important for the success of *in vitro* culture and cryopreservation (figures represent average values for the resp. accessions)

Subgroup	Number of cloves per bulb	Inflorescence formation character	Number of bulbils per inflorescence	Bulbil size (weight in mg)
longicuspis	3–7	complete	20–150	15–80
pekinense	3–7	complete	50–100	500–800
ophioscorodon	5–7	complete	20–80	150–800
sativum	4–25	no, incomplete, or complete*	20–150; 0–50*	15–80; 300–1500*

* Bulbil numbers and sizes rather heterogeneous depending on the completeness of the inflorescence formation, incomplete inflorescences very often consist of only one or few very large bulbils.

Central Asia, the presumable site of origin of garlic. Some forms of the "*longicuspis*-group" preserved a certain degree of fertility that is now under investigation in order to make garlic a seed-propagated crop. So far, progress is rather limited in this direction, and the overwhelming majority of types remain vegetative. This is true for all other groups such as some small groups of East Asia (the "*pekinense*-group") and Southeast Asia as well as the European groups from which the world's most widespread garlic has developed which lost more or less the ability to form flower stalks or produces reduced stalks with some large or very large bulbils (the "*sativum*-group"). In central Europe, a small and relatively uniform group exists with high stalks, which are curled when unripe, possessing large bulbils (the "*ophioscorodon*-group"). As discussed below, the bulbil size is a factor that influences the success in cryopreservation. Therefore, the mean bulbil numbers and sizes for the different groups are given in Table 1 for the Gatersleben garlic collection together with the numbers of cloves per mother bulb, which is determined by the number of developing axillary meristems, another influencing factor when using basal plates as the explant source for cryopreservation.

2 Storage of Germplasm

Allium has been the target of in vitro approaches since the 1970s. First reports came from experiments on callus (Havranek and Novak 1973) and meristem culture (Havranek 1972). A modified B5 medium (Gamborg et al. 1968), called BDS, possessing increased levels of phosphate and less nitrogen (Dunstan and Short, 1977) was useful for many purposes. In later years, many reports cite micropropagation and other subjects of *Allium* in vitro culture (Bhojwani 1980; Novak et al. 1986; Moriconi et al. 1990; Keller 1992; Mohamed-Yasseen et al. 1994; Nagakubo et al. 1997). Studies on in vitro storage are, however, relatively rare. Slow growth conditions have been used for plant cultures (El-Gizawy and Ford-Lloyd 1987, Viterbo et al. 1994). Bulblets are formed under in vitro conditions. Their development is triggered by inductive light conditions, in at least some cases (Debergh and Standaert-de Metsenaere 1976; Keller 1991; Kahane et al. 1992; Mohamed-Yasseen et al. 1995). For *Allium cepa*, a high sucrose content in the medium favors storage of in vitro bulblets to temperatures as low as −1 °C (Keller 1993; Kästner et al. 2001). Thus, it can be assumed that in vitro bulblets could also be used for medium-term storage of garlic, similar to that described for microtubers in potato (Thieme 1988-89). It should be emphasized that an obstacle to permanent in vitro culture and other in vitro methods is the relatively high degree of microbial infection in garlic (Fellner et al. 1996), which can lead to outbreaks of visible contamination, even after several years of culture, and may be exacerbated by severe stress such as would occur during cryopreservation.

3 Cryopreservation

Successful reports for cryopreservation of garlic are rare, despite the increasing demand for garlic and other *Allium* species, not only for food but also for medicinal purposes. All reports so far rely on vitrification as a method of cryopreservation. The first article was published by Niwata (1995). Explants from basal plates of post-dormant cloves were treated by the vitrification method using the cryoprotectant solution PVS 2 (0.4 M sucrose + 30% glycerol + 15% ethylene glycol + 15% DMSO, Sakai et al. 1990; Niwata 2000). The regrowth percentage approached 100% in this study. Hannan and Garoutte (1996, 1998) repeated this approach, mainly experimenting with the pre-conditioning phase, and reached up to 71.4% survival. These authors emphasized the problems caused by the latent infection of the source explants.

In contrast to the experiments published by Niwata (1995) and Hannan and Garoutte (1996, 1998), much better results were obtained in our vitrification experiments using PVS 3 (50% sucrose + 50% glycerol in liquid standard culture medium; Nishizawa et al. 1993) instead of PVS 2. This weaker cryoprotectant does not contain DMSO. With bulbil explants survival after PVS 2 was much lower and reached 80% in only one single genotype, whereas survivals were 85%–95% using PVS 3. Even more extreme was the difference of regrowth, which was, for three compared genotypes 0, 10, and 0% with PVS 2 and for the same material 17, 27, and 60% with PVS 3. The penetration behavior of PVS 3 may be different to that of PVS 2 because for the optimum effect longer pretreatments of about 120-240 min were necessary in comparison to those with PVS 2 (15-20 min; Makowska et al. 1999).

Makowska et al. (1999) summarized the results of comparative series of PVS 3 pretreatment times of 0; 60; 120; 180; 240; and 300 min using cloves and bulbils of four accessions with large or medium bulbils. The best mean regrowth percentages have been found for cloves after 240 min (80%) and for bulbils after 120 min (37%).

Subsequently, in this group, bulbils were predominantly used for cryopreservation. These are formed in the inflorescences and, because these organs are considerably less contaminated than the cloves, this avoids a high percentage loss of explants after cryopreservation due to contamination if the bulbs had come from the soil. Survival and regrowth of bulbil explants depend, however, on the size of the bulbils. This parameter is genotype dependent as described above. Cryopreservation of explants taken from small bulbils is less successful than that of large-bulbil material. In accessions with large bulbils regrowth was 13–100%, with medium bulbils 15–28%, respectively, and in accessions with small bulbils no regrowth was obtained (Makowska et al. 1999; Keller and Senula 2000). Therefore, the best accession (All 290) possessing large bulbils and regrowth of 100% was chosen as standard material for further experiments.

3.1 Effects of Different Parameters on Cryopreservation Success with Standard Material

3.1.1 Material

Bulbils of garlic All 290, belonging to the *ophioscorodon* group, are of the large type (bulbils of 320–570 mg; Fig. 1A) and were harvested at the end of July and stored at 10 °C. Bulbils directly after harvest exhibit dormancy. Therefore, the experiments were performed between December (Fig. 1B) of the harvest year and May (Fig. 1C) of the next one. Because the quality of the bulbils declines towards the end of the storage period, survival and regrowth data are not directly comparable between experiments performed at different times of the year. Two different explant sizes were used (Fig. 1D). The difference consisted in the presence of different numbers of leaf sheaths in the respective explants (Fig. 1E). For comparison, explants from in vitro cultivated shoots (Fig. 1F) of a defined size have been used (Fig. 1G,H). The in vitro donor material is always regenerated via cyclic micropropagation (Keller et al. 1997; Fig. 1I,K).

3.1.2 Methods

The vitrification method was used (Makowska et al. 1999). Bulbils were sterilized by dipping them in 70% ethanol and subsequently placing them in a 3% sodium hypochlorite solution for 15 min. The isolated shoot tips were cultivated overnight on MS medium (Murashige and Skoog 1962) + 0.1 mg/l IAA + 0.1 mg/l kinetin with 0.3 M sucrose at 25 °C in dark. The explants were immersed in a loading solution containing 2 M glycerol + 0.4 M sucrose for 20 min. In the standard technique, apices were then treated with the vitrification solution PVS 3 for 120 min. After pretreatment, the apices were suspended in 1 ml of vitrification solution within a cryotube and were plunged into liquid nitrogen (LN). The control samples were rapidly rewarmed by plunging the samples in a 40 °C water-bath. Explants were then rinsed with liquid medium containing 1.2 M sucrose for 10 min, and placed on solid

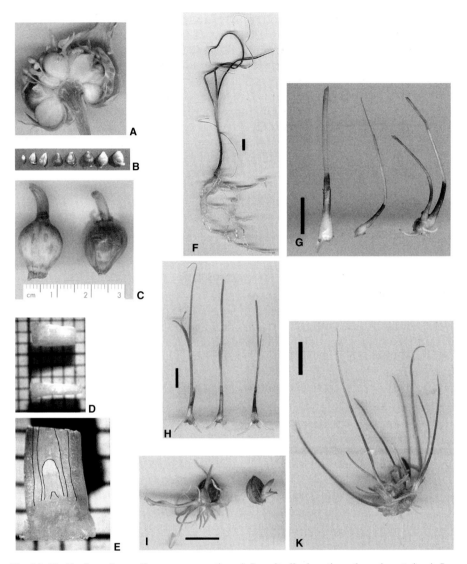

Fig. 1A–H. Explants for garlic cryopreservation. **A** Longitudinal section of an almost ripe inflorescence with bulbils and flower buds; **B** bulbils in the beginning phase of usability (December); **C** bulbils at the end of their usability (May); **D** "large" (*above*) and "small" explants on millimeter paper; **E** scheme of the leaf sheaths in a longitudinally cut "large" explant; **F** in vitro plant in the slow growth phase; **G** early and **H** later phase of in vitro source explants; **I** early and **K** late phase for explant source multiplication

medium containing 0.3 M sucrose in the dark for 1 d. Apices were then transferred to the standard medium MS +80 mg/l adenine sulfate +0.1 mg/l IAA +0.1 mg/l kinetin with 0.1 M sucrose and cultivated at 25 °C in a culture room equipped with fluorescent lamps under a photoperiod of 16 h light/8 h dark with light intensity of 60–80 µmol cm^{-2} s^{-1}, measured with an LI-250 data-logger coupled using a quantum sensor LI-190SZ (LI-COR, Lincoln, Nebraska).

Several parameters were examined. Pre-culture was performed on 0.1 or 0.3 M sucrose. New cryoprotectants were used: PVS 4 (0.6 M sucrose +35% glycerol +20% ethylene glycol; Sakai 2000), a cryoprotectant mixture described by Steponkus (15% sorbitol, 40% ethylene glycol, 60 g/l bovine serum albumin; Langis et al. 1989). Also, the sucrose concentrations before cooling (0.1; 0.3; 0.45 M) and after warming (0.1; 0.3; 0.8; 1.2 M), respectively, were changed or other hormone combinations were applied at that time.

The droplet method described for potato (Schäfer-Menuhr et al. 1994) was modified with respect to the basal medium and compared with the vitrification protocol. The explants were isolated as described above. A preculture phase followed on medium MS + 80 mg/l adenine sulfate + 0.1 mg/l IAA + 0.1 mg/l kinetin on 0.3, 0.6, or 0.9 M sucrose overnight. The explants were then transferred into 10% DMSO solution. After incubation for 2 h at room temperature, 2.5-µl droplets of 10% DMSO were pipetted onto heat-sterilized pieces (0.7×2 cm) of 0.03-mm-thick aluminum foils at six drops per foil. One explant was transferred into each drop and the foils were dropped directly into cryovials filled with LN.

In each experiment, controls of at least five explants per treatment variant were warmed and cultured on standard medium (Fig. 2A,B). The plants derived from the experiments were then compared 4 months later (Fig. 2C), transplanted into pots (Fig. 2D) and characterized for growth. Plants from test experiments were transferred to the field. No changes were observed in comparing treated to non-frozen material (Fig. 2E,F).

3.1.3 Results

The effect of the explant size was compared in both the "traditional" vitrification as well as the droplet methods (Table 2). Percentages showing regrowth in controls with cryoprotectant but without cooling did not differ, whereas, when using small explants, percentages showing regrowth were significantly better after cooling both in the vitrification and in the droplet method.

In experiments with the standard clone, the cryoprotectant solution PVS 2 was previously found to be inferior to PVS 3 (Makowska et al. 1999). The subsequently used cryoprotectant solutions PVS 4 and the solution according to Steponkus did not give better results either (Table 3). For technical reasons, it was not possible to perform these experimental series at the same time. Therefore, statistical treatment was impossible. However, the observed tendencies favor clearly the further use of PVS 3.

In a further experiment, the relationship between the effect of sucrose concentration in the loading phase after explant preparation was tested against the sucrose concentration present in the 1-d recovery phase after cooling, warming, and rinsing in 1.2 M sucrose (Table 4). In the controls, when sucrose was too high after cryoprotecant compared with the concentration before, growth was reduced. In the cooled samples, the best results were also

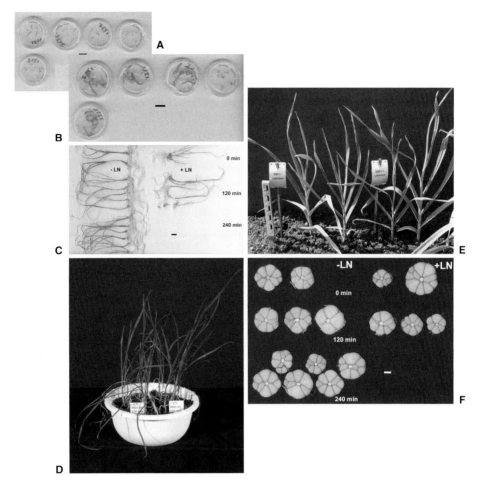

Fig. 2A–F. Plants from garlic cryopreservation. **A** Explants 5 days, **B** explants 8 days after warming (Petri dishes in the lower lines are the controls without LN); **C** plantlets 4 months after warming (three variants with different cryoprotectant pretreatment durations); **D** plantlets in pots 6 months after warming; **E** plantlets in the field 18 months after warming (*left* control plants, *right* plants from cryopreservation); **F** transversal cuts from garlic bulbils derived from experiment in picture c in the same arrangement harvested and scanned 22 months after warming

found when the concentrations were relatively balanced at a high level before and after the treatment.

Similar relationships were examined in another experiment (Table 5) where the sucrose concentration after warming was varied in two steps after the rinsing treatment by 1.2 M sucrose. The recovery culture of 1 day was followed by an intermediate culture phase of 1 week before the explants were finally transferred to the standard medium. In this experiment, reduction of regrowth was observed in both cases (1) when the difference between the sucrose concentration before cryoprotectant/cooling treatment and after

Table 2. Regrowth (%) of several experiments with "large" and "small" explants (for definition see Fig. 1d), cryoprotectant, PVS 3 pretreatment 120 min.

			Large						Small			
		LN	I	II	III	IV	V	Total	I	II	Total
V	EN	–	20	20	20			60	20	20	40
		+	20	20	20			60	20	20	40
	R(%)	–	90.0	100.0	100.0			96.7	95.0	100.0	97.5
		+	5.0	15.0	10.0			<u>10.0</u>	80.0	85.0	<u>67.5</u>
D	EN	–	6	15	5	10	5	41	10	9	19
		+	20	60	20	39	20	159	39	39	78
	R(%)	–	100.0	86.7	100.0	100.0	100.0	95.1	90.0	100.0	94.7
		+	0.0	0.0	0.0	35.9	0.0	<u>8.8</u>	76.9	79.5	<u>78.2</u>

Abbreviations: V vitrification method, D, droplet method, EN explant numbers per variant, R (%) percentage regrowth.
Comparison of the totals between the corresponding variants using "large" and "small" explants, respectively. Underlined figure pairs possess statistically different values, tested by the 2 × 2 contingency tables using χ^2 distribution ($p < 1\%$).

Table 3. Vitrification: Comparison of several experimental series with different cryoprotectant solutions using the standard clone All 290

LN	PVS 3			PVS4			Steponkus		
	T (min)	No.	R (%)	T (min)	No.	R (%)	T (min)	No.	R (%)
–	0	60	96.7	0	40	97.4	0	80	100
	120	60	93.3	120	40	100	10	40	97.5
	240	20	85.0	240	40	85.5	20	40	100
							30	40	100
							60	40	97.5
+	0	60	8.3	0	40	32.5	0	80	13.8
	120	60	45.0	120	40	30.0	10	40	12.5
	240	20	55.0	240	40	32.5	20	40	0
							30	40	2.5
							60	40	0

Abbreviations: T pretreatment time, No. explant numbers, R (%) percentage regrowth. No statistical treatment possible because no time congruence.

Table 4. Vitrification: Interaction between sucrose concentrations before and after cooling/warming. Regrowth (%) after cryoprotectant PVS3 pretreatment of 120 min (40 explants per variant)

	Sucrose concentration after warming					
	–LN			+LN		
	0.3 M	0.8 M	1.2 M	0.3 M	0.8 M	1.2 M
Preculture sucrose	regrowth (%)					
0.1 M	94.1	90.0	***30.3***	0.0	0.0	0.0
0.3 M	95.0	*73.1*	*60.0*	7.5	2.3	3.1
0.44 M	95.0	90.0	60.0	5.1	**17.5**	0.0

Statistical treatment (2 × 2 contingency tables using χ^2 distribution): values in italics – significantly lower than the others in the same line (within –LN), bold italics – significantly lower in line and row (–LN), bold – significantly higher in line and row (+LN) ($p < 5\%$ at least).

Table 5. Vitrification: Regrowth (%) after 0.3 M sucrose after explant preparation in dependence on sucrose concentration after warming (40 explants/variant)

first subculture	0.1 M	0.3 M		0.8 M		1.2 M	
next subculture	0.1 M	0.1 M	0.24 M	0.1 M	0.24 M	0.1 M	0.24 M
		Regrowth (%)					
−LN	76.7	77.5	81.1	57.9	85.0	*50.0*	70.0
+LN	40.0	40.0	35.0	20.0	30.0	*13.1*	27.5

Statistical treatment (2×2 contingency tables using χ^2 distribution): values in italics are significantly lower than the values of the best variants in the same line ($p < 5\%$).

Table 6. Regrowth percentages of bulbil and *in vitro* explants in comparable vitrification/droplet experiments

Method		Vitrification 0.3 M		Droplets 0.6 M		Droplets 0.8 M	
Sucrose in the recovery medium				Regrowth			
		No.	R (%)	No.	R (%)	No.	R (%)
BL	−LN	54	98.1	5	100.0	5	100.0
	+LN	99	72.7	20	79.0	20	80.0
IV	−LN	15	20.0	5	0.0	5	40.0
	+LN	60	13.3	20	15.0	20	20.0

Abbreviations: BL explants from bulbils, IV explants from *in vitro* plantlets, No. explant numbers, R (%), percentage regrowth. Statistical treatment (2×2 contingency tables using χ^2 distribution): underlined figure pairs mark significant differences between ($p < 5\%$ at least) corresponding variants of BL and IV, respectively.

warming was too high and (2) when the difference between the recovery medium and the intermediate phase was too extreme.

Explants from in vitro grown plantlets were compared with explants from bulbils (Table 6). In most controls and all cooled variants for both the vitrification and droplet methods the regrowth percentages of explants from in vitro cultures were significantly lower than those from bulbils.

Finally, the results of the experiments described in Tables 2 and 6 allow a comparison of the "traditional" vitrification with the droplet method. In no case could any statistically significant difference be found. This means that both methods are equally usable both for explants from bulbils and in vitro grown plantlets.

4 Summary and Conclusions

It could be shown, at least for some explant types (cloves) and also for bulbil explants in several genotypes, that very high levels of survival and regrowth were achieved. However, currently there is limited use of cryopreservation for garlic. From the literature survey and our experiments this may be due to the frequent occurrence of contamination. The latter may cause serious problems for the establishment of in vitro cultures, and is a general obstacle for bulbous plants where the meristems or shoot tips are close to the soil. Therefore, we intended to use cleaner plant parts such as bulbils from the aerial portion of the garlic scape. Cryopreservation of explants from bulbils is possible when these are large enough, whereas, presumably, preparation barriers need to be overcome with small bulbils. This technique can be especially useful in cases when plant material is limited and the cloves of the mother plant are, therefore, not available. In the long-term, however, cryopreservation of shoot tips from in vitro cultures offers the chance to use material previously freed of viruses by meristem culture (Senula et al. 2000), such that a virus-free state can be retained. The poor regrowth of in vitro derived explants is still under investigation.

Acknowledgement. The author thanks Mrs. Doris Büchner for the management of the donor plant material and thorough performance of the cryopreservation experiments.

References

Al-Zahim M, Newbury HJ, Ford-Lloyd BV (1997) Classification of genetic variation in garlic (*Allium sativum* L.) revealed by RAPD. HortSci 32:1102–1104

Bhojwani SS (1980) In vitro propagation of garlic by shoot proliferation. Sci Hortic 13:47–52

Dunstan DI, Short KC (1977) Improved growth of tissue cultures of the onion, *Allium cepa*. Physiol Plant 41:70–72

El-Gizawy AM, Ford-Lloyd B (1987) An in vitro method for the conservation and storage of garlic (*Allium sativum*) germplasm. Plant Cell Tissue Organ Cult 9:147–150

Fellner M, Kneifel W, Gregorits D, Leonhardt W (1996) Identification and antibiotic sensitivity of microbial contaminants from callus cultures of garlic *Allium sativum* L. and *Allium longicuspis* Regel. Plant Sci 113:193–201

Gamborg OI, Miller RA, Ojima K (1968) Nutrient requirements of suspension cultures of soybean root cells. Exp Cell Res 50:151–158

Hannan R, Garoutte D (1996) Cryopreservation of garlic germplasm as meristems in LN for long term storage. Proc Natl Onion Res Conf USA, pp 101–103

Hannan R, Garoutte D (1998) Long-term storage of garlic (*Allium sativum* L.) using cryopreservation and regeneration in tissue culture. Proc Natl Onion Res Conf. USA, pp 292–298

Havranek P (1972) Virus-free garlic clones obtained from meristem cultures. Ochr Rostl 8:291–298

Havranek P, Novak FJ (1973) The bud formation in the callus cultures of *Allium sativum* L. Z Pflanzenphysiol 68:308–318

Kahane R, Teyssendier de la Serve B, Rancillac M (1992) Bulbing in long-day onion (*Allium cepa* L.) cultured in vitro: comparison between sugar feeding and light induction. Ann Bot 69:551–555

Kästner U, Klahr A, Keller ERJ, Kahane R (2001) Formation of onion bulblets in vitro and viability in medium-term storage. Plant Cell Rep 20:137–142

Keller ERJ (1993) Sucrose, cytokinin and ethylene influence formation of in vitro bulblets in onion and leek. Genet Resour Crop Evol 40:113–120

Keller ERJ, Lesemann D-E, Lux H, Maass HI, Schubert I (1997) Application of in vitro culture to onion and garlic for the management and use of genetic resources at Gatersleben. Acta Hortic 433:141–150

Keller ERJ, Senula A (2000) Maintenance of the European garlic core collection (Northern part): slow growth culture, phytosanitary aspects and progress in cryopreservation. Proc 3rd Int Symp Edible Alliaceae, Athens, Georgia, USA (in press)

Langis R, Schnabel-Preikstas B, Earle ED, Steponkus PL (1989) Cryopreservation *of Brassica campestris* L. cell suspensions by vitrification. Cryo Lett 10:421–428

Maass H (1996) Morphologische Beobachtungen an Knoblauch. Palmengarten 60:65–69

Maass HI, Klaas M (1995) Infraspecific differentiation of garlic (*Allium sativum* L.) by isozyme and RAPD markers. Theor Appl Genet 91:89–97

Makowska Z, Keller J, Engelmann F (1999) Cryopreservation of apices isolated from garlic (*Allium sativum* L.) bulbils and cloves. Cryo Lett 20:175–182

Mohamed-Yasseen Y, Splittstoesser WE, Litz RE (1994) In vitro shoot proliferation and production of sets from garlic and shallot. Plant Cell Tissue Organ Cult 36:243–247

Mohamed-Yasseen Y, Barringer SA, Splittstoesser WE (1995) In vitro bulb production from *Allium* spp. In Vitro Cell Dev Biol Plant 31:51-52

Moriconi DN, Conci VC, Nome SF (1990) Rapid multiplication of garlic (*Allium sativum* L.) in vitro. Phyton 51:145–151

Murashige T, Skoog F (1962) A revised medium for rapid growth and bioassays with tobacco tissue cultures. Physiol Plant 15:473–487

Nagakubo T, Takaichi M, Oeda K (1997) Micropropagation of *Allium sativum* L. (garlic). In: Bajaj YPS (ed) Biotechnology in agriculture and forestry, vol 39. High-tech and micropropagation V. Springer, Berlin Heidelberg New York, pp 3–19

Nishizawa S, Sakai A, Amano Y, Matsuzawa T (1993) Cryopreservation of asparagus (*Asparagus officinalis* L.) embryogenic suspension cells and subsequent plant regeneration by vitrification. Plant Sci 91:67–73

Niwata E (1995) Cryopreservation of apical meristems of garlic (*Allium sativum* L.) and high subsequent plant regeneration. Cryo Lett 16:102–107

Niwata E (2000) In: Engelmann F, Takagi H (eds.) Cryopreservation of tropical plant germplasm – current research progress and applications. Proc JIRCAS/IPGRI Joint Int Worksh, 20-23 Oct 1998, Tsukuba, Japan, pp 429–430

Novak FJ, Havel L, Dolezel J (1986) Onion, garlic and leek (*Allium* species). In: Bajaj YPS (ed) Biotechnology in agriculture and forestry, vol 2. Crops I. Springer, Berlin Heidelberg New York, pp 387–404

Sakai A (2000) Development of cryopreservation techniques. In: Engelmann F, Takagi H (eds) Cryopreservation of tropical plant germplasm – current research progress and applications. Proc JIRCAS/IPGRI Joint Int Worksh, 20–23 Oct 1998, Tsukuba, Japan, pp 1–7

Sakai A, Kobayashi S, Oiyama I (1990) Cryopreservation of nucellar cells of navel orange (*Citrus sinensis* Osb. var. *brasiliensis* Tanaka) by vitrification. Plant Cell Rep 9:30–33

Schäfer-Menuhr A, Schumacher H-M, Mix-Wagner G (1994) Langzeitlagerung alter Kartoffelsorten durch Kryokonservierung der Meristeme in flüssigem Stickstoff. Landbauforsch Völkenrode 44:301–313

Senula A, Keller ERJ, Lesemann DE (2000) Elimination of viruses through meristem culture and thermotherapy for the establishment of an in vitro collection of garlic (*Allium sativum*). In: Cassels AC, Doyle BM, Curry RF (eds) Proc Int Symp on Methods and Markers for Quality Assurance in Micropropagation. Acta Hortic 530:121–128

Thieme R (1988-89) An in vitro potato cultivar collection: microtuberization and storage of microtubers. FAO/IBPGR Plant Genet Resour Newsl 88/89:17–19

Viterbo A, Altman A, Rabinowitch HD (1994) In vitro propagation and germplasm cold-storage of fertile and male-sterile *Allium trifoliatum* subsp. *hirsutum*. Genet Resour Crop Evol 41: 87–98

II.2 Cryopreservation of *Apium graveolens* L. (Celery) Seeds

M. Elena González-Benito and José María Iriondo

1 Introduction

Celery (*Apium graveolens* L., 2n = 22) is a biennial plant that belongs to the *Apiaceae*. Its center of origin is the Mediterranean basin, with two other secondary centers in the Caucasus and the Himalayan regions. Celery's natural distribution ranges from Europe to India, North and South Africa, and America (Willis 1973; Simon et al. 1984).

1.1 Distribution and Importance

Celery is used both as a cooked and fresh vegetable, and as a condiment. Its cultivation as a vegetable started in Italy in the 16th century, although it had been used earlier because of its medicinal properties (Maroto 1983). Two varieties are cultivated: *Apium graveolens* L. var. *dulce* Pers. (celery), more common and with large petioles that are edible; and *Apium graveolens* L. var. *rapaceum* (Mill.) DC. (celeriac or German celery), with small petioles and a subterranean edible rootstock (partly hypocotyl).

The brown, characteristically aromatic, pungent seed is used in salads, soups, stews, vegetable dishes, meat dishes and celery salt. An essential oil is obtained from seeds or seed chaff by a process of crushing and steam distillation. The essential oil of celery seed includes d-limonene, selinene, sesquiterpene alcohols, sedanolide, and sedanonic anhydride, and it is used as a flavouring or fragrance in liqueurs, perfumes and cosmetics. As a medicinal plant, celery has been used as an aphrodisiac, anthelmintic, antispasmodic, carminative, diuretic, emmenanogue, laxative, sedative, stimulant, and tonic (Simon et al. 1984).

1.2 Germplasm Storage

There are 25 germplasm banks in Europe where old and new cultivars of *Apium graveolens* are conserved (Table 1; Frison and Serwinski 1995). There

Departamento de Biología Vegetal, Escuela Universitaria de Ingeniería Técnica Agrícola, Universidad Politécnica de Madrid, Ciudad Universitaria, 28040 Madrid, Spain

Biotechnology in Agriculture and Forestry, Vol. 50
L.E. Towill and Y.P.S. Bajaj (Eds.) Cryopreservation of Plant Germplasm II
© Springer-Verlag Berlin Heidelberg 2002

Table 1. European countries where *Apium* germplasm is preserved and number of accessions. (Frison and Serwinski 1995)

Country	Number of accessions
Belgium	25
Czech Republic	123
France	77
Germany	201
Greece	25
Hungary	71
Lithuania	10
Nordic Gene Bank	15
Poland	18
Portugal	ND
Russia	250
Slovakia	15
Spain	10
Switzerland	5
Turkey	6
United Kingdom	175

ND, not determined.

are also important collections in the USA (National Seed Storage Laboratory, Fort Collins) and India (National Bureau of Plant Genetic Resources, New Delhi) (Bettencourt and Konopla 1990). Celery germplasm is stored as seeds. These do not lose viability when desiccated (orthodox seeds) and are, in most cases, stored at low temperature (–10 to –20 °C) and low moisture content (generally 5–10%) (Bettencourt and Konopla 1990).

1.3 Need for Conservation and Cryopreservation

When possible, seed storage is the most effective and efficient method for ex situ conservation of plant genetic resources. Conventional seed banks maintain seed viability for decades (Ellis et al. 1993). However, there is evidence that cryopreservation in liquid nitrogen (LN) increases longevity (Stanwood and Sowa 1995; Lakhanpaul et al. 1996). For cultivated species, decreased deterioration in storage could be of importance for rare or scarce genotypes, as could be the case of old celery cultivars no longer in use.

2 Cryopreservation

2.1 General Account

Celery seeds are desiccation-tolerant and liquid nitrogen-tolerant according to the categories established by Stanwood (1985). Cooling and rewarming rates are probably not of great importance provided the seed moisture content

Table 2. Germination percentage of celery cv. Tall Utah 52-70R seeds (15 days after sowing), after different periods of storage in liquid nitrogen, at two moisture contents. Mean values ± standard error with 100 seeds per treatment

Seed moisture content (%, fresh weigh basis)	Control	Storage time in LN		
		1 day	7 days	30 days
9.9	20 ± 6	10 ± 4	21 ± 3	28 ± 2
5.2	29 ± 3	26 ± 3	14 ± 2	14 ± 3

is sufficiently low as to be well under the "high moisture freezing limit". This limit is the seed moisture content above which viability decreases when seeds are cooled to LN and rewarmed to ambient temperature.

Celery seeds survived LN storage for almost 2 and 3 years without losing viability when they were rapidly cooled and either rewarmed rapidly (90 s) or slowly (30 °C/min) (Styles et al. 1982; Stanwood 1985, respectively). In these studies the seed moisture contents were 7.6 and 10.2%, respectively.

Germination of one seed accession of cv. Tall Utah 52-70R was studied after exposure to LN at two moisture contents: 9.9%, and after desiccation in a chamber with silica gel for 45 days (5.2% moisture content, fresh weight basis) (Table 2; Iriondo et al. 1992). No significant differences in the germination percentages were detected at the 0.05 level between seed samples with different moisture contents, nor among seed samples with different LN exposure times (0, 1, 7 and 30 days).

Celery seeds show low and gradual germination (Thomas et al. 1979). From the agronomic point of view, it is important to retain high viability and uniform germination. Several techniques have been assayed to synchronise and reduce the time of germination. Seed priming (the incubation of seeds in high osmotic potential solutions prior to germination) has been successfully applied to celery, allowing a more uniform germination (Salter and Darby 1976). On the other hand, as in many other horticultural crops, celery seeds are often pelleted for mechanical sowing. These factors should also be considered in seed cryopreservation. In the following sections, a study carried out on seed cryopreservation of seven celery cultivars and the possible interaction with priming and pelleting will be detailed (González-Benito et al. 1995).

2.2 Methodology

Cultivars tested are shown in Table 3. Seed moisture content was determined after exposure of seeds to 105 °C for 24 h in an oven, and calculated on a fresh weight basis (Tables 3 and 4).

Seeds were placed in polypropylene cryovials, immersed in liquid nitrogen (cooling rate of approximately 200 °C/min) and maintained there for 1 or 30 days. After removal from the liquid nitrogen, samples were allowed to warm at ambient temperature (ca. 25 °C). The effect of cryopreservation in combination with priming was studied by adding 3 ml of $NaNO_3$ (15.7 g l^{-1}) or KNO_3 (19.3 g l^{-1}) solutions to Petri dishes containing 50 seeds on two sheets of filter

Table 3. Germination percentage (mean values ± SE) and T_{50} values (the number of days needed to reach 50% of the final germination percentage) for seven celery cultivars (30 days after sowing), after two exposure times to liquid nitrogen. (With permission of *Annals of Botany*)

Cultivar (moisture content, %)	Duration of storage in LN					
	Control*		1 day		30 days	
	Germination (%)	T_{50}	Germination (%)	T_{50}	Germination (%)	T_{50}
Florida 683 (8.4%)	74 ± 1	7.5	74 ± 0	8.5	74 ± 0	9.0
Utah (8.9%)	64 ± 3	9.0	64 ± 0	10.5	70 ± 3	9.5
Golden Spartan (7.2%)	93 ± 1	7.0	93 ± 1	7.0	92 ± 1	7.0
Isel (8.6%)	95 ± 1	7.0	96 ± 0	8.0	96 ± 1	7.0
Tall Utah 52-70R (7.9%)	100	5.0	94 ± 0	5.0	99 ± 0	5.0
Istar (5.5%)	93 ± 1	7.0	85 ± 6	7.0	88 ± 1	7.0
Golden Boy (7.2%)	78 ± 1	7.0	86 ± 4	7.0	77 ± 6	7.0

* Control = not exposed to LN.
T_{50} = number of days to reach 50% of the final germination percentage.

Table 4. Germination percentage of non-pelleted and pelleted seeds of three celery cultivars (30 days after sowing), after two exposure times to liquid nitrogen. Mean values ± standard error with 100 seeds per treatment. (With permission of *Annals of Botany*)

Cultivar (moisture content, %)	Duration of storage in LN					
	Control*		1 day		30 days	
	Germination (%)	T_{50}	Germination (%)	T_{50}	Germination (%)	T_{50}
Florida 683						
Non-pelleted (8.4%)	83 ± 3	6	83 ± 1	6	81 ± 1	7
Pelleted (4.4%)	78 ± 3	8	66 ± 3	9	71 ± 2	9
Golden Spartan						
Non-pelleted (7.2%)	88 ± 1	6	94 ± 1	7	97 ± 1	7
Pelleted (1.5%)	97 ± 1	10	98 ± 1	10	97 ± 1	10
Utah						
Non-pelleted (8.9%)	66 ± 1	8	59 ± 1	8	79 ± 4	9
Pelleted (5.7%)	43 ± 2	16	43 ± 3	16	51 ± 4	15

* Control = not exposed to LN.

paper (for both solutions $\Psi = -0.8\,\mathrm{Mpa}$). The seeds were kept at 25 °C in darkness for 7 days and, subsequently, desiccated for 3 days in a chamber with silica gel. Priming was carried out before (priming + LN) or after (LN + priming) cryopreservation (1 day in LN).

For germination tests, two replicates of 50 seeds each were placed in Petri dishes on two sheets of filter paper moistened with 3 ml distilled water, replacing it regularly. Incubation took place at 25/15 °C day/night, with 16/8 h day/night light regime, and an irradiance of $35\,\mu\mathrm{mol\,m^{-2}\,s^{-1}}$, provided by cool white fluorescent tubes. The number of seeds showing radicle emergence was counted every 2 days and they were removed from the Petri dishes. The number of days needed to reach 50% of the final germination percentage (T_{50}) was estimated by median values of the germination time (Georghiou et al. 1987). Analyses of variance of germination percentage after 30 days and of T_{50} were carried out, in the first case after arcsine transformation.

2.3 Results

No significant differences among the germination percentages for the three cryopreservation treatments (including control) were found in seven cultivars (Table 3). No coat damage was detected in observations with a scanning electron microscope (Fig. 1). Overall, T_{50} values significantly increased ($p < 0.05$) for seeds cryopreserved for 1 day with respect to control seeds. However, when results were analysed separately for each cultivar, the differences were not significant.

Both for pelleted and non-pelleted seeds, no significant differences among the three treatments (control, 1 and 30 days in LN) were found in Florida 683 and Golden Spartan (Table 4). However, in Utah, the best response was observed after 30 days in LN. In this cultivar the germination of pelleted seeds was lower than that of non-pelleted ones.

The germination percentages of the three priming treatments (priming alone, and before and after cryopreservation) did not differ significantly using the two priming solutions (Table 5) nor were significant differences found when either control or LN treatments (1 and 30 days) were included. However, significant ($p < 0.001$) differences among treatments and cultivar x treatment interactions were found for T_{50} (Table 5). In all seven cultivars, the mean T_{50} value of the three priming treatments turned out to be lower than that of the control and LN treatments (Tables 4 and 5). This difference was significant in Florida, Utah and Istar both in the $NaNO_3$ and the KNO_3 solutions. The LN + priming treatment gave the lowest T_{50} in Golden Spartan and Tall Utah 52-70R with the $NaNO_3$ solution and in Tall Utah 52-70R and Golden Boy with the KNO_3 solution (Table 5).

2.4 Discussion

Celery seed storage in LN did not reduce germination percentage in any of the cultivars for either periods of time studied, thus confirming the results of others (Styles et al. 1982; Stanwood 1985). Moreover, cryopreservation did not have a negative effect on the germination of pelleted seeds.

The priming (with or without cryopreservation) did not affect final germination percentage. Heydecker and Gibbins (1978) stated that some of the advantages of priming could be lost if seeds are subsequently desiccated. In our experiments, seeds were always desiccated after priming to eliminate the adverse effects of high moisture contents on cryopreservation. In spite of this, the positive effects of priming were still noticeable in some cultivars. In general, the sensitivity of seeds to dehydration increases with the priming

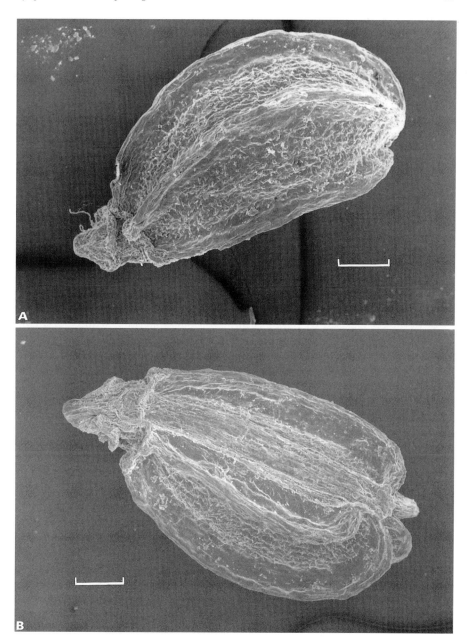

Fig. 1. SEM micrograph of **A** control and **B** frozen seeds of celery cv. Florida 683, *bar* = 200 μm (with permission of Annals of Botany)

Table 5. Germination percentage (mean values ± SE) and T_{50} values (days, in parentheses) of seven celery cultivars (30 days after sowing), after various combinations of priming and freezing treatments. (With permission of *Annals of Botany*)

Cultivar	Treatment					
	NaNO$_3$			KNO$_3$		
	P	P + LN	LN + P	P	P + LN	LN + P
Florida 683	64 ± 3 (5.0)	73 ± 1 (5.0)	70 ± 1 (5.0)	67 ± 1 (5.0)	73 ± 1 (5.0)	67 ± 4 (5.0)
Utah	74 ± 3 (6.0)	61 ± 4 (7.0)	72 ± 1 (5.0)	68 ± 1 (6.0)	71 ± 1 (5.0)	64 ± 3 (6.0)
Golden Spartan	91 ± 1 (7.0)	95 ± 1 (7.0)	92 ± 1 (6.0)	91 ± 2 (6.5)	96 ± 0 (7.0)	94 ± 0 (6.0)
Isel	94 ± 2 (5.5)	98 ± 1 (5.0)	93 ± 1 (6.0)	94 ± 0 (5.0)	92 ± 1 (5.0)	94 ± 1 (6.0)
Tall Utah 52-70R	93 ± 1 (5.0)	95 ± 1 (5.0)	95 ± 1 (3.0)	97 ± 1 (5.0)	98 ± 0 (5.0)	95 ± 1 (3.0)
Istar	94 ± 1 (5.0)	93 ± 1 (5.0)	91 ± 2 (6.0)	89 ± 1 (5.0)	97 ± 0 (5.0)	96 ± 1 (6.0)
Golden Boy	80 ± 1 (6.5)	77 ± 4 (7.0)	86 ± 2 (6.0)	81 ± 1 (7.0)	78 ± 0 (7.0)	82 ± 1 (6.0)

P = Priming; LN = 1 day in LN; P + LN = priming before LN; LN + P = priming after LN.

incubation time (Heydecker and Gibbins 1978). Under our conditions, the time of priming incubation was short (7 days) when compared with periods of 21 days (at 10 °C) used by Singh et al. (1985), also working with celery. Thus, a 7-day pretreatment appears to be insufficient to activate celery seed desiccation sensitivity.

Thus, the feasibility of cryopreservation on celery and its compatibility with other treatments, such as priming and pelleting, make it an interesting tool for celery germplasm storage. Current research in the development of synthetic seeds from celery somatic embryos (Kim and Janick 1990; Saranga et al. 1992; Nadel et al. 1995) can greatly benefit from existing cryopreservation protocols for celery seeds.

Cryopreservation of celery is also of great interest in several other areas of applied biotechnology research where significant advances are taking place. For instance, celery somaclonal variants highly resistant to several pathogens have been obtained (Wright and Lacy 1988; Heath-Pagliuso et al. 1988; Toth and Lacy 1992). Moreover, cell suspension cultures (either undifferentiated or differentiated) have also been explored for the production of secondary products (Collin and Isaac 1991). Thus, the development of cryopreservation protocols for cell and tissue cultures would be of great interest for the maintenance of their genetic stability and the avoidance of secondary product yield decline due to subculture.

3 Conclusions

From studies carried out on celery seeds it can be concluded that cryopreservation is a feasible technique for germplasm conservation, as has been shown for other horticultural species (Stanwood and Roos 1979; Styles et al. 1982).

In celery this technique is compatible with other seed pretreatments of great relevance in horticulture, such as priming and pelleting. Cryopreservation is an interesting alternative tool for celery germplasm storage. The development of cryopreservation protocols for cell and tissue cultures would be very useful in the existing areas of research on celery biotechnology.

References

Bettencourt E, Konopla J (1990) Directory of germplasm collections, 4 vegetables. IBPGR, Rome
Collin HA, Isaac S (1991) *Apium graveolens* L. (celery): In vitro culture and the production of flavors. In: Bajaj YPS (ed) Biotechnology in agriculture and forestry, vol 15. Medicinal and aromatic plants III. Springer, Berlin Heidelberg New York, pp 73–94
Ellis RH, Hong TD, Martín MC, Pérez-García F, Gómez-Campo C (1993) The long-term storage of seeds of seventeen crucifers at very low moisture content. Plant Varieties Seeds 6:75–81
Frison EA, Serwinski J (eds) (1995) Directory of European institutions holding crop genetic resources collections, vols 1 and 2, 4th edn. IPGRI, Rome
Georghiou K, Thanos CA, Passam HC (1987) Osmoconditioning as a means of counteracting the ageing of pepper seeds during high-temperature storage. Ann Bot 60:279–285
González-Benito ME, Iriondo JM, Pita JM, Pérez-García F (1995) Effects of seed cryopreservation and priming on germination in several cultivars of *Apium graveolens*. Ann Bot 75: 1–4
Heath-Pagliuso S, Pullman J, Rappaport L (1988) Somaclonal variation in celery: screening for resistance to *Fusarium oxysporum* f. sp. *apii* Theor Appl Genet 75:446–451
Heydecker W, Gibbins BM (1978) The priming of seeds. Acta Hortic 83:213–223
Iriondo JM, Pérez C, Pérez-García F (1992) Effect of seed storage in liquid nitrogen on germination of several crop and wild species. Seed Sci Technol 20:165–171
Kim YH, Janick J (1990) Synthetic seed technology: improving desiccation tolerance of somatic embryos of celery. Acta Hortic 280:23–28
Lakhanpaul S, Babrekar PP, Chandel KPS (1996) Monitoring studies in onion (*Allium cepa* L.) seeds retrieved from storage at −20 °C and −180 °C. Cryo Lett 17:219–232
Maroto JV (1983) Horticultura herbácea especial. Mundi-Prensa, Madrid
Nadel BL, Altman A, Ziv M (1995) Somatic embryogenesis and synthetic seed in *Apium graveolens* (celery). In: Bajaj YPS (ed) Biotechnology in agriculture and forestry, vol 31. Somatic embryogenesis and synthetic seed. Springer, Berlin Heidelberg New York, pp 306–322
Salter PJ, Darby RJ (1976) Synchronisation of germination of celery seeds. Ann Appl Biol 84:415–424
Saranga Y, Kim KH, Janick J (1992) Changes in tolerance to partial desiccation and in metabolite content of celery somatic embryos induced by reduced osmotic potential. J Am Soc Hortic Sci 117:342–345
Simon JE, Chadwick AF, Craker LE (1984) Herbs: an indexed bibliography, 1971–1980. The scientific literature on selected herbs, and aromatic and medicinal plants of the temperate zone. Archon Books, Hamden, CT
Singh H, Morss S, Orton TJ (1985) Effect of osmotic pretreatment and storage on germination of celery seed. Seed Sci Technol 13:551–558
Stanwood PC (1985) Cryopreservation of seed germplasm for genetic conservation. In: Kartha KK (ed) Cryopreservation of plant cells and organs. CRC Press, Boca Raton, pp 199–226
Stanwood PC, Roos EE (1979) Seed storage of several horticultural species in liquid nitrogen (−196 °C). Hortscience 14:628–630
Stanwood PC, Sowa S (1995) Evaluation of onion (*Allium cepa* L.) seed after 10 years of storage at 5, −18, and −196 °C. Crop Sci 35:852–856
Styles ED, Burgess JM, Manson C, Huber BM (1982) Storage of seed in liquid nitrogen. Cryobiology 19:195–199

Thomas TH, Biddington NL, O'Toole DF (1979) Relationship between position of the parent plant and dormancy characteristics of seeds of three cultivars of celery (*Apium graveolens*). Physiol Plant 45:492–496

Toth KF, Lacy ML (1992) Micropropagation of celery (*Apium graveolens* var. *dulce*). In: Bajaj YPS (ed) Biotechnology in agriculture and forestry, vol 19. High-tech and micropropagation III. Springer, Berlin Heidelberg New York, pp 218–229

Willis JC (1973) A dictionary of the flowering plants and ferns. Cambridge Univ Press, Cambridge

Wright JC, Lacy ML (1988) Increase of disease resistance in celery cultivars by regeneration of whole plants from cell suspension cultures. Plant Dis 72:256–259

II.3 Cryopreservation of *Armoracia rusticana* P. Gaert., B. Mey. et Scherb. (Horseradish) Hairy Root Cultures

Kazumasa Hirata[1], Phunchindawan Monthana[1], Akira Sakai,[2] and Kazuhisa Miyamoto[1]

1 Introduction

Cryopreservation has become important as a means of ensuring the long-term preservation of plant germplasm. It is particularly useful for the conservation of precious and rare species on the verge of extinction due to global environmental problems. In the case of higher plants, roots are considered to be appropriate material for cryopreservation since many uniform root samples can be taken from such plants growing in the field without inflicting lethal damage. In vitro transformed hairy root culture is a very useful model material for investigating cryopreservation of roots because of the high levels of uniformity of their morphology and rapid growth. Their property of high plantlet regeneration frequency allows clonal propagation of elite plants. Furthermore, hairy root cultures have been shown to be very useful material for the production of valuable secondary metabolites. Therefore, cryopreservation of hairy root cultures is also important to establish safe patent repositories of lines capable of producing products of commercial interest. However, there are only a few reports in the literature on the cryopreservation of root cultures by conventional slow-freezing (Benson and Hamil 1991; Bajaj 1995; Teoh et al. 1996) or vitrification (Yoshimatsu et al. 1996) techniques, indicating that these methods are not applicable to roots in general, and that alternative, more reliable techniques are needed.

Recently, cryopreservation by encapsulation-dehydration has been applied to meristematic tissues of various plants (Dereuddre et al. 1990; Bajaj 1995; Vandenbussche and De Proft 1998). In contrast to slow-freezing or vitrification techniques, this technique does not require toxic cryoprotective additives, like dimethyl sulfoxide and ethylene glycol, and the encapsulation of the fine hairy roots in gel beads makes their handling much easier. In the present study, the utility of the encapsulation-dehydration technique for the cryopreservation of plant roots was investigated using horseradish hairy root cultures as a model material.

---

[1] Environmental Bioengineering Laboratory, Graduate School of Pharmaceutical Sciences, Osaka University, 1-6 Yamada-oka, Suita, Osaka 565-0871, Japan
[2] 1-5-23 Asabu-cho, Kita-ku, Sapporo 001–0045, Japan

Biotechnology in Agriculture and Forestry, Vol. 50
L.E. Towill and Y.P.S. Bajaj (Eds.) Cryopreservation of Plant Germplasm II
© Springer-Verlag Berlin Heidelberg 2002

2 Cryopreservation

2.1 Materials

Horseradish (*Armoracia rusticana* P. Gaert., B. Mey. et Scherb.) of the family Cruciferae is a native of southeastern Europe. It is well known that roots of this plant are a raw material for commercial production of peroxidase (EC 1.11.1.7), which is widely used as an important reagent for colorimetric analyses of biological materials. Taya et al. (1989) have investigated effective production procedures of peroxidase from hairy root cultures. They also reported that the hairy roots could be used to produce artificial seeds because encapsulated root tips in alginate beads regenerated whole plants with high frequency (Uozumi et al. 1992).

Hairy root cultures were subcultured every 4 weeks in a hormone-free Murashige-Skoog (MS) liquid medium (Murashige and Skoog 1962) containing 90 mM sucrose at 25 °C in the dark with rotary shaking at 80 rpm. Shoot primordia, which were used for cryopreservation, were induced from 2-week-old hairy roots by cultivation in the medium supplemented with 1 µM naphthaleneacetic acid under continuous illumination with $60 \mu E m^{-2} s^{-1}$ of white fluorescent light. Root tips were sampled for cryopreservation from 2-week-old hairy root cultures.

2.2 Encapsulation-Dehydration Technique

Shoot primordia (1–2 mm in diameter) or root tips (2–3 mm length) were suspended in a 2% sodium alginate (500 centi-poise) solution supplemented with various concentrations of sucrose and/or glycerol, and then immediately dropped into 100 ml of a 50 mM calcium chloride solution containing the same concentrations of sucrose and/or glycerol. After holding for 5 min to allow the formation of Ca-alginate beads (ca. 5 mm in diameter), the beads were transferred onto solid MS medium containing the same concentrations of sucrose and/or glycerol and precultured for 1 day. Then the beads were transferred into a Petri dish (3.5 cm in diameter) and slowly dehydrated by placing the dish in a desiccator over silica gel to reduce the water content of the beads to various values. In case of encapsulated shoot primordia, dehydration was carried out by a two-step procedure (Hirata et al. 1995). To simplify dehydration, a one-step procedure was employed, in which encapsulated root tips were slowly dehydrated at 0.8% reduction in the water content per hour (average rate) (Hirata et al. 1998). Dehydrated beads were placed in 1-ml cryotubes and plunged into liquid nitrogen (LN). After 3 days the beads were rapidly rewarmed in a water bath at 40 °C and then transferred onto the medium without washing.

To test the effect of ABA treatment on survival of root tips after dehydration and cooling, excised root tips were precultured in MS medium containing 2 µM ABA for 1 day in the dark without shaking. The root tips precultured were encapsulated and dehydrated in the same manner described above (Hirata et al. 1998).

2.3 Encapsulation-Vitrification Technique

The encapsulation-vitrification technique was also applied to the cryopreservation of shoot primordia (Phunchindawan et al. 1997). The precultured and encapsulated primordia described above were dehydrated with a vitrification solution (PVS2, Sakai et al. 1990) in a 100-ml Erlenmeyer flask with occasional shaking at 25 or 0 °C for different lengths of time. The PVS2 solution contained 30% (w/v) glycerol, 15% (w/v) ethylene glycol, and 15% (w/v) dimethyl sulfoxide in 0.5 M sucrose. The ratio of the number of beads to volume of PVS2 was 1 to 1 ml. The solution was completely exchanged to the new solution every 30 min. Cryotubes containing 15 beads dehydrated as above were plunged into LN and kept there for 3 days. After rewarming, 30 ml of liquid MS medium containing 1.2 M sucrose were added to 15 beads in a 50-ml test tube and held for 40 min. Then, the beads were placed onto filter paper on solidified MS medium. After overnight cultivation, they were transferred onto the fresh medium and cultured under the conditions described above. After 2 days, the beads were transferred onto fresh medium again and cultured for 4 weeks.

2.4 Measurement of Viability

Survival after cooling in LN was recorded as the percentage of the total number of shoot primordia forming normal shoots, or root tips forming new roots, 4 weeks after cultivating on the medium containing 4.4 μM benzyladenine or 2.7 μM naphthaleneacetic acid, respectively.

3 Results and Discussion

3.1 Encapsulation-Dehydration

Encapsulation-dehydration can be easily applied to various plant materials. In this technique, desiccation tolerance, which is necessary to retain a high level of viability after dehydration, was induced by treatment with sucrose. By optimizing the dehydration conditions, i.e., encapsulation in Ca-alginate containing 0.5 M sucrose, preculturing for 1 day in the presence of the same concentration of sucrose, slow dehydration to 73% of the water content, holding for 2 days and finally rapid dehydration to 24% of the water content, a survival rate of 46% was obtained with shoot primordia induced from horseradish hairy root cultures (Table 1). However, this level of survival is proba-

Table 1. Effect of dehydration conditions on survival of encapsulated horseradish shoot primordia after cooling in liquid nitrogen

Dehydration condition	Survival (%)	
	LN (−)[a]	LN (+)[b]
Rapid[c]	90	0
Slow-rapid[d]	76 ± 11	16 ± 6
Slow-holding-rapid[e]	92 ± 6	46 ± 8

Values indicate means ± S.D. (n = 7)
[a] Survival of the shoot primordia cultivated for 4 weeks after dehydration without cooling in liquid nitrogen (control).
[b] Survival of the primordia cultivated for 4 weeks after dehydration and cooling in liquid nitrogen.
[c] Dehydrated rapidly until 24% water content (average rate was 10% of water content loss per hour).
[d] Dehydrated slowly until 73% water content (average rate was 0.17% of water content loss per hour) and rapidly until 24% water content.
[e] Dehydrated slowly until 73% water content, held for 2 days, and then dehydrated rapidly until 24% water content.

Table 2. Effects of water content and concentrations of sucrose and glycerol on survival of encapsulated horseradish shoot primordia after cooling in liquid nitrogen

Sucrose (M)	Glycerol (M)	Bead weight (% of initial wt)	Water content (%)	Survival (%)	
				LN (−)	LN (+)
0.5	0	30	37	93 ± 5	404 ± 28
0.5	0	25	24	85 ± 5	53 ± 29
0	1.0	25	56	90	0
0	1.0	20	45	69 ± 26	0
0.5	0.5	35	33	95 ± 4	28 ± 8
0.5	0.5	30	21	77 ± 4	73 ± 10
0.5	1.0	40	30	94 ± 4	94 ± 7
0.5	1.0	35	20	86 ± 9	84 ± 8
0.5	1.5	45	27	93 ± 4	91 ± 4
0.5	1.5	40	18	87 ± 7	83 ± 6

Values indicate means ± S.D. (n = 4–6).

bly not sufficient for germplasm preservation. Since extending the applicability of this technique is contingent upon improving the survival rate, attempts were made to ensure higher levels of survival by treating encapsulated primordia with 0.5 M sucrose in combination with various cryoprotective solutes. Among the solutes tested, only glycerol was effective in increasing survival: rates of 94 and 91% were obtained after cooling in LN when glycerol at concentrations of 1 and 1.5 M, respectively, was combined with 0.5 M sucrose (Table 2). The exact mechanism by which a mixture of sucrose and glycerol increases desiccation tolerance is unknown. It may be postulated that glycerol contributes to minimizing the injurious membrane changes resulting from severe dehydration by placing the water molecules at the charged exterior surface of the membrane. Under the treatment with glycerol and sucrose, the rate of plantlet formation from shoot primordia was significantly faster than that of primordia treated with 0.5 M sucrose alone, and was similar to that of treated control primordia without cooling (control). No irregular morphology was evident in the regrown plantlets. Furthermore, such rapid regeneration at a high rate was observed after storing beads in LN for 1 year.

Encapsulation-dehydration with glycerol and sucrose was applied to the cryopreservation of horseradish hairy root tips. When root tips were encapsulated with 0.5 M sucrose and 1 M glycerol, concentrations that gave a high level of survival in the case of shoot primordia, the survival rate of the root tips after dehydration was less than 5%, and dropped to zero after cooling in LN. At high concentrations of glycerol and sucrose, root tips were exposed to serious osmotic stress during dehydration, which is thought to be lethal to them. Hence, combinatorial optimization of the concentrations of glycerol and sucrose to increase survival rates after dehydration and cooling was performed. Addition of 0.5 M glycerol and 0.3 M sucrose in the preculture and encapsulation media gave the highest survival, 33%, among the various conditions tested (Fig. 1).

This was the first time hairy roots were successfully cryopreserved by the encapsulation-dehydration technique. However, survival after dehydration was around 50% even at reduced solute concentrations, indicating that the concentrations of glycerol and sucrose were still high. To obtain a higher survival level, the effect of abscisic acid (ABA) on survival was investigated. ABA is a well-known phytohormone that plays an important role in the acquisition of tolerance to desiccation and osmotic stress (Chandler and Robertson 1994; Ingram and Bartels 1996). The hormone has also been found to be effective

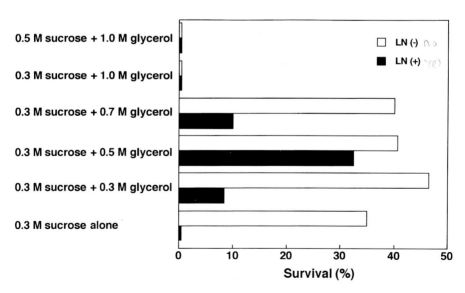

Fig. 1. Effects of concentrations of sucrose and glycerol on survival rates of encapsulated horseradish hairy root tips after cooling in liquid nitrogen

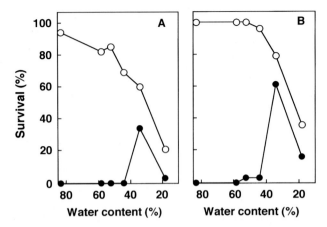

Fig. 2. Effect of 2 μM abscisic acid on survival of encapsulated horseradish hairy root tips with various water contents after dehydration and after cooling in liquid nitrogen. The root tips were precultured and encapsulated **A** without **B** with ABA before dehydration and cooling in liquid nitrogen

in conferring tolerance against freezing (Keith and McKersie 1986) and for obtaining high survival levels in the cryopreservation of several plant materials (Reaney and Gusta 1987; Kendall et al. 1993; Vandenbussche and De Proft 1998). Figure 2 shows the effect of ABA (2 μM) on the survival of root tips at different water contents after dehydration and cooling in LN. As the water content was reduced, survival of ABA-treated dehydrated root tips (open circles in Fig. 2B) declined more gently in comparison with that in the non-ABA-treated control (open circles in Fig. 2A). This result indicates that the root tips acquired desiccation tolerance as a consequence of preculture and dehydration in the presence of ABA. Compared at 33% water content, a survival rate of about 60% was obtained in ABA-treated cryopreserved root tips (closed circles in Fig. 2B), higher than that in non-ABA-treated cryopreserved root tips (closed circles in Fig. 2A).

When the period required for detectable elongation was compared between ABA-treated and non-treated cryopreserved root tips, the former showed significantly earlier elongation than the latter. Figure 3 shows hairy root regeneration after a 4-week cultivation of dehydrated ABA-treated (ABA [+]) and non-treated (ABA [−]) root tips before cooling (LN [−]) or after cooling (LN [+]) in LN. The ABA-treated root tips before (ABA [+], LN [−]) and after cooling (ABA [+], LN [+]) exhibited markedly faster regeneration of hairy roots from survivors compared with their non-treated counterparts (ABA [−], LN [−] and (ABA [−], LN [+]).

The treatment with ABA was applicable to increasing the desiccation tolerance and survival after cooling in liquid nitrogen in encapsulated hairy root tips of two other plants (*Ajuga reptans* L. var. *atropurpurea* and *Vinca minor* L.; data not shown).

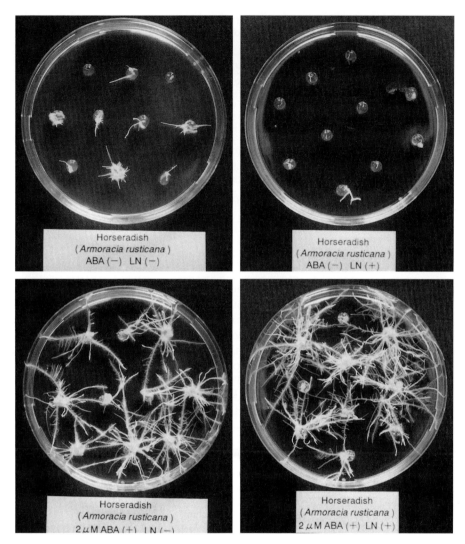

Fig. 3. Effect of abscisic acid on regeneration of hairy roots from encapsulated horseradish hairy root tips after dehydration and after cooling in liquid nitrogen. The root tips were precultured and encapsulated in Ca-alginate beads containing 0.3 M sucrose and 0.5 M glycerol in the absence (ABA[–]) or in the presence (ABA[+]) of $2\,\mu$M abscisic acid. They were dehydrated by air-drying until the water content reached 33%, and then cultivated for 4 weeks without cooling (LN[–]), or after cooling in liquid nitrogen (LN[+])

3.2 Encapsulation-Vitrification Technique

Cryopreservation of encapsulated horseradish shoot primordia was investigated using a vitrification technique. Considerable numbers of plants have been successfully cryopreserved using this technique (Langis et al. 1989; Sakai

Table 3. Effects of temperature and duration of exposure to PVS2 on survival of encapsulated horseradish shoot primordia after cooling in liquid nitrogen

Temperature (°C)	Duration (h)	Survival (%)	
		LN (−)	LN (+)
25	1	87 ± 23	0
25	2	80 ± 17	40 ± 12
25	3	57 ± 15	15 ± 4
0	1	98 ± 5	0
0	2	90 ± 14	14 ± 9
0	3	90 ± 14	56 ± 17
0	4	88 ± 11	69 ± 10
0	5	85 ± 17	33 ± 21

Values indicate means ± S.D. (n = 3–5).

and Kobayashi 1990; Bajaj 1995). Although the technique has so far been used mainly for meristematic tissue without encapsulation, its application to encapsulated materials has recently been reported (Matsumoto et al. 1995). In this technique, dehydration is achieved by treatment with high concentrations of cryoprotective solutes. To obtain the optimal vitrification conditions, the effects of the temperature and duration for treatment on survival were investigated. The highest survival rate, 69%, was attained when primordia were dehydrated with PVS2, a solution widely used for the vitrification technique, for 4h at 0°C (Table 3). The regrowth rate of shoot primordia into plantlets was slightly lower than that of untreated primordia on a control experiment. This technique thus appears to be applicable to the cryopreservation of shoot primordia. Yoshimatsu et al. (1996) reported the successful cryopreservation of ginseng hairy root cultures by vitrification. In the case of encapsulated horseradish root tips, no survival has yet been obtained after cooling in LN.

4 Summary and Conclusions

Horseradish hairy root tips and shoot primordia induced from the roots were used as model materials for cryopreservation of plant roots by encapsulation-dehydration. The survival rates after dehydration and after cooling in LN were markedly increased by addition of glycerol combined with sucrose in a preculture medium and encapsulation beads. In the case of hairy root tips, ABA treatment was effective in further increasing survival after dehydration and cooling. Thus, the revised encapsulation-dehydration technique presented here is promising as a practical procedure for cryopreserving plant roots.

References

Bajaj YPS (1995) Cryopreservation of germplasm of potato (*Solanum tuberosum* L.) and cassava (*Manihot esculenta* Crantz). In: Bajaj YPS (ed) Biotechnology in agriculture and forestry, vol 32. Cryopreservation of plant germplasm I. Springer, Berlin Heidelberg New York, pp 398–416

Benson EE, Hamil JD (1991) Cryopreservation and post-freeze molecular and biosynthetic stability in transformed roots of *Beta vulgaris* and *Nicotiana rustica*. Plant Cell Tissue Org Cult 24:163–172

Chandler DM, Robertson M (1994) Gene expression regulated by abscisic acid and its relation to stress tolerance. Annu Rev Plant Physiol Plant Mol Biol 45:113–141

Dereuddre JC, Scottez C, Arnaud Y, Duron M (1990) Resistance of alginate-coated axillary shoot tips of pear tree (*Pyrus communis* L. Beurre Hardy) in vitro plantlets to dehydration and subsequent freezing in liquid nitrogen: effect of previous cold hardening. CR Acad Sci Paris Ser III 310:317–325

Hirata K, Phunchindawan M, Knemoto M, Miyamoto K, Sakai A (1995) Cryopreservation of shoot primordia induced from horseradish hairy root cultures by encapsulation and two-step dehydration. Cryo Lett 16:122–127

Hirata K, Goda S, Phunchindawan M, Du D, Ishio M, Sakai A, Miyamoto K (1998) Cryopreservation of horseradish hairy root cultures by encapsulation-dehydration. J Ferment Bioeng 86: 418–420

Ingram J, Bartels D (1996) The molecular basis of dehydration tolerance in plants. Annu Rev Plant Physiol Plant Mol Biol 47:377–403

Keith CN, McKersie BD (1986) The effect of abscisic acid on the freezing tolerance of callus cultures of *Lotus corniculatus* L. Plant Physiol 80:766–770

Kendall EJ, Kartha KK, Qureshi JA, Chermak P (1993) Cryopreservation of immature spring wheat zygotic embryos using an abscisic acid pretreatment. Plant Cell Rep 12:89–94

Langis R, Schnebel B, Earle ED, Steponkus PL (1989) Cryopreservation of *Brassica compestris* L. cell suspensions by vitrification. Cryo Lett 10:421–428

Matumoto T, Sakai A, Takahashi C, Yamada K (1995) Cryopreservation of in vitro grown apical meristems of wasabi (*Wasabia japonica*) by encapsulation-vitrification method. Cryo Lett 16: 189–196

Murashige T, Skoog F (1962) A revised medium for rapid growth and bioassays with tobacco tissue cultures. Physiol Plant 15:473–497

Phunchindawan M, Hirata K, Sakai A, Miyamoto K (1997) Cryopreservation of encapsulated shoot primordia induced in horseradish (*Armoracia rusticana*) hairy root cultures. Plant Cell Rep 16:469–473

Reaney MJT, Gusta LV (1987) Factors influencing the induction of freezing tolerance by abscisic acid in cell suspension cultures of *Bromus inermis* and *Medicago sativa*. Plant Physiol 83: 423–427

Sakai A, Kobayashi S (1990) A simple and effective procedure for cryopreservation of nucellar cells of navel orange by vitrification. Cryobiology 27:657

Sakai A, Kobayashi S, Oiyama I (1990) Cryopreservation of nucellar cells of navel orange (*Citrus sinensis* Osb. var 'brasiliensis' Tanaka) by vitrification. Plant Cell Rep 9:30–33

Taya M, Yoyama A, Nomura R, Kondo O, Matsui C, Kobayashi T (1989) Production of peroxidase with horseradish hairy root cells in a two step culture system. J Ferment Bioeng 67:31–34

Teoh KH, Weathers PJ, Cheetham RD, Walcerz DB (1996) Cryopreservation of transformed (hairy) roots of *Artemisia annua*. Cryobiology 33:106–117

Uozumi N, Nakashimada Y, Kato Y, Kobayashi T (1992) Production of artificial seed from horseradish hairy root. J Ferment Bioeng 74:21–26

Vandenbussche B, De Proft MP (1998) Cryopreservation of in vitro sugar beet shoot tips using the encapsulation-dehydration technique: influence of abscisic acid and cold acclimation. Plant Cell Rep 17:791–793

Yoshimatsu K, Yamaguchi H, Shimomura K (1996) Traits of *Panax ginseng* hairy roots after cold storage and cryopreservation. Plant Cell Rep 15:555–560

II.4 Cryopreservation of *Chamomilla recutita* L. Rauschert (Chamomile) Callus

Eva Čellárová, Katarína Kimáková, and Martina Urbanová

1 Introduction

1.1 Plant Distribution and Important Species

Chamomile is one of oldest known medicinal herbs and was used in ancient Egypt, Greece and other Mediterranean countries. There are two herbs commonly called chamomile: German chamomile (Hungarian chamomile or wild chamomile) – (*Chamomilla recutita* L. Rauschert, syn. *Matricaria recutita* L.) and Roman chamomile (*Chamaemelum nobile* L., syn. *Anthemis nobilis* L.). *Chamomilla recutita* was originally native to southeastern and southern Europe but is now commonly distributed all over the world, except in tropical and arctic regions. Chamomile grows in fields and many other habitats throughout Europe, Russia and Asia, and is naturalised in Australia and the United States; however, it prefers cultivated ground. *Chamaemelum nobile* is native to southwestern and northwestern regions of Europe (Spain, France, England) and is scattered around the eastern Mediterranean, the Balkans and Crimea. Both German and Roman chamomiles are cultivated but German chamomile is the principle article of commerce, both on the European continent and in the United States. Some 4000 to 5000 tons of chamomiles are produced annually throughout the world (Mann and Staba 1986). Both chamomiles are used in traditional herbalism and medicine. However, German chamomile is more frequently preferred for medicinal use. In Europe today, a large number of pharmaceutical preparations are available containing either extracts of chamomile or volatile oil. The volatile oil content in German chamomile varies from 0.14%–1.9% and is composed of (−)-α-bisabolol, (−)-α-bisabololoxids A and B, (−)-α-bisabolonoxide, chamazulene, farnesene and spiro-ether. The volatile oil constituents have anti-inflammatory and anti-spasmodic effects, reduce inflammation and act as mild antibacterials. Other important compounds identified in German chamomile are flavonoids including apigenin, luteolin and quercetin. These display inhibitory effects in vitro on the proliferation of certain malignant cells (Agullo et al. 1997). Some alkylated flavonoids contribute to the anti-inflammatory and anti-spasmodic properties. The anti-inflammatory effect is characteristic for other classes

Department of Experimental Botany and Genetics, Faculty of Science, P. J. Šafárik University, Mánesova 23, 041 67 Košice, Slovakia

Biotechnology in Agriculture and Forestry, Vol. 50
L.E. Towill and Y.P.S. Bajaj (Eds.) Cryopreservation of Plant Germplasm II
© Springer-Verlag Berlin Heidelberg 2002

of compounds like coumarins, herniarin and umbelliferone. The essential oil of the Roman chamomile contains less chamazulene and contains mainly esters of angelic and tiglic acid. It also contains farnesene and α-pinene. Some important flavonoids found in German chamomile are also present in Roman chamomile.

Recently found activities of chamomile as immunomodulating and antimutagenic agents (Birt et al. 1986; Huang et al. 1996) contribute to its diverse therapeutic uses.

1.2 Methods for Storage of Germplasm

Chamomilla recutita is an annual herb which is easily propagated by seeds. Fresh seeds have very low germination (>1%). Aging of seeds for 200–300 days significantly increases germination to 80%–100%. There are minor differences between germination values for diploid and tetraploid seeds when the seeds are stored in glass containers at 10°C and the relative humidity does not exceed 30%. Storage at −18°C had no effect on germination of aged seeds (Carle et al. 1991). Seeds persisting in the soil retain the ability to germinate for 10–15 years (Bernáth 1993). However, under laboratory conditions with low relative humidity and the temperature between 20–25°C, the germination ability significantly decreases after 2 or 3 years.

Chamomilla recutita shoot tips can be successfully stored in liquid nitrogen after both slow and ultrarapid cooling procedures with about 65% survival and 40% regeneration ability (Diettrich et al. 1990).

1.3 Need for Cryopreservation

Breeding of *Chamomilla recutita* is still of great interest in many countries. As an alternative to the conventional breeding methods, shoot tip or meristem culture and callus culture capable of regeneration of the whole plants is a suitable tool for establishment of a gene bank of elite genotypes. The former allows plants with desirable properties to be cloned that may contribute to the rationalisation of the breeding process, and the latter may widen the genetic variability of plants regenerated from different cell populations.

2 Cryopreservation

2.1 General Account

Different protocols are available for in vitro proliferation of *Chamomilla recutita* shoots which produce plant material suitable for cryopreservation. Direct or indirect shoot differentiation can be induced in different types of explants, including leaves, stems, apical meristems, roots, inflorescences or seeds (Čellárová et al. 1982; Diettrich et al. 1990; Sakr et al. 1991; Tavoletti et al. 1994; Menghini et al. 1996).

Chamomile shoot tips of up to 1 mm length were isolated from in vitro regenerated axenic shoots and successfully cryopreserved by Diettrich et al. (1990). Shoot tips were hardened at 4 °C for 8 weeks, treated with a cryoprotective solution consisting of dimethyl sulfoxide (DMSO), glycerol and sucrose for 16 h, and then cooled either slowly at 0.5 °C/min to –40 °C or ultrarapidly by submerging the samples in liquid nitrogen. Different methods of cooling did not significantly affect the survival and regeneration capacity of frozen shoot tips; however, cold hardening prior to cryoprotection was essential for a high percentage of post-freezing survival and regeneration ability of chamomile shoot tips. This method is suitable for cryostorage of elite, genetically and biochemically determined genotypes for breeding purposes.

Callus cultures of chamomile consist of different cell populations and are very heterogeneous cell systems. Thus, several steps in the cryopreservation process need to be examined. To avoid the possible interaction between plant growth regulators present in the culture medium and different additives used during the preculture process, habituated chamomile callus cultures were used for cryopreservation studies (Čellárová et al. 1992a,b; Kimáková 1995).

2.2 Methodology/Protocol

2.2.1 Plant Material and Culture Conditions

Callus cultures of tetraploid *Chamomilla recutita* plants originated from leaves cultured on the basal Linsmaier-Skoog medium (Linsmaier and Skoog 1965) containing Linsmaier-Skoog basal salt mixture, Gamborg's B5 vitamins (Gamborg et al. 1968), 30 g/l sucrose, and 2 mg/l glycine (RM medium). This basal medium was supplemented with 3 mg/l 2,4-dichlorophenoxyacetic acid (RM3 medium). From these cultures, a habituated cell line was isolated. Callus cultures were maintained under artificial illumination ($7.5 \mu E m^{-2} s^{-1}$), 16 h photoperiod, 26/22 °C temperature and a 70% relative humidity.

2.2.2 Pregrowth

Habituated callus cultures for cryopreservation experiments were precultured as follows:

1. 3-month preculture on the RM3 medium with 0.5 mg/l abscisic acid or 0.1 mg/l gibberellic acid
2. Preculture on the basal medium with an increasing sucrose content ($3 \rightarrow 5 \rightarrow 7 \rightarrow 10\%$, 1 month each)
3. 3-day preculture on the basal medium supplemented with 0.5 M mannitol.

2.2.3 Cryoprotection

Cryoprotectants (PEG MW 400, 1000 and 6000, glucose, glycerol, DMSO and sucrose in concentrations of 5, 10 and 20%; Table 1) and their mixtures (glycerol/sucrose/DMSO and PEG/glucose/DMSO; Table 2) were tested on the cells at the beginning of the exponential growth phase. Callus, 0.5 g, was re-suspended in 0.5 ml of the basal nutrient medium in Eppendorf tubes and cooled on ice. Cryoprotectant solutions (0.5 ml) were added drop by drop and the samples were allowed to equilibrate at 0 °C for 10 or 60 min.

2.2.4 Cooling and Thawing

Cooling was performed in a freezer (POLAR) constructed at the Institute of Experimental Physics of the Slovak Academy of Science, Košice, Slovakia. The cooling rate varied between 0.1–0.5 °C/min to −40 °C, at that temperature the samples were kept for 30 min and then subsequently immersed directly in liquid nitrogen where they were stored for 1 to 14 days. Samples were warmed rapidly in a 40 °C water bath.

2.2.5 Viability and Regrowth

The viability of cryoprotected and cryopreserved cells was determined after cryoprotection and immediately after cryopreservation and at the end of the first post-cryopreservation subculture by FDA (Widholm 1972, modified) and TTC tests (Steponkus and Lanphear 1967, modified) and expressed as percent survival of the non-treated and uncooled control. Growth of callus cultures was ascertained at regular time intervals during the first two post-cryopreservation subcultures.

2.3 Results

2.3.1 Effect of Pregrowth on Viability of Chamomilla recutita Cells

The use of abscisic and gibberellic acid in the preculture period followed by cryoprotection with a mixture of glycerol/sucrose/DMSO (5:10:10, v/w/v) resulted in a prolonged phase of the rapid cell growth. However, it did not enhance the post-thaw viability of cells (Čellárová et al. 1992a). Similarly, an increased sucrose level during the pregrowth period for 1–3 months, depending on the sucrose concentration, caused a decrease in fresh and dry weight of cells and a reduced cell volume (Fig. 1). On the contrary, after a 3-day preculture with 0.5 M mannitol, neither changes in cell volume nor prolongation of exponential growth phase was observed and this process gave the highest post-thaw viability (Kimáková 1995).

Table 1. Comparison of cell viability of habituated Chamomilla recutita cells after 10 and 60 min equilibration on ice in different cryoprotectants

Cryoprotectant/ content	10 min equilibration					60 min equilibration				
	FDA		TTC			FDA		TTC		
	M	viability [%]	x	s	%	M	viability [%]	x	s	%
1. PEG 400 5%	N	90	0.90	0.13	9	N	90	3.06	0.46	89
2. PEG 400 10%	N	90	1.14	0.22	17	N	90	0.89	0.19	8
3. PEG 1000 10%	N	95	1.70	0.50	28	N	95	3.84	1.20	119
4. PEG 1000 20%	N	95	5.30	0.89	172	N	95	3.64	0.72	111
5. PEG 6000 5%	N	95	4.92	0.66	159	N	95	3.43	0.28	102
6. PEG 6000 10%	N/S	45	2.19	0.25	50	S	0	9.89	0.70	340
7. PEG 6000 20%	N/S	0	1.83	0.15	43	S	0	7.16	2.21	244
8. Glu 5%	N	100	1.69	0.28	37	N	90	5.68	1.32	188
9. Glu 10%	N	100	2.00	0.20	49	N	80	4.20	0.60	138
10. Glu 20%	N/S	80	5.96	0.68	198	N/P	0	6.74	2.56	228
11. Gly 5%	N	50	3.72	0.33	114	N	50	6.08	0.34	203
12. Gly 10%	N	50	2.60	0.60	72	N	50	4.94	0.56	160
13. Gly 20%	N	100	2.92	0.34	84	N/S	100	2.06	0.28	49
14. DMSO 5%	N	100	7.85	1.30	258	N	60	8.26	0.52	284
15. DMSO 10%	N	100	2.56	1.10	40	N	10	1.90	0.40	46
16. Sucrose 5%	N	90	5.24	0.98	172	N	30	3.66	0.52	112
17. Sucrose 10%	N	20	3.99	1.35	113	N	10	2.21	0.48	57
18. Sucrose 20%	N/P	10	2.66	0.92	73	N/S	10	1.50	0.48	27

FDA　fluorescein diacetate
PEG　polyethylene glycol
Glu　glucose
Gly　glycerol
DMSO　dimethyl sulfoxide
M　microscopy
N　normal appearance
P　plasmolysis
S　shrunken cells

TTC　triphenyltetrazolium chloride
control without cryoprotective treatment (100%): 3.46 ± 0.70
dead cells (0%): 0.80 ± 0.10
x　mean value
s　standard deviation

Table 2. Comparison of cell viability of habituated Chamomilla recutita cells after 10 and 60 min equilibration on ice in different cryoprotective mixtures

Cryoprotective mixture/content	10 min equilibration					60 min equilibration				
	FDA		TTC			FDA		TTC		
	M	viability [%]	x	s	%	M	viability [%]	x	s	%
1. Gly-sucrose-DMSO 5:10:10	N	100	4.00	0.30	125	N	50	3.20	0.42	95
2. Gly-sucrose-DMSO 5:5:5	N	100	5.08	0.74	190	N	100	3.90	1.00	125
3. Gly-sucrose-DMSO 10:10:10	N	100	2.60	0.36	72	N/S	50	3.46	0.34	104
4. Gly-sucrose-DMSO 10:5:10	N	60	4.34	0.88	136	N/S	50	3.96	1.14	122
5. Gly-sucrose-DMSO 10:10:5	N	60	1.78	0.32	42	N/S	30	2.90	0.70	82
6. Gly-sucrose-DMSO 5:5:10	N	70	7.00	1.44	236	N	30	3.36	0.80	101
7. Gly-sucrose-DMSO 5:10:5	N	50	3.14	0.60	75	N	25	4.68	0.62	140
8. Gly-sucrose-DMSO 10:5:5	N/S	50	5.40	1.12	178	N/S	10	4.72	1.08	145
9. PEG4000-Glu-DMSO 10:5:10	N	20	3.60	1.48	113	S	0	2.10	0.36	53
10. PEG1000-Glu-DMSO 10:5:10	N	40	6.40	0.48	214	P	0	5.05	0.19	164
11. PEG6000-Glu-DMSO 10:5:10	N	0	5.60	0.62	187	S/P	0	3.36	0.90	101

FDA   fluorescein diacetate
PEG   polyethylene glycol
Glu   glucose
Gly   glycerol
DMSO dimethyl sulfoxide
M   microscopy;
N   normal appearance
P   plasmolysis
S   shrunken cells

TTC   triphenyltetrazolium chloride
      control without cryoprotective
      treatment (100%): 3.46 ± 0.70
      dead cells (0%) : 0.80 ± 0.10
x   mean value
s   standard deviation

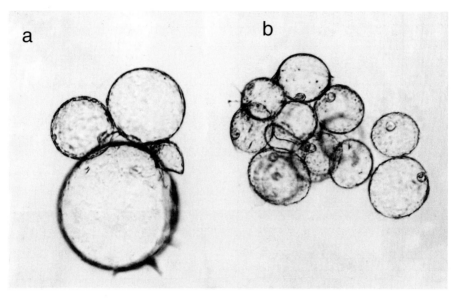

Fig. 1. Chamomile cells precultured on **a** 3% sucrose and **b** 7% sucrose

2.3.2 Effect of Cryoprotectants on Cell Viability

To determine the effect of single cryoprotectants and their mixtures on the survival of chamomile cells, viability tests of both fluorescein diacetate (FDA) and triphenyltetrazolium chloride (TTC) were applied after 10 and 60 min equilibration on ice prior to (Tables 1 and 2) and after cryopreservation (data not shown) was used. The application of a dilute solution of fluorescein diacetate differentiated between live and dead cells by conversion of FDA to fluorescein as a result of esterase activity. Cells with an intact plasmalemma fluoresce in ultraviolet light. The triphenyltetrazolium chloride reduction is based on the mitochondrial respiratory efficiency of cells that converts the tetrazolium salt to insoluble formazan, which is extracted and measured spectrophotometrically. In general, the use of cryoprotective mixtures did not show significant differences between 10 and 60 min equilibration. These results were compared with the microscopic picture of the treated cells.

2.3.3 Effect of the Cooling Rate on Cell Viability

The habituated chamomile cells were treated with a cryoprotective mixture consisting of glycerol/sucrose/DMSO (10:10:10, v/w/v) and cooled at 0.1, 0.2, 0.3, 0.4 and 0.5 °C/min to −10 °C and then at a rate of 1 °C/min to −40 °C. The two slowest cooling rates were lethal and gave extensive plasmolysis. Cooling at the rates of 0.4 and 0.5 °C/min caused damage to cells. Under given

Table 3. Cell viability of habituated callus pre-cultured on the basal RM medium with 0.5 M mannitol and cryoprotected with a mixture of glycerol (10%): sucrose (10, 20, or 30%): DMSO (10%) assessed by TTC treatment

Sucrose content	1 days after warming			After 1 month re-growth		
	x	s	%	x	s	%
10%	4.54	1.21	13.5	0.50	0.02	0.00
20%	10.25	1.62	35.6	0.55	0.02	0.00
30%	7.76	0.62	27.9	0.45	0.02	0.00

Unfrozen control: 28.68 ± 2.23 100%
Dead cells   0.49 ± 0.02 0%
x   mean value
s   standard deviation

Table 4. Comparison of cell viability of habituated Chamomilla recutita cells assessed by TTC test immediately after cryopreservation and after 1 month re-culture on the medium with plant growth substances

Immediately after cryopreservation			After one month re-culture		
$\bar{x}$	s	%	$\bar{x}$	s	%
10.26	1.62	35.60	23.90	7.30	79.00

Unfrozen control: 28.68 ± 2.23 100%
Dead cells: 0.49 ± 0.02 0%
x   Mean value
s   standard deviation

conditions of cryoprotection, cooling at 0.3 °C/min resulted in unchanged microscopic appearance of cells and was chosen for further experiments.

Since regrowth was not satisfactory under these conditions, the sucrose content in the cryoprotective mixture was increased to 20 and 30%, respectively. The highest viability occurred with 20% sucrose content in the cryoprotective mixture (Table 3). Increasing the sucrose content also required modifying the cooling rate to 0.5 °C/min. However, these habituated cultures still did not grow when placed on the basal medium without plant growth substances; this indicates a significant decrease or loss of capacity for phytohormone biosynthesis. When the cryopreserved cultures were placed on a medium supplemented with 0.5 mg/l benzylaminopurine (BAP) and 0.5 mg/l naphthaleneacetic acid (NAA), they started to divide after a 2-week lag phase (Table 4, Fig. 2). During the next subculture on the growth regulator medium, growth characteristics of cryopreserved cultures were comparable with the control unfrozen cells.

2.3.4 Protocol for Cryopreservation of Chamomile Habituated Cells

Plant material:	Habituated *Chamomilla recutita* callus cultures.
Preculture:	Cells in exponential growth phase on the basal medium supplemented with 0.5 M mannitol for 2 or 3 days.

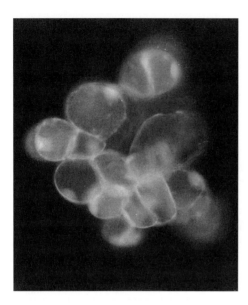

Fig. 2. Cell division in recovered chamomile cells after cryopreservation observed under a fluorescent microscope (FDA staining)

Cryoprotection:	0.5 g of callus placed in 1.5 ml of the liquid basal medium with a mixture of glycerol/sucrose/DMSO (10:20:10, v/w/v) on ice for 10 or 60 min.
Cooling rate:	0.5 °C/min to −10 °C and 1 °C/min to −40 °C.
Cryostorage:	After 30 min at −40 °C the samples are immersed in liquid nitrogen.
Warming:	In a water bath at 40 °C for 1 min.
Viability tests:	After warming the cells were placed on filter paper and subjected to FDA and TTC staining immediately and after 1 month re-growth on medium supplemented with 0.5 mg/l BAP and 0.5 mg/l NAA.

2.4 Discussion

The pregrowth phase of plant cells is considered as very important for attaining cryopreservation. During this period, various changes may occur on the cellular level including a decrease in cell and vacuole size, changes in the flexibility and thickness of cell walls, and alteration of metabolic activities (Withers 1978).

Addition of ABA and GA_3 to the pregrowth medium for chamomile habituated callus cultures stimulated callus growth and prolonged the exponential growth phase but did not increase survival of cryopreserved cells.

Long-term preculture on the basal medium with an increased sucrose content of 5, 7 and 10% resulted in a decreased cell volume and water content, but no regrowth after cryopreservation. In accordance with the results of Withers (1985) and Gnanapragasam and Vasil (1992), mannitol was the most effective agent during a short-term preculture for chamomile cells.

Information from the literature on the toxicity of cryoprotectants prior to the cooling process is limited. In some species cryoprotectants cause a temporary loss of semipermeability of membranes (McLellan et al. 1990). Pushkar et al. (1976) and Moiseyev et al. (1982) found in the presence of low molecular weight PEGs, in accordance with our results, a decreased activity of enzyme systems. In our experiments addition of PEG 400 as a single cryoprotectant or as a component of a cryoprotective mixture resulted in a significant loss of cell viability. According to Chen et al. (1984), DMSO concentrations higher than 5% significantly decreased the viability of *Catharanthus* cells. The same tendency was evident in our experiments with chamomile. Mixtures of DMSO, glycerol and sucrose, which were effective for cell suspensions of many plant species (Withers and King 1979), not only in a nutrient medium but also in water, were also the most appropriate for chamomile.

Chamomile cells were very sensitive to the cooling rate. Cooling rates lower than 0.3 °C/min caused plasmolysis of cells and those higher than 0.3 °C/min resulted, in accordance with the results of Demeulemeester et al. (1992), in mechanical damage of cells. A rate of 0.3 °C/min damaged the larger and non-dividing cells; however, as ascertained by FDA staining, the clusters of small isodiametric cells remained alive. Comparable results have been reported by Panis et al. (1990).

The ability to assess both quantitatively and qualitatively the condition of a specimen after the various stages in the cryopreservation procedure is one of the most important aids to the development of a freeze-preservation protocol. To provide a rapid and quite objective assessment, several tests are available. Quantification of the dehydrogenase-reductase activity in the callus cells by TTC reduction and its comparison in the control and treated samples may lead to inadequate results in several respects. Firstly, callus cultures are very heterogeneous systems. Secondly, retaining TTC reduction is not a definite indicator of cellular integrity and may even occur when the cells are extensively damaged. We performed two different viability tests that were supplemented with direct observation of cells in the light microscope and, on the basis of these observations, we optimised the individual steps of the procedure.

The most valuable viability response of recovering cells is evaluation of growth during several post-cryopreservation subcultures. The initial estimate of 35% viability of habituated chamomile cells after cryopreservation subsequently decreased which indicated loss of habituation on the hormone-free medium. When these cultures were transferred to a medium with plant growth regulators, cells started to divide after a 2-week lag phase. Similarly, Panis et al. (1990) showed that cytokinins stimulate the cell growth of cryopreserved cells but not the non-frozen cells.

3 Summary and Conclusions

Habituated *Chamomilla recutita* callus cultures were subjected to cryopreservation. The cryopreservation protocol presented is a result of an assessment of some necessary steps prior to and after cryopreservation. Some observations were made on the duration and conditions of the pregrowth period, effect of single cryoprotectants and their mixtures, cooling rate and recovery nutrient conditions. Cell viability was assessed by two tests. On the basis of these observations, a protocol for the cryopreservation of chamomile callus cultures was designed. It is hoped that with slight modifications this procedure can be applied for the storage of callus and cell cultures of chamomile which represent a potential source of secondary compounds of medical significance. The results of Diettrich et al. (1990), who successfully stored chamomile meristems that were capable of differentiation into plants, have an impact for the establishment of a gene bank for use in breeding programs.

References

Agullo G, Gamet-Payrastre L, Manenti S, Viala C, Remesy C, Chap H, Payrastre B (1997) Relationship between flavonoid structure and inhibition of phosphatidylinositol 3-kinase: a comparison with tyrosine kinase and protein kinase C inhibition. Biochem Pharmacol 53: 1649–1657

Bernáth J (ed) (1993) Vadon termö és termesztett gyógynövények. Mezögazda Kiadó, Budapest, p 355

Birt DF, Walker B, Tibbels MG, Bresnick E (1986) Anti-mutagenesis and anti-promotion by apigenin, robinetin and indole-3-carbinol. Carcinogenesis 7:959–963

Carle R, Seidel F, Franz C (1991) Investigation into seed germination of *Chamomilla recutita* (L.) Rauschert. Angew Bot 65:1–8

Čellárová E, Greláková K, Repčák M, Hončariv R (1982) Morphogenesis in callus tissue cultures of some *Matricaria* and *Achillea* species. Biol Plant 24:430–433

Čellárová E, Černická T, Vranová E, Brutovská R, Lapár M (1992a) Viability of *Chamomilla recutita* (L.) Rauschert cells after cryopreservation. Cryo Lett 13:37–42

Čellárová E, Kimáková K, Brutovská R (1992b) Effect of cryoprotectants on the viability of plant cells. In: Theoretical basis of cryopreservation. Abstracts of the international conference, Hradec Králové, Czechoslovakia, 7–9 Oct 1992, p 26

Chen THH, Kartha KK, Constabel F, Gusta LV (1984) Freezing characteristics of cultured *Catharanthus roseus* (L.) G. Don cells treated with dimethylsulfoxide and sorbitol in relation to cryopreservation. Plant Physiol 75:720–725

Demeulemeester MAC, Panis BJ, De Proft MP (1992) Cryopreservation of in vitro shoot tips of chicory (*Cichorium intybus* L.). Cryo Lett 13:165–174

Diettrich B, Donath P, Popov AS, Butenko RG, Luckner M (1990) Cryopreservation of *Chamomilla recutita* shoot tips. Biochem Physiol Pflanzen 186:63–67

Gamborg OL, Miller RA, Ojima K (1968) Nutrient requirements of suspension cultures of soybean root cells. Exp Cell Res 50:148–151

Gnanapragasam S, Vasil IK (1992) Ultrastructural changes in suspension culture cells of *Panicum maximum* during cryopreservation. Plant Cell Rep 11:169–174

Huang YT, Kuo ML, Liu JY, Huang SY, Lin JK (1996) Inhibition of protein kinase C and proto-oncogene expression in NIH 3T3 cells by apigenin. Eur J Cancer 32A:146–151

Kimáková K (1995) Vplyv kryokonzervácie na zmeny bunkových funkcií kalusových kultúr *Matricaria recutita* (L.) Rauschert a na genetickú stabilitu *Hypericum perforatum* L. (The effect of cryopreservation on the changes of the *Matricaria recutita* (L.) Rauschert cell functions and on genetic stability of *Hypericum perforatum* L.). PhD Thesis, PJ Šafárik University, Košice, pp 117

Linsmaier EM, Skoog F (1965) Organic growth factor requirements of tobacco tissue cultures. Physiol Plant 18:100–127

Mann C, Staba EJ (1986) The chemistry, pharmacology, and commercial formulations of chamomile. In: Cracker LE, Simon JE (eds) Herbs, spices and medicinal plants. Recent advances in botany, horticulture, and pharmacology, vol 1. Oryx Press, Phoenix, pp 237–280

McLellan MR, Schrijnemakers EWM, Iren FV (1990) The responses of four cultured plant cell lines to freezing and thawing in the presence or absence of cryoprotectant mixtures. Cryo Lett 11:189–204

Menghini A, Standardi A, Tavoletti S, Veronesi F (1996) In vitro culture technique for chamomile propagation. Atti convegno internazionale: Coltivazione e miglioramento di piante officinali, Trento, Italy, 2–3 July 1996, pp 375–379

Moiseyev VA, Nardid OA, Belous AM (1982) On a possible mechanism of the protective action of cryoprotectants. Cryo Lett 3:17

Panis BJ, Withers LA, De Langhe EAL (1990) Cryopreservation of musa suspension cultures and subsequent regeneration of plants. Cryo Lett 11:337–350

Pushkar NS, Sheenberg MG, Oboznaya EI (1976) On the mechanism of cryoprotection by polyethylene oxide. Cryobiology 13:142–146

Sakr SS, Badawy EM, Morsi HA, El-Bahr MK, Taha HS (1991) Regeneration of chamomile plant (*Chamomilla recutita* L.). Bull Fac Agr Univ Cairo 42:1461–1483

Steponkus PL, Lanphear O (1967) Refinement of the triphenyltetrazolium chloride method of determining cold injury. Plant Physiol 42:1423–1426

Tavoletti S, Standardi A, Veronesi F (1994) Response to in vitro culture of two chamomile (*Matricaria chamomilla* L.) populations with different ploidy levels. J Gen Breeding 48:125–130

Widholm JM (1972) The use of fluorescein diacetate and phenosafranine for determining viability of cultured plant cells. Stain Technol 47:189–194

Withers LA (1978) A fine-structural study of the freeze preservation of plant tissue cultures. Protoplasma 94:235–247

Withers LA (1985) Cryopreservation of cultured plant cells and protoplasts. In: Kartha KK (ed) Cryopreservation of plant cells and organs. CRC Press, Boca Raton, pp 243–268

Withers LA, King PJ (1979) Proline: a novel cryoprotectant for the freeze preservation of cultured cells of *Zea mays* L. Plant Physiol 64:675–678

II.5 Cryopreservation of *Cichorium intybus* L. var. *foliosum* (Chicory)

B. Vandenbussche, M. Demeulemeester, and M. De Proft

1 Introduction

1.1 Taxonomy

Chicory is a member of the *Compositae* (or *Asteraceae*) family and belongs to the genus *Cichorium*. The genus consists of six different species, but only *Cichorium intybus* L. and *Cichorium endivia* L. are economically important. The first species is subdivided in three main groups: *C. intybus* var. *intybus* (wild form), *C. intybus* var. *foliosum* (leaf chicory) and var. *sativum* (root chicory). Within *C. intybus* var. *foliosum*, four different cultivar groups are distinguished, among them Belgian endive or witloof chicory (Vermeulen et al. 1994). Chicory originated from the Mediterranean area and is now widespread over Western, Central and Southern Europe, Northern Africa and temperate regions of Asia.

1.2 Commercial Production of Belgian Endive

During the first year of growth, plants remain vegetative and rosettes of green leaves and large taproots are formed. These roots are harvested in autumn, stripped of their green leaves and stored at −1 or +1 °C (90–95% RH) for up to 1 year. Finally, roots are forced in complete darkness at a temperature of 15 °C and an air humidity of 85–90%. After approximately 3 weeks a white, yellowish, etiolated chicory head is harvested. The vegetable has a cylindrical shape with a conical tip. Chicory heads are produced in both hydroponics and heated soil (Schoofs and De Langhe 1988).

1.3 Storage of Chicory Germplasm

In most chicory collections germplasm is stored as seed at either room temperature or at 4 °C in a cold room. Seeds can also be conserved in freezers at −18 °C. The maintenance of these collections is very laborious since accessions

Laboratory of Plant Culture, Faculty of Agricultural and Applied Biological Science, Katholieke Universiteit Leuven, Willem de Croylaan 42, 3001 Heverlee, Belgium

Biotechnology in Agriculture and Forestry, Vol. 50
L.E. Towill and Y.P.S. Bajaj (Eds.) Cryopreservation of Plant Germplasm II
© Springer-Verlag Berlin Heidelberg 2002

have to be regenerated every 4 to 10 years. Because chicory is a cross-fertilizing and insect-pollinated species there is a risk of losing accessions due to unwanted cross-pollination.

However, not all chicory accessions are stored as seeds. With chicory, breeders select individual mother plants and maintain them in in vitro culture until their breeding value is tested (Desprez et al. 1994). This allows breeders to improve chicory populations and thus increase response to selection in a polycross-breeding scheme (Frese 1996). The use of in vitro germplasm collections has some important disadvantages. The different genotypes have to be propagated regularly to keep plant material in a healthy status. This is not only time consuming, but also increases the risk of loosing valuable resources due to infections and/or human errors (Towill 1988). Cryopreservation helps to overcome some of these problems.

1.4 Need for Conservation of Chicory Germplasm

In 1996 a meeting was held at Braunschweig (Germany) to start an international program to safeguard genetic resources of chicory. The aim was to construct a network to help collect, regenerate and conserve different accessions.

Until now, only small collections have existed in different European countries. In France, a national chicory collection (GEVES, Brion) was founded in 1995 containing more than 600 accessions of *Cichorium intybus* L. (root and leaf chicory) and *C. endivia* L. (Kelechian-Cadot and Boulineau 1996). Frese and Dambroth (1987) described a smaller collection of 50 accessions (root and leaf chicory) at Braunschweig (Germany). Also in Sweden and Italy different accessions of *Cichorium intybus* L. (root and leaf chicory) are conserved. In Belgium, more than 800 accessions of *Cichorium intybus* var. *foliosum* (Belgian endive) were collected and are currently stored at the Laboratory of Plant Culture, Katholieke Universiteit, Leuven. The greater part of this collection consists of ancient varieties obtained by farmers through mass selection and which only produce chicory heads when forced in heated soil. Nowadays, these varieties no longer have a direct economic value since F1 hybrids are used and chicory heads are mostly produced in hydroponics. Nevertheless, these accessions are very valuable to enlarge the genetic base for future breeding of Belgian endive.

2 Cryopreservation

2.1 Introduction

Several techniques have been optimized to cryopreserve in vitro chicory shoot tips. In 1992, Demeulemeester and colleagues reported the first successful cryopreservation of chicory shoot tips. Plants were regenerated, their morphology screened and seed set evaluated (Demeulemeester et al. 1993). Later that year a different approach was tested to cryopreserve chicory shoot tips. This resulted in the optimization of an encapsulation-dehydration technique (Vandenbussche et al. 1993). Recently, seeds of chicory have been cryopreserved using a rapid freezing procedure.

2.2 Materials and Methods

2.2.1 Plant Material and Seeds

The in vitro experiments used an early cultivar, Flash (INRA, France), and two late cultivars, Rumba (Clause, France) and Carolus (Bucomat, Belgium), of *Cichorium intybus* var. *foliosum* (Belgian endive). Plants were sown in spring at the Research Station for chicory in Herent, Belgium, on a sandy-loam soil. In autumn, roots were harvested and stored at −1 or 1 °C depending on the cultivar used. In vitro cultures were initiated using these roots.

Seeds of two early cultivars, Venus (Clause, France) and Focus (Nunhems, The Netherlands), and one late cultivar, Tabor (Nunhems, The Netherlands), of *Cichorium intybus* var. *foliosum* (Belgian endive) were cryopreserved.

2.2.2 In Vitro Culture Technique

Roots were washed, peeled and disinfected by immersion in ethanol (96% for 20s) and NaOCl (4% for 15min). After rinsing three times (10, 5, 5min) with sterile distilled water, cylindrical fragments with a diamctcr of 10mm were excised from roots and placed on a nutrient medium. A modified MS medium (Murashige and Skoog 1962) without NH_4NO_3, $ZnSO_4$ and $CuSO_4$ and with a reduced $Na_2 \cdot EDTA$ concentration was used. NH_4^+ was omitted from the medium because it evoked hyperhydricity (Vasseur et al. 1986). The medium was supplemented with $10\,gl^{-1}$ sucrose (Fluka, biochemical grade), $6\,gl^{-1}$ agar (Gibco BRL), $10^{-5}\,M$ BAP and $10^{-6}\,M$ IAA. This medium is referred to as MS1. The pH was adjusted to 6.0 before autoclaving for 15min at 121 °C. Four root fragments were placed in a 360-ml glass culture container sealed with a transparent polycarbonate lid (De Proft et al. 1985). Cultures were maintained in an air-conditioned room with an air humidity of 60(±5)% and a temperature of 23(±2) °C. Plants grew under a light intensity of 80–90 μmol m^{-2} s^{-1} (white fluorescent lamps Philips 36 W, TLD/54) with a 16-h photoperiod.

2.2.3 Cryopreservation of Shoot Tips

2.2.3.1 Two-Step Cooling Method (Demeulemeester et al. 1992)

Shoot tips (1–2mm in length) were dissected from 14-day-old, in vitro initiated root fragments using a stereomicroscope (Nikon, magnification 8–10).

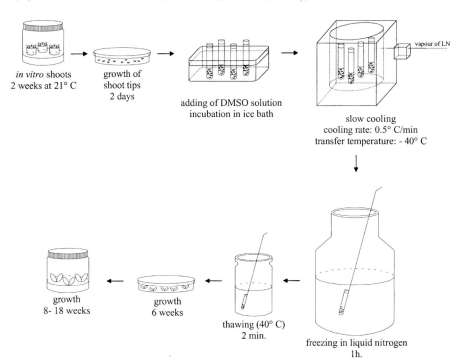

in vitro shoots
2 weeks at 21° C

growth of
shoot tips
2 days

adding of DMSO solution
incubation in ice bath

vapour of LN

slow cooling
cooling rate: 0.5° C/min
transfer temperature: - 40° C

growth
8- 18 weeks

growth
6 weeks

thawing (40° C)
2 min.

freezing in liquid nitrogen
1h.

Fig. 1. Scheme of the different manipulations involved in cryopreservation of *Cichorium intybus* L. shoot tips when using a two-step cooling method

An overview of the applied procedure is shown in Fig. 1. Dissected meristems were grown for 2–7 days in Petri dishes on a filter paper (Schleicher and Schuell NR 515, 9 cm) soaked with 2.5 ml of liquid medium before use. A modified MS medium without plant growth regulators (MS2) was used; however, it was supplemented with different DMSO concentrations (Janssen Chimica, 99.9%).

For cryoprotection, shoot tips were transferred to 5-ml cryotubes containing 2 ml of liquid MS2 medium. Cryotubes were chilled in an ice bath and 2.0 ml of cryoprotectant was gradually added over a period of 30 min. Cryovials were kept for another 1–2 h in the same ice bath. Chilled MS2 medium mixed with DMSO at twice the chosen concentration was added to each cryotube.

Shoot tips (ten per cryotube) were cooled at different rates (0.1–0.7 °C/min) to a transfer temperature of −30, −40 or −50 °C. The applied cooling device was controlled by a Gulton West 2050 unit (L'air Liquide) and used liquid nitrogen as cooling agent. Cryotubes were held at a transfer temperature for 10 min and were subsequently plunged into liquid nitrogen for at least 1 h. Shoot tips were warmed by agitating the cryovials for 2–3 min in sterile water at 40 °C. Shoot tips were drained of superficial cryoprotectant solution and placed in Petri dishes (diameter 9 cm) on a solidified MS1 medium. During preculture and growth, shoot tips were maintained under similar conditions as described earlier (Sect. 2.2.2). After 4 weeks, growth (direct shoot formation or callusing) was evaluated.

For each experiment or treatment approximately 30 shoot tips were used. Data were analyzed using a logistic regression model, allowing comparison of results with a χ^2-test. A significance level of P_r: $\chi^2 > 0.05$ was used.

2.2.3.2 Encapsulation-Dehydration Technique (Vandenbussche et al. 1993)

Shoot tips (1–2 mm in size) were dissected from 14-day-old root fragments and grown for 2 days in Petri dishes on a sterile filter paper (Schleicher and Schuell NR 515, 9 cm) soaked with 2.5 ml of liquid MS2 medium (Fig. 2). Shoot tips were subsequently suspended in Ca^{2+}-free MS1 medium supplemented with 30 g l^{-1} Na-alginate (Fluka, Biochemika). This mixture was dropped from a pipette with a sterile tip into MS1 medium containing 100 mM $CaCl_2$ (Dereuddre et al. 1990). After 30 min, beads of approximately 4 mm and containing one or two shoot tips were selected and precultured for 24 h in liquid MS1 medium enriched with different sucrose concentrations. Sucrose levels from 0.3–1.0 M were tested. When applying higher sucrose levels, the concentration was sometimes increased stepwise. With a two-step preculture, encapsulated shoot tips were first cultured for 24 h in liquid MS1 medium enriched with 0.3 M sucrose and were subsequently placed for another 24 h in liquid MS1 medium supplemented with 0.75 M sucrose. In a following step, beads

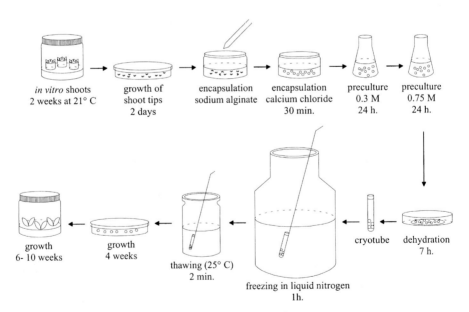

Fig. 2. Scheme of the different manipulations involved in cryopreservation of *Cichorium intybus* L. shoot tips when using an encapsulation-dehydration technique

were placed on a sterile filter paper (Schleicher and Schuell NR 515, 9 cm) in an open Petri dish and dried at room temperature for 4–10 h using a sterile air flow in a laminar air flow cabinet. The water content of 30 beads was measured. Each bead was weighed immediately after dehydration and after drying in an oven at 70 °C. To test the impact of preculture and dehydration, control beads were placed in Petri dishes (diameter 9 cm) on solidified MS1 medium. For each combination of preculture and dehydration, 30 encapsulated shoot tips were used as a control treatment. To test the influence of cryopreservation, 50 dehydrated beads were placed into 5-ml cryotubes (ten per tube) and the tubes were plunged immediately into liquid nitrogen. After 1-h storage, cryotubes were agitated in sterile water at 25 °C (2 min) to warm shoot tips. Beads were then placed in Petri dishes (diameter 9 cm) on solidified MS1 medium. Throughout the different manipulations and growth, similar culture conditions were applied as described in Section 2.2.2. After 2 weeks, growth (direct shoot formation or callusing) was evaluated. Data were analyzed as described in Section 2.2.3.1.

2.2.4 Cryopreservation of Seeds

Seeds were obtained from different breeding companies. Without a prior treatment, 100 seeds were placed into 5-ml cryotubes and plunged immediately into liquid nitrogen and held for 1 h. Each tube contained 20 seeds. To warm seeds, cryovials were agitated in sterile water at 38 °C for 2 min. Finally, seeds were placed in Petri dishes (20 per dish) on a filter paper (Schleicher and Schuell NR 515, 9 cm) soaked with 2.5 ml of sterile water. The same culture conditions were applied as described earlier (Sect. 2.2.2). As a control treatment, 100 seeds were immediately germinated. To calculate the average water content, 100 seeds were weighed before and after drying in an oven at 70 °C. Germination was evaluated after 2 weeks. Each experiment was repeated once and results were compared using Duncan's multiple range test ($\alpha = 0.05$).

2.3 Results

2.3.1 Two-Step Cooling (Demeulemeester et al. 1992, 1993)

With the three cultivars approximately 90% of the shoot tips resumed growth after dissection. To test the influence of DMSO on growth, dissected shoot tips of 'Flash' were placed for 2–7 days in Petri dishes on MS1 medium supplemented with different DMSO concentrations (Table 1). Within the exposure time tested, growth was strongly reduced when applying 10% DMSO (significant at a 5% level). Shoot tips turned brown or formed callus. Levels of up to 5% DMSO during preculture did not influence growth.

To optimize the cryoprotectant solution, different DMSO concentrations and exposure times were tested (Table 2). Without preculture, shoot tips were cooled (0.5 °C/min) to a transfer temperature of –40 °C. Growth of

Table 1. Influence of exposure time to different DMSO concentrations during preculture on survival [%] of chicory shoot tips (cv Flash)

DMSO concentration [%]	duration of preculture [days]			
	2	3	4	7
0	100	100	100	100
1	100	100	100	100
3	100	97 ± 3	100	100
5	94 ± 4	100	100	97 ± 3
10	27 ± 8	8 ± 5	0	0

Table 2. Influence of exposure time to different DMSO concentrations during cryoprotection on survival [%] of uncooled and cryopreserved chicory shoot tips (cv Flash)

DMSO concentration [%]	duration of cryoprotection (h)			
	1		2	
	control	cryopreserved	control	cryopreserved
0	100	0	100	0
5	100	0	100	0
10	100	0	100	45 ± 9
15	100	53 ± 9	95 ± 5	–
20	94 ± 4	40 ± 10	86 ± 8	–

–: not tested, cooling rate: 0.5 °C. min⁻¹, transfer temperature: –40 °C.

Table 3. Influence of different preculture and cryoprotection treatments on survival [%] of uncooled and cryopreserved chicory shoot tips (cv Flash)

DMSO concentration during preculture for 2d [%]	cryoprotection			
	15% DMSO for 1h		10% DMSO for 2h	
	control	cryopreserved	control	cryopreserved
0	100	83 ± 7	100	22 ± 7
1	100	50 ± 9	100	26 ± 8
3	100	32 ± 8	100	10 ± 5
5	100	16 ± 7	100	3 ± 3
10	13 ± 6	0	53 ± 9	0

Temperature during preculture: 23(± 2) °C, cooling rate: 0.5 °C.min⁻¹, transfer temperature: –40 °C.

control shoot tips of 'Flash' was not influenced after incubation in a cryoprotectant solution for 1 or 2 h. Growth was slightly reduced only when incubated for 2 h in a 20% DMSO solution. However, shoot tips of 'Flash' only survived cryopreservation with a 15 or 20% DMSO solution as cryoprotectant (significant at a 5% level). After incubation for 2 h, a lower concentration (10%) was sufficient to protect shoot tips.

To further increase survival after freezing, shoot tips of 'Flash' were precultured on media supplemented with different DMSO concentrations (Table 3). Especially after cryopreservation, survival decreased with higher

Table 4. Influence of different preculture conditions, cooling rates and transfer temperatures on survival [%] of cryopreserved chicory shoot tips (cv Flash)

transfer temperture [°C]	DMSO concentration during preculture [%]	cooling rate [°C.min^{-1}]			
		0.1	0.3	0.5	0.7
−40	0	2 ± 2	4 ± 3	83 ± 7	7 ± 5
	1	4 ± 3	2 ± 2	50 ± 9	0
	3	0	0	30 ± 8	0
	5	0	2 ± 2	17 ± 7	0
−50	0	–	43 ± 9	23 ± 8	–
	1	–	26 ± 8	23 ± 9	–
	3	–	0	20 ± 7	–
	5	–	3 ± 3	12 ± 6	–

–, not tested.

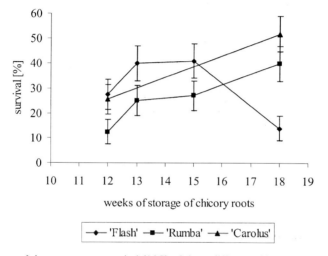

Fig. 3. Influence of the root storage period (1 °C) of three different chicory cultivars on survival (%) of shoot tips after cryopreservation

DMSO concentrations during preculture. Best results were obtained on MS1 medium lacking DMSO and after incubation for 1 h in a 15% DMSO solution (significant at a 5% level).

The cooling protocol was optimized by testing different cooling rates and transfer temperatures (Table 4). Shoot tips of 'Flash' were grown for 2 days on MS1 medium without DMSO and were incubated in 15% DMSO for 1 h. Both transfer temperatures and cooling rates affected survival after cryopreservation. The highest survival was obtained with a cooling rate of 0.5 °C/min and a transfer temperature of −40 °C (significant at a 5% level). At a transfer temperature of −30 °C, no survival was observed.

Finally, shoot tips of two other cultivars were cryopreserved using the optimized protocol (Fig. 3). The storage period of roots had a distinct influence on survival of shoot tips after cryopreservation. With the early cultivar 'Flash',

Fig. 4. Regenerated plants of *Cichorium intybus* L. (cv. Flash) after cryopreservation with a two-step cooling method (cryoprotectant: 15% DMSO solution; cooling rate: 0.5 °C/min; transfer temperature: –40 °C)

survival decreased after 15 weeks of storage, while the opposite was true for the late cultivars Rumba and Carolus.

Callus formation was a major problem during growth of cryopreserved chicory shoot tips. Moreover, shoots were often hyperhydric. Callus formation was reduced by placing shoot tips for 1 week on solidified MS1 medium, followed by a transfer to medium lacking plant growth regulators (MS2, results not shown).

Regenerated shoots were rooted either on a half-strength MS2 medium (5–7 weeks) or by treating shoots as cuttings (4–5 weeks). With the latter method shoots were dipped into commercial rooting powder (Rhizopon B) containing 0.2% α-NAA, placed in potting soil and kept under a plastic cover. To acclimate rooted shoots, the vessel cover was gradually opened to reduce humidity. Finally, plants were transferred to a greenhouse for further growth. No morphological differences were observed in treated plants when compared with control plants (Fig. 4). Taproots of regenerated plants were harvested. After vernalisation, taproots were planted in a greenhouse to produce seeds (Fig. 5). Again, plant morphology was not changed and obtained seeds had a germination rate of 80–93% depending on the tested cultivar.

Fig. 5. Regenerated flowering plants of *Cichorium intybus* L. (cv. Flash) after cryopreservation with a two-step cooling method (cryoprotectant: 15% DMSO solution; cooling rate: 0.5 °C/min; transfer temperature: –40 °C)

2.3.2 Encapsulation-Dehydration Technique (Vandenbussche et al. 1993)

Excised shoot tips of 'Flash' were grown for 2 days before encapsulation and placed immediately on solidified MS1 medium. Encapsulation did not influence survival (100%), although growth was slower.

Alginate-coated shoot tips of 'Flash' were precultured in liquid MS1 medium supplemented with different sucrose concentrations. Preculture for 24 h in a 0.3 M sucrose solution did not influence survival (100%). When higher sucrose concentrations were applied, survival gradually decreased (results not shown). By using a two-step preculture this problem was overcome. After preculture in a 0.75 M sucrose solution only 54(±9)% of shoot tips survived. However, when the sucose concentration was stepwise increased (0.3, 0.75 M), 100% of the encapsulated shoot tips resumed growth.

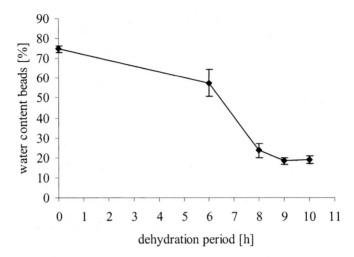

Fig. 6. Time course of dehydration of encapsulated chicory shoot tips (cv. Flash) after a two-step preculture (0.3, 0.75 M)

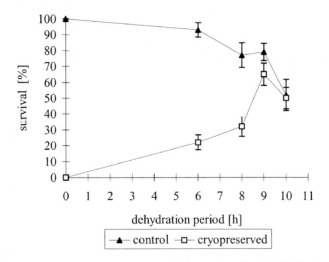

Fig. 7. Survival (%) of control and cryopreserved chicory shoot tips (cv. Flash) after a two-step preculture (0.3, 0.75 M) and different dehydration times

To obtain a vitrified state during cooling precultured beads were dehydrated prior to plunging into liquid nitrogen (Fig. 6). After preculture in a 0.3 or 0.5 M sucrose solution and 6 h of dehydration, no shoot tips survived cryopreservation. However, when beads were stepwise precultured (0.3, 0.75 M) and dehydrated for 6 h, shoot tips survived cryopreservation (22 ± 5%). Further dehydration (9 h) of alginate-coated shoot tips to a water content of 18.3 ± 1.9% clearly increased survival after cryopreservation (65 ± 7%, significant at 5% level). A longer dehydration period did not improve survival after cooling (Fig. 7). A three-step preculture (0.3, 0.75, 1.0 M), and dehydration to

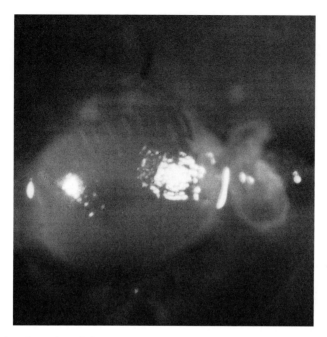

Fig. 8. Surviving shoot tips of *Cichorium intybus* L. (cv. Flash) after cryopreservation with an encapsulation-dehydration technique (two-step preculture, dehydration of beads to a water content of around 20%)

a water content of $21.8 \pm 1.9\%$, did not increase survival after cryopreservation ($51 \pm 7\%$).

To test whether the medium influenced survival after cooling, the sucrose concentration was elevated (0.3 M) during the first week of growth. This higher sucrose concentration during early growth did not enhance survival after cryopreservation ($57 \pm 13\%$).

During growth of cryopreserved shoot tips of 'Flash' two different patterns could be distinguished. Most of the shoot tips (79%) resumed growth by direct shoot formation, while only 21% formed a callus (Fig. 8). Sometimes shoots were slightly hyperhydric. Four weeks after cryopreservation, surviving shoot tips were transferred to MS2 to inhibit new or further callusing. Finally, shoots were rooted, acclimated and planted in a greenhouse. No morphological changes were observed compared with control plants.

2.3.3 Cryopreservation of Chicory Seeds

Chicory seeds of three commercial cultivars were, without further dehydration, placed into cryotubes and plunged immediately into liquid nitrogen (Table 5). No differences in germination were observed between control and cryopreserved seeds (Fig. 9). Although the water content of seeds from three

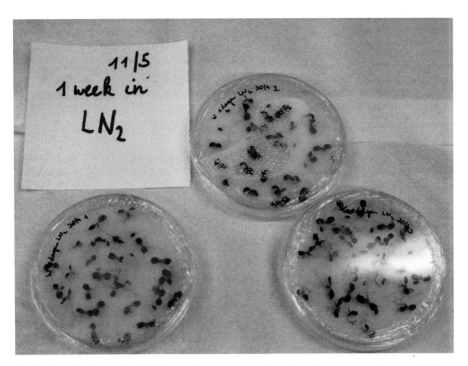

Fig. 9. Germination test of *Cichorium intybus* L. (cv. Focus) seeds when applying a rapid freezing technique. Dry seeds were plunged into liquid nitrogen

Table 5. Germination [%] of control and cryopreserved seeds of three chicory cultivars. Dry seeds were plunged into liquid nitrogen

cultivar	water content of seeds [%]	survival [%]	
		control	cryopreserved
'Focus'	12.79 ± 1.3	86.0 ± 2.0	85.0 ± 3.0
'Tabor'	8.33 ± 0.8	87.0 ± 2.0	88.0 ± 2.0
'Venus'	10.95 ± 1.5	89.0 ± 2.0	94.0 ± 2.0

chicory cultivars varied, no significant differences in survival (germination) after cooling were observed when comparing cultivars.

2.4 Discussion

Shoot tips of chicory can be cryopreserved using both a two-step cooling method (Demeulemeester et al. 1992) and an encapsulation-dehydration technique (Vandenbussche et al. 1993).

When applying a slow cooling method, plant tissues are cooled to −35 or −40 °C at 0.1–10 °C/min prior to a transfer to liquid nitrogen (Withers 1980).

Hence controlled freezing devices are necessary. With these cooling rates extracellular ice crystals are formed, which dehydrate the surrounding cells and help to induce intracellular vitrification (Grout and Morris 1987). To optimize a two-step freezing method the most important aspects are the applied cryoprotectant, cooling rate and transfer temperature (Towill 1983).

With chicory, best results were obtained using a 15% DMSO solution as cryoprotectant. At lower DMSO concentrations no shoot tips survived freezing, unless the incubation time was doubled. Doubling of the DMSO concentration did not further increase survival after cryopreservation.

As demonstrated by other authors, the cooling rate can be very decisive for survival after cryopreservation (Reed 1990). With chicory, cooling rates lower or higher than 0.5 °C/min decreased survival after cryopreservation. When the applied cooling rate is too slow, cells are strongly dehydrated resulting in an increase of the intracellular solute concentration. As a consequence, enzyme activity is reduced and proteins are denaturated. Furthermore, a transition of membrane phospholipids to a hexagonal II phase may occur which disrupts functioning of membranes (Grout and Morris 1987). When applying faster cooling rates, cells contain too much water and intracellular ice crystals can be formed upon freezing in liquid nitrogen (Kartha 1987).

A transfer temperature of at least −40 °C was needed to obtain survival after cryopreservation. An interaction between cooling rate and transfer temperature was observed. With a cooling rate of 0.5 °C/min, best results were obtained when a transfer temperature of −40 °C was used. However, at a cooling rate of 0.3 °C/min, results were improved by lowering the transfer temperature to −50 °C.

To further increase survival after cryopreservation, chicory shoot tips were precultured on media supplemented with different DMSO concentrations. Adding DMSO to the media reduced survival after cryopreservation. Moreover, when meristems were grown for 2 days on MS1 medium lacking DMSO, survival after cryopreservation improved. This beneficial effect could be explained as a recovery of shoot tips from dissection. A similar healing was observed with meristems of potato (Benson et al. 1989), *Phoenix dactylifera* L. (Bagniol and Engelmann 1991) and *Beta vulgaris* L. (Vandenbussche and De Proft 1996).

Fabre and Dereuddre (1990) first applied the encapsulation-dehydration technique. This methodology can only be used if plant cells tolerate severe dehydration. Sucrose is mostly used as cryoprotectant to induce a vitrification state during cooling (Dereuddre 1992; Sakai 1993). Moreover, sucrose enhances the dehydration tolerance by stabilizing membranes through interactions with membrane phospholipids (Crowe et al. 1984).

Alginate-coated shoot tips of chicory survived cryopreservation after dehydration to a water content between 18 and 22%. Survival only occurred if a two-step preculture was used. Similar results were observed with shoot tips of *Vitis vinifera* L. (Plessis et al. 1993) and *Beta vulgaris* L. (Vandenbussche and De Proft 1996). A further increase of the sucrose concentration during preculture (up to 1.0 M) did not increase survival after cryopreservation.

Unlike walnut embryos (de Boucaud et al. 1994) and shoot tips of *Salix* spp. (Blakesley et al. 1996), survival and growth did not improve with a higher sucrose concentration during regrowth (0.3 M). After cryopreservation of chicory shoot tips with a two-step cooling or an encapsulation-dehydration technique, callus was formed during growth. Especially when using a two-step cooling method callus was often formed prior to shoot formation. A similar growth pattern was observed with shoot tips of *Mentha* (Towill 1988). However, a transfer of chicory shoot tips, after 1 week of growth, to a medium lacking plant growth regulators reduced callusing. Also with shoot tips of *Brassica napus* the growth pattern after cryopreservation was influenced by the applied culture medium (Withers et al. 1988). With an encapsulation-dehydration technique callus formation was strongly reduced compared to a two-step cooling method. Around 80% of the surviving shoot tips regrew by direct shoot formation. This could indicate that shoot tips were less damaged after cryopreservation with an encapsulation-dehydration technique compared to two-step cooling. Furthermore, encapsulated chicory shoot tips resumed growth faster than shoot tips that had been cryopreserved with a two-step cooling method. Regenerated plants were planted in a greenhouse after 10–12 weeks when applying an encapsulation-dehydration method, compared to 18–20 weeks after a two-step cooling technique. Similar findings were reported with shoot tips of *Solanum* sp. (Fabre and Dereuddre 1990). With both cooling methods no macromorphological abnormalities were observed in regenerated plants.

Cryopreservation is often indicated as a powerful method for long-term conservation of seeds (Stanwood 1985). To survive cooling, seeds need to be sufficiently dehydrated to vitrify (Sun et al. 1994). The optimal water content differs among species but usually varies between 5 and 10% (Withers 1987; Salomão 1995; Lakhanpaul et al. 1996). However, with some species, higher survival percentages were obtained at higher water contents (Wang et al. 1998). Orthodox seeds are tolerant to this severe dehydration because they accumulate different kinds of sugars (Amuti and Pollard 1977; Sun et al. 1994) and proteins (Blackman et al. 1992) that help to stabilize membranes (Crowe et al. 1984). Recalcitrant and imbided seeds cannot be dehydrated that strongly (Withers 1987). If they are strongly dehydrated, their cell solutions will vitrify but membranes change to a gel phase that hinders normal functioning (Sun et al. 1994). Sometimes the use of a slow cooling rate (Vertucci 1989; Dussert et al. 1997), addition of a cryoprotectant (de Boucaud et al. 1991) or extraction of embryonic axes (Berjak and Dumet 1996) can improve results. The applied drying rates can influence survival as well (Vertucci et al. 1991; Potts and Lumpkin 1997).

Commercially available chicory seeds have water contents of around 10%, indicating that this species is orthodox. When seeds, without additional drying, were plunged into liquid nitrogen high germination was obtained. Although the water content of seeds from different cultivars varied, no influence on germination after cryopreservation was observed.

3 Conclusions

In vitro shoot tips of chicory can be cryopreserved using either a two-step cooling method or an encapsulation-dehydration technique.

With a two-step cooling method, shoot tips were grown for 2 days on MS2 medium lacking DMSO and cryoprotected for 1 h with a 15% DMSO solution. Best results were obtained at a transfer temperature of −40 °C and a cooling rate of 0.5 °C/min. With the encapsulation-dehydration technique shoot tips were grown for 2 days on MS2 medium. After encapsulation, beads were subsequently precultured stepwise in a sucrose solution (0.3 M, 0.75 M), dehydrated to water contents of around 20% and plunged into liquid nitrogen. To store in vitro chicory germplasm, an encapsulation-dehydration technique is most appropriate since the use of a freezing device and DMSO is avoided. Furthermore, growth is not only improved but also accelerated.

Chicory appeared to be an orthodox species and seeds with water contents of about 10.0% survived a direct plunge into liquid nitrogen. Therefore, this method is recommended for the long-term safe storage of seeds.

Acknowledgements. The first author is recipient of an IWT grant. The second author was recipient of an NFWO accomplishment during this study.

References

Amuti KS, Pollard CJ (1977) Soluble carbohydrates of dry and developing seeds. Phytochemistry 16:529–532

Bagniol S, Engelmann F (1991) Effects of pregrowth and freezing conditions on the resistance of meristems of date palm (*Phoenix dactylifera* L var. Bou Sthammi Noir) to freezing in liquid nitrogen. Cryo Lett 12:279–286

Benson EE, Harding K, Smith H (1989) Variation in recovery of cryopreserved shoot tips of *Solanum tuberosum* exposed to different pre- and post-freeze light regimes. Cryo Lett 10: 323–344

Berjak P, Dumet D (1996) Cryopreservation of seeds and isolated embryonic axes of neem (*Azadirachta indica*). Cryo Lett 17:99–104

Blackman SA, Obendorf RL, Leopold A (1992) Maturation proteins and sugars in desiccation tolerance of developing soybean seeds. Plant Physiol 100:225–230

Blakesley D, Pask N, Henshaw GG, Fay MF (1996) Biotechnology and the conservation of forest genetic resources: in vitro strategies and cryopreservation. Plant Growth Regul 20: 11–16

Crowe LM, Mouradian R, Crowe JH, Jackson SA, Womersley C (1984) Effects of carbohydrates on membrane stability at low water activities. Biochim Biophys Acta 769:141–150

de Boucaud MT, Brison M, Ledoux C, Germain E, Lutz A (1991) Cryopreservation of embryonic axes of recalcitrant seed: Jug*lans regia* L. cv. Franquette. Cryo Lett 12:163–166

de Boucaud MT, Brison M, Negrier P (1994) Cryopreservation of walnut somatic embryos. Cryo Lett 15:151–160

Demeulemeester MAC, Panis BJ, De Proft MP (1992) Cryopreservation of in vitro shoot tips of chicory (*Cichorium intybus* L.). Cryo Lett 13:165–174

Demeulemeester MAC, Vandenbussche B, De Proft MP (1993) Regeneration of chicory plants from cryopreserved in vitro shoot tips. Cryo Lett 14:57–64

De Proft MP, Maene ML, Debergh P (1985) Carbon dioxide and ethylene evolution in the culture atmosphere of M*agnolia* cultured in vitro. Physiol Plant 65:375–379

Dereuddre J (1992) Cryopreservation of in vitro cultures of plant cells and organs by vitrification and dehydration. In: Dattée Y, Dumas C, Gallais A (eds) Reproductive biology and plant breeding. Springer, Berlin Heidelberg New York, pp 291–300

Dereuddre J, Scottez C, Arnoud Y, Duran M (1990) Resistance of alginate-coated axillary shoot tips of pear tree (P*yrus communis* L) in vitro plantlets to dehydration and subsequent freezing in liquid nitrogen: effects of previous cold hardening. CR Acad Sci Paris 310 (Ser III):317–325

Desprez BF, Delesalle L, Dhellemmes C, Desprez MF (1994) Génétique et amélioration de la chicorée industrielle. CR Acad Agric Fr 80:47–62

Dussert S, Chabrillange N, Engelmann F, Anthony F, Hamon S (1997) Cryopreservation of coffee (C*offea arabica* L) seeds: importance of the precooling temperature. Cryo Lett 18:269–276

Fabre J, Dereuddre J (1990) Encapsulation-dehydration: a new approach to cryopreservation of *Solanum* shoot tips. Cryo Lett 11:413–426

Frese L (1996) Breeding of root chicory and Jerusalem artichoke – the state of the art. In: Fuchs A, Schittenhelm S, Frese L (eds) Proceedings of the 6th seminar on inulin, 14–15 Nov, Braunschweig, Germany, pp 41–50

Frese L, Dambroth M (1987) Research on the genetic resources of inulin-containing chicory (Cich*orium intybus*). Plant Breed 99:308–317

Grout BWW, Morris GJ (1987) Freezing and cellular organisation. In: Grout BWW, Morris GJ (eds) The effect of low temperature on biological systems. Arnold, London, pp 147–173

Kartha KK (1987) Advances in the cryopreservation technology of plant cells and organs. In: Green CE, Somers DA, Hackett WP, Biesboer DD (eds) Plant tissue and cell culture, vol 3, Alan R Liss, New York, pp 447–458

Kelechian-Cadot V, Boulineau F (1996) Setting up a French network for the management of genetic resources – *Cichorium endivia* L., *Cichorium intybus* L. In: Fuchs A, Schittenhelm S, Frese L (eds) Proceedings of the 6th seminar on inulin, 14–15 Nov, Braunschweig, Germany, pp 155–157

Lakhanpaul S, Babrekar PP, Chandel KPS (1996) Monitoring studies in onion (Al*lium cepa* L.) seeds retrieved from storage at –20 °C and –180 °C. Cryo Lett 17:219–232

Murashige T, Skoog F (1962) A revised medium for rapid growth and bioassays with tobacco tissue cultures. Physiol Plant 15:473–497

Potts SE, Lumpkin TA (1997) Cryopreservation of *Wasabia* spp. seeds. Cryo Lett 18:185–190

Plessis P, Leddet C, Collas A, Dereuddre J (1993) Cryopreservation of *Vitis vinifera* L. cv. Chardonnay shoot tips by encapsulation-dehydration: effects of pretreatment, cooling and postculture conditions. Cryo Lett 14:309–320

Reed BM (1990) Survival of in vitro-grown apical meristems of *Pyrus* following cryopreservation. HortScience 25:111–113

Sakai A (1993) Cryogenic strategies for survival of plant cultured cells and meristems cooled to –196 °C. In: Cryopreservation of plant genetic resources, ref no 6. Japanese International Cooperation Agency, Tokyo, Japan, pp 5–26

Salomão AN (1995) Effects of liquid nitrogen storage on *Zizyphus joazeiro* seeds. Cryo Lett 16:85–90

Schoofs J, De Langhe E (1988) Chicory (*Cichorium intybus* L.). In: Bajaj YPS (ed) Biotechnology in agriculture and forestry, vol 6. Crops II. Springer, Berlin Heidelberg New York, pp 294–321

Stanwood PC (1985) Cryopreservation of seed germplasm for genetic conservation. In: Kartha KK (ed) Cryopreservation of plant cells and organs. CRC Press, Boca Raton, pp 199–226

Sun WQ, Irving TC, Leopold AC (1994) The role of sugar, vitrification and membrane phase transition in seeds desiccation tolerance. Physiol Plant 90:621–628

Towill LE (1983) Improved survival after cryogenic exposure of shoots derived from in vitro plantlet cultures of potato. Cryobiology 20:567–573

Towill LE (1988) Genetic considerations for germplasm preservation of clonal materials. HortScience 23:91–94

Vandenbussche B, De Proft MP (1996) Cryopreservation of in vitro sugar beet shoot tips using the encapsulation-dehydration technique: development of a basic protocol. Cryo Lett 17:137–140

Vandenbussche B, Demeulemeester MAC, De Proft MP (1993) Cryopreservation of alginate-coated in vitro grown shoot tips of chicory (*Cichorium intybus* L.) using rapid freezing. Cryo Lett 14:259–266

Vasseur J, Lefebvre R, Backoula E (1986) Sur la variabilité de la capacité rhizogène d'explantats racinaires de *Cichorium intybus* L. (var witloof) cultivés in vitro: influence de la dimension des explantats initiaux et de la durée de la conservation des racines au froid. Can J Bot 64:242–246

Vermeulen A, Desprez B, Lancelin D, Bannerot H (1994) Relationships among *Cichorium* species and related genera as determined by analysis of mitochondrial RFLPs. Theor Appl Genet 88:159–166

Vertucci CW (1989) Effects of cooling rate on seeds exposed to liquid nitrogen temperature. Plant Physiol 90:1478–1485

Vertucci CW, Berjak B, Pammenter NW, Crane J (1991) Cryopreservation of embryonic axes of an homeohydrous (recalcitrant) seed in relation to calorimetric properties of tissue water. Cryo Lett 12:339–350

Wang JH, Ge JG, Liu F, Bian HW, Huang CN (1998) Cryopreservation of seeds and protocorms of *Dendrobium candidum*. Cryo Lett 19:123–128

Withers LA (1980) Low temperature storage of plant tissue cultures. In: Fietcher A (ed) Advances in biochemical engineering. Plant cell cultures II, vol 18. Springer, Berlin Heidelberg New York, pp 101–149

Withers LA (1987) The low temperature preservation of plant cell, tissue and organ cultures and seed for genetic conservation and improved agricultural practice. In: Grout BWW, Morris GJ (eds) The effects of low temperature on biological systems. Arnold, London, pp 389–409

Withers LA, Benson EE, Martin M (1988) Cooling rate/culture medium interactions in the survival and structural stability of cryopreserved shoot tips of *Brassica napus*. Cryo Lett 9:114–119

II.6 Cryopreservation of *Colocasia esculenta* L. Schott (Taro)

Nguyen Tien Thinh[1] and Hiroko Takagi[2]

1 Introduction

1.1 Plant Distribution and Species Importance

Taro [*Colocasia esculenta* L. Schott] is one of the three important root crops of the monocotyledonous family Araceae. The other two comprise species of the genera *Alocasia* and *Xanthosoma*. The Greek name *colocasia* was derived from the Arabic "qolquas" and the Malay name "tallas" gave rise to the Polynesian *taro* (Porteres 1960). There are confusions in the taxonomy of different edible *Colocasia* cultivars. Based on the morphology of the sterile appendage within the spathe, some researchers recognize two species, namely, *C. esculenta* and *C. antiquorum*. Others consider only one species, *C. esculenta*. Currently, many workers recognize *C. esculenta* but with two varieties, var. *esculenta* (syn. var. *typica* A. F. Hill) or "dasheen" type and var. *antiquorum* (Schott) Hubbard and Rehder (syn. var. *globulifera* Engl. and Krause) or "eddoe" type (Purseglove 1975). Generally, the main difference between these two varieties lies in the form of their corm ("tuber"). Variety *esculenta* has a main large corm with few small side cormels while var. *antiquorum* produces a relatively small central corm but with many side cormels.

Almost all parts of a taro plant can be used for food or for other purposes, of which the corm is the most delicious and nutritive. Edible carbohydrate, protein and fat are present in the corm (Table 1). In some cultivars, leaves and petioles can be used as vegetables. Waste leaves and corms can be cooked or fermented into silage for animal feeding. A root extract of taro is used as a traditional medicine to cure rheumatism and acne, and a leaf extract is used for clotting blood, neutralizing snake poison and as a purgative medicine (Winarno 1990).

Taro is commercially grown throughout the humid and semi-humid tropics and even in some warm regions of the temperate areas, where irrigation or rain-fed condition is possible. Today, it is cultivated throughout South-East Asia, the Pacific Islands (including Papua New Guinea), south and central

[1] Department of Biotechnology and Nuclear Techniques, Nuclear Research Institute, 1 Nguyen Tu Luc St., Dalat City, Vietnam
[2] Japan International Research Center for Agricultural Sciences, 1–2 Ohwashi, Tsukuba, Ibaraki 305, Japan

Biotechnology in Agriculture and Forestry, Vol. 50
L.E. Towill and Y.P.S. Bajaj (Eds.) Cryopreservation of Plant Germplasm II
© Springer-Verlag Berlin Heidelberg 2002

Table 1. The approximate nutrient and vitamin composition of fresh taro corm (After Djazuli 1996)

Approximate/vitamins/nutrient	Unit/100 g	1	2	3
Calori	cal	–	153	98.0
Water	g	75.1	–	73.0
Carbohydrate	g	18.2	37.0	23.7
Protein	g	2.00	1.00	1.90
Sugar	g	1.42	–	–
Ash	g	1.17	–	–
Crude fiber	g	0.80	–	–
Fat	g	0.20	–	–
Phosphor	g	–	0.051	0.061
Calcium	g	–	0.026	0.028
Fe	g	–	0.001	0.001
Vitamin C	mg	–	–	4.0
Vitamin B1	mg	–	0.092	0.13
Riboflavin	mg	–	0.030	–
Niacin	mg	–	0.85	–
Thiamin	mg	–	–	4.0
Vitamin A	I.U.	–	–	20.0

1: Payne 1941. In Danumihardja 1978.
2: Murai 1956. In Plucknett 1976.
3: Anon 1989.

China, the southern parts of Japan, India, the West Indies and in West and North Africa. There are, however, no reliable data available on world and national production and price. FAO data in 1992 estimated a world production of 5.6 million tons from 1 million ha. In the Philippines, 112,000 tons were produced from 33,000 ha in 1992 and in Papua New Guinea, 438,000 tons were harvested from 77,000 ha in 1993 (Wilson and Siemonsma 1996). In general, as a minor food crop, taro has received little attention from both policy makers and scientific researchers. Yet, as pointed out by Jackson (1994), the importance of this crop is far greater than its contribution to nutrition and revenue; it is very much a part of people's custom, and assumes high cultural status.

1.2 Methods for the Storage of Taro Germplasm

Taro genetic resources comprise hundreds of edible cultivars selected and cultivated by farmers over generations plus diverse wild species occurring in the areas of taro origin and domestication. The wealth of germplasm is unfortunately threatened. The replacement of traditional cultivars with a smaller number of species selected or bred for high yield (monoculture), the deforestation in various developing countries, and the lesser attention given to taro (a minor food crop) from the policy makers and researchers are causes of widespread genetic erosion of taro (Jackson 1994). To prevent this erosion, exploration, collection and particularly conservation of taro genetic resources is the only dependable strategy (Velayudhan et al. 1991). Activities for the conservation of taro genetic resources are described below.

Table 2. Losses of taro germplasm collections in Pacific Island countries (modified after Jackson 1994)

Country	Year*	Grade of loss	Reason for loss
FSM (Pohnpei)	1994	all?	Lack of staff
Papua New Guinea	1980	partial	Alomae disease
Solomon Islands	1974	all?	Alomae disease
	1991	all?	Cost of maintenance
Tonga	1985	all?	Drought
USA (Hawaii)	1988	most	Unknown
Vanuatu	1976	all?	Drought
	1994	partial	Lack of staff
Western Samoa	1986	all?	Cost of maintenance

* Dates are indicative only. In many cases, losses occurred over several years.

1.2.1 Field Gene Banks

Field gene banks are the most common way to maintain taro as well as other vegetatively propagated species in many countries. The collected plants are grown in the field, thus readily providing samples for research. However, plants collected from the field can be infected with diseases, threatened by natural disasters and continuously exposed to natural selection. A field gene bank is also expensive due to high demands of land, labor and maintenance. These disadvantages can cause a serious loss of collected germplasm, as has been the case for taro collections in Pacific Island countries (Table 2).

1.2.2 Corm Storage

The easiest way to conserve taro is by storage of corms (tubers), and this method is commonly used by farmers to store the best corms in cool conditions. The storage period, however, is only a couple of months as taro corms can quickly perish. Thus, for the purpose of long-term, safe conservation of germplasm, corm storage is clearly not a desirable method.

1.2.3 Seed Conservation

Some wild species and cultivars of taro can flower and set seeds naturally (Jackson 1994), but many others must be induced to flower by using gibberellic acid (Katsura et al. 1986; Miyazaki et al. 1986). Even so, in term of genetically true-to-type maintenance of germplasm, seeds are not the material of choice as they will not reproduce plants with the same genotype as the collected mother clones. Little information on the storage of taro seeds is recorded and reports are not promising. As reviewed by Jackson (1994), Strauss (1979, 1980) kept air-dried seeds of taro for up to 81 days at 22, 4 and −5 °C. Seed viability declined at 4 °C and fell most sharply at −5 °C. Wilson (1990) reported that taro seeds could be stored for at least 2 years in a desiccator kept in a refrigerator. However, there were no data to support this statement.

1.2.4 In Vitro Shoot Cultures

Plant germplasm, especially those of vegetatively propagated species, can be conserved in vitro (Engelmann 1991b). This storage often uses slow growth procedures for the cultured accessions (plantlets or shoots), such that the time between successive subcultures can be prolonged (Ng et al. 1991). Compared to field collection, conservation of germplasm by in vitro shoot cultures has the following advantages: (1) accession is kept free from pathogens and natural disasters, (2) less space is required, (3) deterioration of accessions can be detected visually, and (4) propagation potential of cultures can be very high. Attempts to preserve germplasm of taro and a related genus, *Xanthosoma*, by tissue culture in vitro have been carried out at the Agricultural University, Wageningen, Netherlands, and at the University of the South Pacific, western Samoa. The results showed that both the reduction of incubation temperature (down to 9 °C for taro and 13 °C for *Xanthosoma*) and the enhancement of culture medium osmolarity (addition of 60 g/l mannitol to culture medium) can slow the growth of plantlets. As such, germplasm of some clones of taro and *Xanthosoma* could be stored for more than 1 year under in vitro condition (Staritsky et al. 1986; Zandvoort 1987; Bessembinder et al. 1993). There are also several tissue culture laboratories in Malaysia, Japan, Vietnam, Fiji, and Papua New Guinea, where taro collections are kept and propagated in vitro. Yet, according to Jackson (1994), no attempts have been made to monitor the genetic stability of the conserved taro germplasm.

1.3 Need for Cryopreservation

As described above, traditional methods for conserving taro germplasm are risky and costly. In vitro shoot culture partly reduces such risks, but currently the technology of slow growth induction is available for only a few taro clones. Moreover, in various tropical developing countries (where a wealth of taro genetic resources exists), the unstable supply of electricity and the cost of this kind of energy make storage of in vitro shoot cultures in cold rooms (to induce slow growth) unsafe and somewhat expensive.

With the above situation in mind, cryopreservation currently appears to be more promising as it is the safest and most economical means for long-term conservation. Indeed, in cryopreservation, accessions are compactly kept in liquid nitrogen (LN), i.e., at −196 °C, and maintenance is just periodically refilling the tank with LN. Thus, cryopreservation does not require as much space, manpower and electricity as other methods of conservation and results in a lower cost (a reliably cheap source of LN is, however, necessary). In cryopreservation, risks of losses of taro germplasm due to disease, natural disasters, human error, lack of manpower and accidental interruption of the energy source (as pointed out in Table 2) can also be minimized. It is therefore useful to develop efficient cryoprocedures for the safe and long-term storage of taro germplasm.

2 Cryopreservation

2.1 Brief Review of Cryopreservation Work with Taro

Since the pioneering studies of Sakai (reviewed by Sakai 1995) on hardy twigs tissues frozen in LN, the number of successfully cryopreserved plant species has greatly increased (Engelmann 1991a, 1993; Sakai 1995). However, most of the successfully reported cases have dealt with cold-tolerant or temperate species. Information on effective cryopreservation of tropical species, especially of shoot tips/meristems of vegetatively propagated tropical plants (such as taro), is still very limited (Engelmann 1991b, 1993). Indeed, in the case of taro, there have been only three attempts dealing with cryopreservation of axillary buds and embryogenic calli. The earliest study was by Zandvoort (1987) and this had little success. Later, Shimonishi and coworkers (1993) were able to preserve embryogenic calli of Japanese taro cv. Eguimo in LN with about 70% post-warm survival. Yet, callus is not an ideal explant for genetically true-to-type conservation of germplasm due to potential somaclonal variation (Scowcroft 1984). The most promising information was provided by Takagi et al. (1994). About 75% of taro axillary buds could withstand LN cooling. However, this work was preliminary, with modest data. Thus, for long-term conservation of taro germplasm by cryopreservation, a reliable, efficient cryoprocedure is required.

In this chapter, we present our research on the development of simple, but efficient and replicable, cryoprocedures for shoot tips of taro, the ideal material for the in vitro conservation of germplasm of vegetatively propagated plants (Kartha 1985).

2.2 Cryopreservation of Taro Shoot Tips

2.2.1 Vitrification Technique

2.2.1.1 Materials and Methods

2.2.1.1.1 Materials
Shoot tips were excised from 2-month-old in vitro propagated plants. Two types of shoot tips were dissected. Large tips contained 3–4 leaf primordia (ca. 2–3 mm long) and the smaller contained only two leaf primordia surrounding the apical dome (ca. 0.8 mm long; Fig. 1A). Taro cv. Eguimo (*Colocasia escu-*

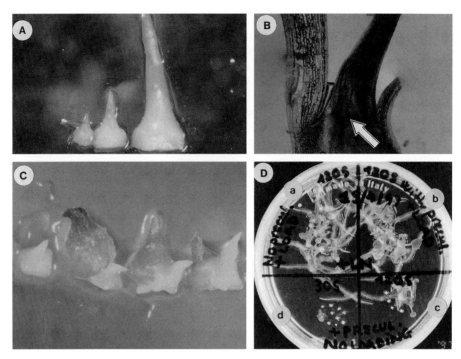

Fig. 1.A From *left* to *right*, taro shoot tips bearing 2, 3 and 4–5 leaf primordia, respectively. **B** Vertical hand section of a shoot tip showing the well-covered apical meristem (*arrow*). **C** Regrowth of loaded, LN-treated shoot tips of cv. Kabira, 1 week after warming. The shoot tip on the far left was dead. **D** *a* High levels of survival of nonprecultured, loaded, LN-treated S12 shoot tips; *b* high levels of survival of precultured, loaded, LN-treated S12 shoot tips; *c* medium levels of survival of precultured, nonloaded, LN-treated S12 shoot tips; *d* low levels of survival of precultured, nonloaded, LN-treated S3 shoot tips (material: taro cv. Eguimo, picture was taken 20 days after warming)

lenta L. Schott var *antiquorum*) was used for experiments. The protocols were applied to other *Colocasia* genotypes as well as to species of *Xanthosoma*, a *Colocasia*-related genus.

2.2.1.1.2 Methods
Basic Vitrification Procedure. Seven to ten shoot tips were wrapped together in a piece of tissue paper (1.5×1.5 cm). For dehydration, the wrapped tips were immersed for 0–60 min at 25 °C in 7 ml of a solution of 30% (w/v) glycerol + 15% (w/v) ethylene glycol + 15% (w/v) dimethyl sulfoxide + 0.4 M sucrose in MS medium. This solution is referred to as PVS2 and was developed by Sakai et al. (1990). For LN cooling, the dehydrated tips in tissue paper were placed into a 0.7-ml plastic cryotube, filled with fresh PVS2 and the tube was rapidly plunged into LN. For warming, after at least 1 h in LN, the cryotube was rapidly transferred to sterile distilled water kept at 40 °C in a water bath and vigor-

ously shaken for 90 s. For removing cryoprotectants, the warmed tips were transferred into 7 ml of liquid MS medium supplemented with 1.2 M sucrose and held for 15 min. For plant recovery, tips were then removed from the tissue paper and cultured overnight in darkness on a filter paper on basic MS medium supplemented with 0.3 M sucrose. The tips were then transferred to MS basic medium supplemented with 0.1 M sucrose and kept in dim light for 10 days. Next, they were placed under a light intensity of $90 \mu E\,m^{-2}s^{-1}$, with a photoperiod of 16/8 h and at 25 °C. Survival was defined as the percentage of treated shoot tips showing new shoot growth.

Pretreatments. Effects of preconditioning, preculture and loading on the post-thaw survival of shoot tips were investigated. These treatments were carried out as described below:

Preconditioning:	The preconditioning was carried out by transferring intact, 1-month-old in vitro plants onto the following modified MS media for an additional 1 or 2 months: (a) MS (hereafter referred to as S3), (b) MS with triple concentrations of macrosalts (hereafter referred to as 3MS), (c) MS with 0.5 or 1 g/l L-proline (hereafter referred to as P0.5 and P1), (d) MS with 5 or 10 mg/l ABA (hereafter referred to as A5 and A10) and (e) MS with either 6, 9 or 12% sucrose (hereafter referred to as S6, S9 and S12, respectively). Cultures were incubated with the light and temperature regimes described above. Shoot tips were dissected from these pre-conditioned shoot tip donor plants (hereafter referred to as SDPs) and subjected to either the PVS2 dehydration, the preculture or the loading step in the vitrification process.
Preculture:	Freshly isolated shoot tips were grown in darkness at 25 °C for 1–2 nights on MS basic medium supplemented with either 5–10% (v/v) dimethyl sulfoxide, 5–10% (v/v) ethylene glycol, 1–2 mg/l ABA or 0.1–0.7 M sucrose (hereafter referred to as S0.1-S0.7). Precultured shoot tips were then subjected to either the loading or the PVS2 dehydration step in the vitrification process.
	All media in the above steps of preconditioning and preculture, except S6-S12 and S0.1-S0.7, were supplemented with 30 g/l sucrose and 2 g/l gellan gum. The pH of the medium was adjusted to 5.7–5.8 with a solution of 0.1 N KOH and/or HCl prior to autoclaving at 121 °C for 15 min.
Loading:	Loading treatments were carried out by immersing the isolated shoot tips for 20 min at 25 °C in different loading solutions. Four loading solutions were tested: (a) 1.5 M glycerol+5% DMSO+0.4 M sucrose (hereafter referred to as L1), (b) 2 M glycerol+0.4 M sucrose (hereafter referred to as L2), (c) PVS2 solution at 20% strength (hereafter

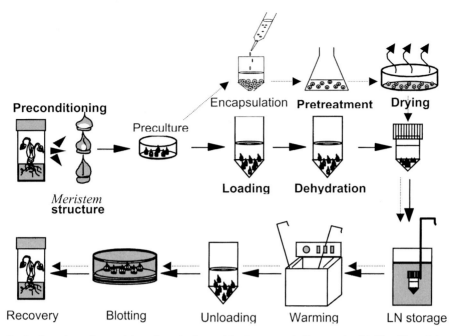

Fig. 2. Overview of materials and methods used in the research. Factors that highly affected the success of cryopreservation are written in **bold**. *Dotted arrows* indicate encapsulation/drying procedure. *Full arrows* indicate vitrification procedure

referred to as L3), and (d) PVS2 "20-40-60-80"%, i.e., for each PVS2 concentration, the shoot tips were treated for 5 min, commencing with 20% PVS2 (hereafter referred to as L4). Loaded shoot tips were then subjected to the PVS2 dehydration step in the vitrification process.

Figure 2 shows a general overview of the materials and methods applied in the research.

2.2.1.2 Results

2.2.1.2.1 Sensitivity of Shoot Tips to Dehydration by PVS2

Large shoot tips were more tolerant to PVS2 dehydration than small ones (Fig. 3). Frequencies of shoot regrowth from these two kinds of shoot tips after being treated with PVS2 for 60 min were 80 and 64.5%, respectively. When the shoot tips were dehydrated without being covered in tissue paper, it was microscopically observed that the large shoot tips appeared to be waterproof and often floated on the surface of PVS2, whereas the small ones became shrunken, slightly transparent, then sank down to the bottom of the PVS2-containing Petri dish after about 10–15 min of treatment. This suggested

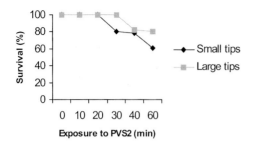

Fig. 3. Effects of duration of PVS2 dehydration on survival of shoot tips of taro cv. Eguimo. Each data point is the average mean of two replications, each with approximately ten explants

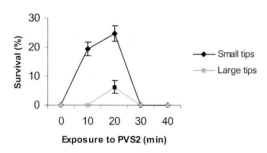

Fig. 4. Effects of duration of PVS2 dehydration on post-warm survival of LN-treated shoot tips of taro cv. Eguimo. Each data point is the mean of three replications, each with approximately ten explants. *Vertical bars* represent standard errors

that the small tips were more readily dehydrated by PVS2 than the large ones.

2.2.1.2.2 Post-Warm Survival of Shoot Tips as Affected by Different PVS2 Dehydration Times

Effective dehydration times for obtaining shoot regrowth from LN-treated shoot tips ranged between 10–20 min (Fig. 4). In these dehydration treatments, in contrast to their high tolerance to PVS2 mentioned above, only 0–5.4% of the large shoot tips showed shoot recovery compared to 19.5–25.9% of the small ones. Moreover, after about 1 month of post-warm culture, while almost all of the surviving small shoot tips formed plantlets (small shoots with roots), most of the surviving large ones produced only a green leaf growing out from the inside. The outer tissues of these large tips gradually turned brown and died.

2.2.1.2.3 Effects of Loading the Shoot Tips

Practical conservation of germplasm by cryopreservation requires fairly high levels of post-warm regrowth. The best survival obtained above (25.9%) was too low to meet this requirement. To enhance survival, a similar loading treatment to that reported by Matsumoto et al. (1994) in wasabi was used. Results of treating the shoot tips with L1 for 20 min prior to PVS2 dehydration and LN cooling are shown in Fig. 5. As such, no improvement in survival was recorded for loaded large shoot tips, but dramatic enhancements occurred for small ones. Indeed, 77.2% of small shoot tips, which were loaded with L1 for

Fig. 5. Effects of duration of PVS2 dehydration on post-warm survival of shoot tips of taro cv. Eguimo exposed to L1 solution for 20 min prior to PVS2 dehydration and LN treatment. Each data point is the mean of three replications, each with approximately ten explants. *Vertical bars* represent standard errors

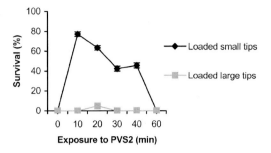

Table 3. Effects of different loading solutions on survival of LN-treated Eguimo shoot tips

Loading solution[a]	Survival (% ± SE)[b]
Not loaded	23.3 ± 1.5
1.5 M glycerol + 5% DMSO + 0.4 M sucrose (L1)	63.3 ± 1.3
2 M glycerol + 0.4 M sucrose (L2)	69.5 ± 2
PVS2 20% (L3)	40.0 ± 0
PVS2 20–40–60–80% (L4)	32.5 ± 1.4

[a] Shoot tips were loaded with L1, L2 or L3 for 20 min at 25 °C, followed by dehydration with PVS2 for 10 min at 25 °C before LN cooling. For L4, each concentration of PVS2 was applied for 5 min.
[b] Data are means plus standard errors of three replications, each with approximately 7–10 shoot tips per treatment.

20 min, followed by dehydration with PVS2 for 10 min then rapidly cooled in LN, showed post-warm shoot growth (Fig. 1C). The range of effective dehydration times to obtain post-warm survival was increased from 10–20 min to 10–40 min. At the same dehydration treatment of 20 min, 63.3% of loaded small shoot tips survived LN cooling compared to the 29.5% mentioned above. Based on these results, a loading treatment was considered essential for vitrification of taro shoot tips, and small shoot tips (hereafter just referred to as shoot tips) were selected as explants for further experiments.

To determine an optimal loading solution, four different loading solutions (see "Methods" for details) were examined. Results presented in Table 3 strongly confirmed the effectiveness of a loading treatment and indicated either L1 or L2 could be used to pretreat the shoot tips. We selected L2 as loading solution for use in the next experiment on the effect of loading time on post-warm survival of shoot tips.

Reducing or prolonging the loading time did not significantly improve the post-warm survival of cryopreserved shoot tips (Fig. 6). Compared to the post-warming survival of 17.5% for the non-loaded treatment, 55, 70 and 39% of shoot tips survived after loading with L2 for 10, 20 and 40 min, respectively, prior to dehydration with PVS2 for 10 min and rapid cooling in LN. Thus, at present, the 20-min loading time is still the best to cryoprotect shoot tips prior to PVS2 dehydration.

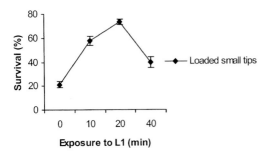

Fig. 6. Effects of loading time in L1 solution on survival of LN-treated shoot tips of taro cv. Eguimo prior to PVS2 dehydration for 10 min. Each data point is the mean of three replications, each with approximately ten explants. *Vertical bars* represent standard errors

Table 4. Effect of sucrose preculture on survival of LN-treated Eguimo shoot tips

Sucrose concentration[a]	Survival (% ± SE)[b]
No preculture	66.6 ± 2.3
0.3 M	83.1 ± 2.3
0.5 M	13.3 ± 0.9
0.7 M	0.0 ± 0.0
0.3–0.5–0.7 M*	56.6 ± 6.7

[a] Shoot tips were precultured for one night (or one night interval, i.e.*) on MS medium supplemented with the above listed sucrose concentrations, loaded with L1 solution for 20 min and dehydrated with PVS2 for 10 min prior to LN cooling.
[b] Data are means plus standard errors of three replications, each with approximately ten shoot tips per treatment.

2.2.1.2.4 Effects of Preculturing Shoot Tips

The beneficial effects of preculture of shoot tips for enhancing post-warming survival have been well documented in different species (Fabre and Dereuddre 1990; Matsumoto et al. 1994, 1995; Panis 1995). Our trials of preculturing taro shoot tips on media supplemented with ethylene glycol (5–10%), dimethyl sulfoxide (5–10%), or ABA (1–2 mg/l) did not significantly improve post-warm survival (data not shown). As shown in Table 4, the highest survival (83.3%) was obtained when shoot tips had been precultured overnight on MS medium supplemented with 0.3 M sucrose prior to loading, dehydrating and LN cooling. Similar preculture of shoot tips with 0.5 M sucrose, however, reduced survival to just 13.3%. More severely, all LN-treated explants died if precultured with 0.7 M sucrose. Interestingly, if shoot tips were steadily precultured on elevated concentrations of sucrose from 0.3 to 0.5 then 0.7 M, with 1-day intervals, 56.6% of tips could withstand the LN cooling. Thus, culturing shoot tips overnight on 0.3 M sucrose appeared to be the most suitable preculture treatment.

Table 5. Effects of preconditioning on survival of LN-treated shoot tips of cv Eguimo

Medium[a]	Survival (%)[b]	
	one month	two months[c]
MS control	65.0	67.5
3MS	40 ns	25.0**
P1	70 ns	66.3 ns
A10	53.3 ns	60 ns
S6	100**	65 ns
S9	97.2*	100**
S12	100**	100**
CV%	21.4	27.8
LSD0.05	26.1	–
LSD0.01	–	32.9

Approximately 10 shoot tips were tested for each medium for each of four replications.
[a] Abbreviations used for medium are described in "methods".
[b] In the same column, means are compared with that of MS control using LSD test, ns = not significantly different, * & ** = significantly different at 5 & 1% level, respectively.
[c] Shoot tips were harvested from plants preconditioned for one or two months.

2.2.1.2.5 Effects of Preconditioning the Shoot Tip Donor Plants (SDPs)

Shoot tips of some species, like apple, lily, pear, potato, *Ribes*, and wasabi, showed higher levels of post-warm survival if their SDPs had been cold hardened at low temperatures (Dereuddre et al. 1990; Niino et al. 1992; Reed 1992; Matsumoto et al. 1995). Yet, our previous trials applying low temperatures (10–15 °C) to harden the SDPs of taro prior to dissection and cryopreservation of shoot tips were unsuccessful. Indeed, growth of taro SDPs was severely inhibited by these low temperatures (data not shown). Since taro is an established tropical species, cold hardening is not a suitable way to condition this plant.

We have found that the culture of taro SDPs for 1 month on S6, S9 and S12 media (see "Methods" for details), i.e., sucrose preconditioning, resulted in high levels of survival of cryopreserved shoot tips, i.e., >90% (Table 5). Sucrose preconditioning, in particular that performed with the S12 medium, significantly decreased water content and promoted the accumulation of free proline as well as soluble sugars (Fig. 7). However, preconditioning the SDPs on media supplemented with proline (P1), ABA (A5) and an elevated concentration of MS macrosalts (3MS) did not improve survival. Interestingly, data presented in Table 6 indicated that when S12 sucrose preconditioning was applied, it was possible to eliminate the step of sucrose preculture in the cryopreservation process without a significant reduction in survival (Fig. 1D, a). This shortened the time needed for successful cryopreservation. Thus, preconditioning taro SDPs for 1 month with 12% sucrose (hereafter referred to

Table 6. Summary of effects of additional pretreatments on survival of LN-treated shoot tips of cv Eguimo

Treatment			Survival (%)
Preconditioning	Preculture	Loading	
–	–	–	25.9
+	–	–	34.6
+	–	+	94.7
+	+	+	97.3
–	+	–	24.5
–	+	+	81.3
–	–	+	77.2

Data are summarized from different experiments. For each treatment, data are mean percentage calculated from 30–40 tested shoot tips. A same basic vitrification protocol was applied in all experiments.

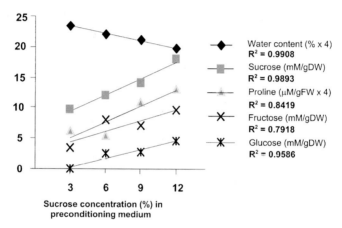

Fig. 7. Effects of sucrose concentration in preconditioning medium on the accumulation of solutes in cultured shoot tip donor plants of taro cv. Eguimo. Note: to scale the graph, actual values of water content and proline were divided by 4

as preconditioning) was included in the cryopreservation process of shoot tips of taro.

2.2.1.2.6 Miscellaneous Technical Experiments

Experiments were carried out to answer technical questions such as how many taro tips can be cryopreserved in one cryotube, how long can taro tips be stored in LN, and can the modified vitrification procedure be applied to other taro genotypes? According to the data, cryopreservation of 3, 5, 7, 10 or even 25 tips in the same cryotube did not significantly affect survival. Within the time of our research project, we observed no difference in growth rates of taro

Table 7. Survival of LN-treated shoot tips of different taro cultivars and taro-related species cryopreserved by vitrification

Species	Cultivar	Common name	Origin	Survival (% ± SE)
C. esculenta	Eguimo	Taro	Japan	100 ± 0.0
C. esculenta	Dotare	Taro	Japan	100 ± 0.0
C. esculenta	Kabira	Taro	Japan	83.3 ± 2.7
C. esculenta	Ginowan	Taro	Japan	95.8 ± 3.2
C. esculenta	So	Taro	Viet Nam	66.6 ± 5.4
C. esculenta	Sap	Taro	Viet Nam	75.0 ± 4.2
X. sagitteafolia	Alotau	Tannia	Japan	66.6 ± 5.4
X. nigra	Mo	Tannia	Viet Nam	61.9 ± 8.3

Approximately ten shoot tips were used for each tested genotype for each of 3–4 replications.

tips that had been stored in LN for 30 min or for 6 months. In addition, several taro genotypes belonging to two different varieties, *esculenta* and *antiquorum*, as well as *Colocasia*-related species – *Xanthosoma sagittaefolia and X. nigra* – were successfully cryopreserved using the described vitrification procedure (Table 7).

2.2.2 Encapsulation/Drying Technique

2.2.2.1 Materials and Methods

2.2.2.1.1 Materials
Based on the results of the above research with vitrification, only small shoot tips, i.e., shoot tips bearing a maximum of two leaf primordia, were used as cryogenic explants in this investigation.

2.2.2.1.2 Methods
Encapsulation of Shoot Tips. Excised shoot tips were first cultured in darkness overnight on plain MS medium supplemented with 0.3 M sucrose. Cultured shoot tips were then suspended in Ca-free liquid MS medium supplemented with 3% Na-alginate and 0.4 M sucrose. This mixture was dropped, from a sterile disposable plastic pipette, into liquid MS medium containing 100 mM $CaCl_2 \bullet 2H_2O$ and 0.4 M sucrose to form beads. The beads, each containing one shoot tip and 3.5±0.4 mm in diameter, were held in this Ca solution for 30 min before being transferred into different liquid media for pretreatment.

Pretreatment. Encapsulated shoot tips were placed into liquid MS medium supplemented with either 0.8 M sucrose (hereafter referred to as 0.8S), 0.8S + 0.5 M glycerol (hereafter referred to as 0.8S0.5G), 0.8S + 1 M glycerol (hereafter referred to as 0.8S1G) or 0.8S + 2 M glycerol (hereafter referred to as 0.8S2G). Pretreatment was carried out by shaking the beads for 16 h at 80 rpm at 25 °C in darkness in a 200-ml flask containing 60 ml of each of the above described medium and with a ratio of 2 ml medium per bead. Pretreated beads were dried over silica gel.

Silica Gel Drying and Cryopreservation Procedure. For drying, 20 pretreated beads were placed on a sterile filter paper laid over 40 g dry silica gel contained in a closed Petri dish (90 × 15 mm). Drying was carried out for 1, 2, 3, 5, 7 or 9 h; the closed Petri dishes were kept in a laminar flow cabinet to prevent exogenous contamination. Seven to ten beads derived from each drying time were placed into a 0.7-ml plastic cryotube and the tube was rapidly plunged into LN. Warming was similarly carried out in water kept at 40 °C, as described in Section 2.2.1.1. Then the beads were cultured overnight on MS medium supplemented with 0.3 M sucrose; after that, they were transferred onto MS medium supplemented with 0.1 M sucrose for shoot recovery. All cultures were similarly incubated under the light and temperature regimes described in Sect. 2.2.1.1. Survival was defined as the ability of an encapsulated shoot tip to form a shoot.

2.2.2.2 Results

2.2.2.2.1 Uptake of Solutes into Beads
The applied pretreatments affected the fresh (FW) and dry weight (DW) of the treated Ca-alginate beads. As shown in Table 8, the 0.8S pretreatment resulted in the greater enhancement of bead FW (3.9 mg), followed by the 0.8S0.5G (3 mg), the 0.8S1G (2.5 mg) and the 0.8S2G (1.8 mg). In contrast, beads pretreated with a combination of sucrose and glycerol, i.e., 0.8S0.5G, 0.8S1G and 0.8S2G, were significantly heavier in DW than those pretreated with only 0.8S. Pretreatment was also observed to significantly reduce bead water content (WC), with beads pretreated with 0.8S1G and 0.8S2G having lowest WC of 69.9 and 71%, respectively.

Table 8. Effects of pretreatment on bead weight and water content

Pretreatment[a]	KFW[b] (mg)	DW[c] (mg)	WC[d] (%)
No	0.0a	6.1a	85.3d
0.8S	3.9d	10.0b	75.7c
0.8S0.5G	3.0c	11.4c	72.2b
0.8S1G	2.6c	11.4c	69.9a
0.8S2G	1.8b	10.9c	71.0ab

[a] Pretreatment was carried out as described in 'Method'.
[b] Increment of bead fresh weight, with KFW = bead FW after pretreatment minus bead FW before pretreatment.
[c] Average bead dry weight.
[d] Bead water content. In the same column, data followed by the same letter are not significantly different at 5% level by Duncan's multiple Range test.

Fig. 8. Effects of silica gel drying on water content (WC, %) of beads derived from different pretreatment regimes. On each line each point is the mean WC % of three replications, each with ten beads

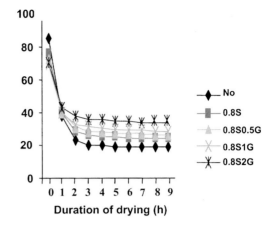

Duration of drying (h)

Upon silica gel drying, pretreated beads, especially those pretreated with mixtures of sucrose and glycerol, lost WC at much slower rates than the non-pretreated counterparts. For instance, after drying for 3 h, the WCs of non-, 0.8S-, 0.8S0.5G-, 0.8S1G- and 0.8S2G-pretreated beads were 20.3, 26.3, 28.9, 31.1 and 36.2%, respectively (Fig. 8). These results indicate that during pretreatment there was an uptake of solutes (sucrose and glycerol) into the beads, with a concomitant reduction in water content. This enhanced the vitrification ability of beads upon cooling in LN.

2.2.2.2.2 Post-Warming Survival of the LN-Treated, Encapsulated Shoot Tips

Based on the above, shoot tips of cv. Eguimo were encapsulated in Ca-alginate beads and pretreated for 16 h in 0.8S, 0.8S0.5G, 0.8S1G and 0.8S2G. Pretreated encapsulated shoot tips (hereafter referred to as ESTs) were then dried over silica gel for 1, 2, 3, 5 or 7 h prior to LN cooling. As shown in Fig. 9, survival was observed mainly in ESTs that had been dried for 2, 3 and 5 h. All ESTs without drying or with drying for 1 or 7 h turned white and died after warming. Among those that survived, ESTs pretreated in 0.8S and dried for 3 h gave the greatest survival (65.5%). Incorporation of glycerol (0.5–2 M) into 0.8S pretreatment solution reduced survival; at the highest concentration (2 M) of glycerol, no ESTs survived LN cooling. All ESTs processed in the same way but without LN cooling gave rise to normal plantlets, though (depending on pretreatment and drying regimes) with different rates of growth (data not shown).

A remarkable improvement in the survival of LN-treated ESTs was obtained when shoot tips were derived from SDPs preconditioned for 1 month on S12 medium (hereafter referred to as S12 shoot tips; see Sect. 2.2.1.1 for details; Figs. 10 and 11). In all treatments, the S12-ESTs always had better post-warming survival percentages than those derived from SDPs grown with 3%

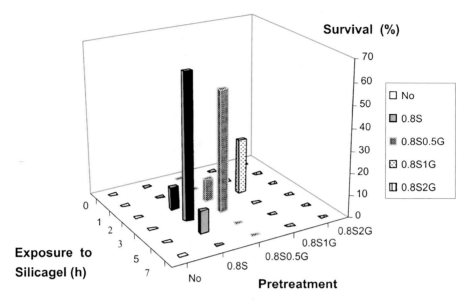

Fig. 9. Effects of pretreatment and drying time on post-warm survival of nonpreconditioned, encapsulated, LN-treated shoot tips of taro cv. Eguimo. See 'Methods' for abbreviations of pretreatment solutions. Each *bar* is the mean of three replications, each with approximately ten encapsulated shoot tips

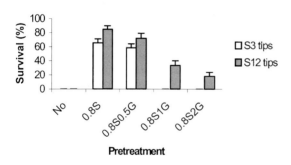

Fig. 10. Comparison of post-warm survival of encapsulated shoot tips of taro cv. Eguimo derived from donor plants grown for 1 month on MS medium supplemented with 3% (S3 tips) or 12% (S12 tips) sucrose. Pretreated, encapsulated tips were dried over silica gel for 3 h before cooling in LN. *Vertical bars* represent standard errors ($n = 3$)

sucrose. Similarly, the S12-ESTs also possessed the highest survival (85.5%) when they had been pretreated with 0.8S medium and dried for 3 h. Increased glycerol concentrations in the pretreatment solution (i.e., the use of 0.8S0.5G, 0.8S1G and 0.8S2G) also reduced survival of the S12-ESTs, but considerable numbers of these still survived LN cooling (black bars in Fig. 10).

2.2.2.2.3 Application of the Developed Encapsulation/Drying
Procedure to S12 Shoot Tips of Other Taro Genotypes
Optimal conditions determined from the above experiments were applied to S12 shoot tips of other taro genotypes, namely cvs. So (Vietnamese taro) and Ginowan (Okinawan taro) which belong to *C. esculenta* var. *esculenta*. As such,

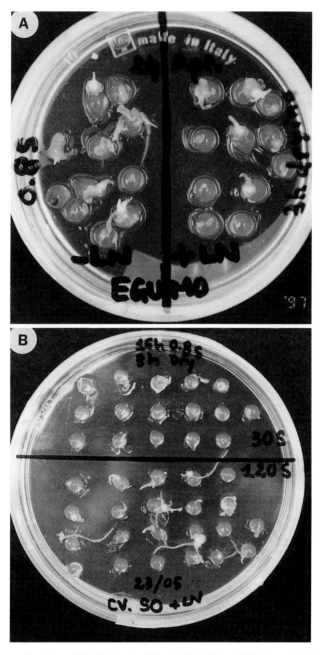

Fig. 11. A Regrowth of non-LN- (*left* part of the Petri dish) and LN-stored (*right* part), encapsulated S12 shoot tips of cv. Eguimo. The encapsulated shoot tips were pretreated for 16 h in 0.8S medium, silica gel dried for 3 h then cooled in LN for 1 week. Picture was taken 20 days after warming. **B** Difference in frequency and speed of regrowth of LN-stored, encapsulated shoot tips of cv. So derived from donor plants grown with 3% (*top* part of the Petri dish, 30S) or with 12% sucrose (*bottom* part, 120S). Picture was taken 1 month after warming

Table 9. Comparison of cryopreservation of shoot tips of taro cv Eguimo using vitrification and encapsulation methods

Cryogenic procedure[a]	Post-warm survival (% ± SE)	Operational time (min)	Regrowth time[b] (day)	Callus formation (%)
Vitrification	94.7 ± 2.8	30	2–4	< 1
Encapsulation/drying	85.5 ± 5.7	2130	8–10	ca. 25

[a] The S12 shoot tips of cv Eguimo were used as materials. For vitrification, shoot tips were loaded with L2 for 20 min, dehydrated with PVS2 for 10 min then rapidly cooled in LN. For encapsulation/drying, shoot tips were precultured for 16 h (960 min) on 0.3 M sucrose MS medium, encapsulated in Ca-alginate beads for 0.5 h (30 min), pretreated for 16 h (960 min) with 0.8S medium, dried with silicagel for 3 h (180 min) then rapidly cooled in LN.
[b] Post-warm time needed for explants to turn green and swollen.

ESTs of these cultivars were pretreated for 16 h with 0.8S medium, followed by drying for 3 h with silica gel then rapid cooling in LN. Respectively, 66.6 and 86.6% of the ESTs of cvs. Ginowan and So showed shoot growth after warming. Thus, like the vitrification procedure described above, encapsulation/drying was also promising for the cryostorage of taro shoot tips. However, as shown in Table 9, compared to the vitrification technique, the encapsulation/drying procedure possesses several drawbacks: (1) post-warm survival is lower, (2) post-warm shoot growth is slower, (3) the overall process is longer, and (4) some ESTs produced callus, which is not desirable for preserving clones true-to-type.

3 Discussion

During rapid cooling in LN, ice crystals are formed in cells within the shoot tip. Thus, prior to freezing, free water in cells needs to be reduced to avoid this undesired crystallization and to make cells capable of vitrifying upon LN treatment (Sakai 1993). In vitrification and encapsulation/drying methods, reduction of cellular water is carried out through dehydration of explants using highly concentrated vitrification solutions or air/silica gel drying, respectively. These treatments, however, can impose severe dehydration stress on cells which itself can be damaging. In the present research, successful cryopreservation was obtained when additional steps such as preconditioning, preculture, pretreatment with concentrated sucrose solutions, and especially loading were applied. Based on the data, these pretreatments enhance tolerance of dehydration. Indeed, preconditioning reduced the water content of taro SDPs and also promoted the accumulation of water-deficiency-responsive solutes like soluble sugars and free proline. These are considered to be protective substances against abiotic stresses (Greenway and Munns 1980; Levitt 1980; Guy et al. 1992; Larher et al. 1993) and should make shoot tips more tolerant to osmotic dehydration. The S12 tips (i.e., preconditioned tips) were more tolerant to pretreatment with mixtures of sucrose and glycerol than those dissected from SDPs grown with 3% sucrose. As documented in other species (Uragami et al. 1990; Panis 1995; Reinhoud 1996), reduction of WC and accumulation of solutes may also have occurred in isolated shoot tips of taro precultured with high concentrations of sucrose. Yet, a sudden change in osmotic pressure due to direct preculture of freshly isolated shoot tips onto sucrose concentrations of 0.5–0.7 M was not beneficial for taro. Although the mechanism of loading is not well understood, this treatment was believed to protect plant cells from sudden osmotic shock caused by a concentrated vitrification solution (PVS2) (Matsumoto et al. 1994). Among pretreatments applied in this research, the most interesting was sucrose preconditioning in place of cold temperatures to harden the SDPs of taro. Preconditioning would be useful in the cryopreservation process of tropical species that commonly cannot tolerate low temperatures.

It was observed, however, that the benefits of the pretreatments described above were not obtained unless shoot tips with a suitable structure were used as cryogenic explants. Small shoot tips, consisting of the apical dome plus a maximum of two leaf primordia, survived LN cooling much more successfully than larger ones. Microscope observations of the cross sections of taro shoot tips revealed that apical meristems of this monocot are well covered inside the overlapped, tubular and thick petiole bases (Fig. 1B). Thus, excessive layers of these petioles, as in the case of large shoot tips, would prevent the penetration of either PVS2 or loading solution into the dome area (in vitrification) or would prevent the domes from being appropriately dried by silica gel (in encapsulation/drying). This probably leads to insufficient dehydration of dome cells or, in other words, the meristem cannot be vitrified upon cooling. Indeed, shoot tips bearing greater numbers of leaf primordia showed lower post-warm survival (data not shown).

Here the vitrification procedure was superior compared with encapsulation/drying for cryopreservation. Thus, based on the developed vitrification procedure and in a descending order of importance, factors affecting the success of cryopreservation of taro shoot tips are: shoot tip structure, loading treatment, PVS2 dehydration time, sucrose preconditioning and sucrose preculture. With these considerations in mind, we also successfully cryopreserved shoot tips of other vegetatively propagated tropical monocots like banana, *Cymbopogon*, *Cymbidium* (terrestrial orchid) and pineapple by vitrification. The successful vitrification procedures for shoot tips of these plants were quite similar to that of taro (Thinh 1997; Thinh and Takagi 2000).

4. Summary and Conclusions

We have shown that in vitro grown shoot tips of taro can be successfully cryopreserved using vitrification and encapsulation/drying methods. Both procedures appear to be simple, easy-to-perform and economical, as they do not require sophisticated equipment. In the present investigation, compared with encapsulation/dehydration, vitrification was the method of choice because it gave higher levels of post-warm survival, required a shorter time for the procedure and led to no callus formation from the cryopreserved shoot tip. Successful vitrification requires small shoot tips with no more than two leaf primordia, and additional pretreatments such as loading and preconditioning to enhance shoot tip tolerance to osmotic shock. The described vitrification technique can be used for the long-term conservation of taro genetic resources since all genotypes tested so far with this procedure had high levels of survival (60–100%). However, prior to such an application, it is necessary to determine if genetic changes occur in germplasm that is processed by vitrification.

Acknowledgements. This research was sponsored by Japan International Research Center for Agricultural Sciences (JIRCAS) through a visiting research fellowship granted to Nguyen Tien Thinh.

References

Bessembinder JJE, Staritsky G, Zandvoort EA (1993) Long-term in vitro storage of *Colocasia esculenta* under minimal growth conditions. Plant Cell Tissue Org Cult 33:121–127
Dereuddre J, Scottez C, Arnaud Y, Duron M (1990) Resistance of alginate-coated axillary shoot tips of pear tree (*Pyrus communis* L. cv. Beurre Hardy) in vitro plantlets to dehydration and subsequent freezing in liquid nitrogen: effects of previous cold hardening. CR Acad Sci Paris 310:317–323
Djazuli M (1994) Taro genetic resources and utilization in Indonesia. In: MAFF international workshop on genetic resources, root and tuber crops. MAFF/AFFRC/NIAR. 15–17 March 1994. Tsukuba, Japan, pp 159–166
Engelmann F (1991a) In vitro conservation of horticultural species. Acta Hortic 298:327–334
Engelmann F (1991b) In vitro conservation of tropical plant germplasm – a review. Euphytica 57:227–243
Engelmann F (1993) Cryopreservation for the long-term conservation of tropical crops of commercial importance. In: Shamaan NA (ed) Applications of plant in vitro technology. Proc Int Sym, UPM Serdang, Selangor, Malaysia
Fabre J, Dereuddre J (1990) Encapsulation/dehydration: a new approach to cryopreservation of *Solanum* shoot tips. Cryo Lett 11:413–426
Greenway H, Munns R (1980) Mechanism of salt tolerance in nonhalophytes. Annu Rev Plant Physiol 31:149–190
Guy LC, Huber JLA, Huber SC (1992) Sucrose phosphate synthase and sucrose accumulation at low temperature. Plant Physiol 100:502–508
Jackson JVH (1994) Taro and yam genetic resources in the Pacific and Asia. A report prepared for the ACIAR and IPGRI, 62p. Anutech Pty, Canberra ACT 0200
Kartha KK (1985) Meristem culture and germplasm preservation. In: Kartha KK (ed) Cryopreservation of plant cells and organs. CRC Press, Boca Raton, pp 115–134
Katsura N, Takayanagi K, Sato T (1986) Gibberellic acid induced flowering in cultivars of Japanese taro. J Jpn Soc Hortic Sci 55:69–74
Larher F, Leport L, Petrivalsky M, Chappart M (1993) Effectors for the osmoinduced proline response in higher plants. Plant Physiol Biochem 31:911–922

Levitt J (1980) Chilling, freezing and high temperature stresses. In: Kozlowski TT (ed) Responses of plants to environmental stresses, vol I. Academic Press, New York, 497 pp

Matsumoto T, Sakai A, Yamada K (1994) Cryopreservation of in vitro-grown apical meristems of wasabi (*Wasabi japonica*) by vitrification and subsequent high plant regeneration. Plant Cell Rep 13:442–446

Matsumoto T, Sakai A, Yamada K (1995a) Cryopreservation of in vitro-grown apical meristems of lily by vitrification. Plant Cell Tissue Org Cult 41:237–241

Miyazaki S, Tashiro Y, Kanazawa K, Yanagawa M, Tabaru M (1986) Promotion of flowering by the treatment of seed corms and young plants with gibberellic acid in taro (*Colocasia esculenta* Schott). J Jpn Soc Hortic Sci 54:450–459

Ng SYC, Ng NQ (1991) Reduced-growth storage of germplasm. In: Dodds JD (ed) In vitro methods for conservation of plant genetic resources. Chapman and Hall, London, pp 11–39

Niino T, Sakai A, Yakuwa H, Nojiri K (1992) Cryopreservation of in vitro-grown shoot tips of apple and pear by vitrification. Plant Cell Tissue Org Cult 28:261–266

Panis B (1995) Cryopreservation of banana germplasm. PhD Diss no 207, Katholieke Universiteit Leuven, Belgium

Porteres R (1960) La sombre aroidee cultivee: *Colocasia antiquorum* Schott ou taro de Polynesie: essai d'etymologie sematique. J Agric Trop Bot Appl 7:92–109

Purseglove JW (1975) Tropical crops – monocotyledons. Longman Group, London, 607 pp

Reed B (1992) Cryopreservation of *Ribes* apical meristems. Cryobiology 29:740

Reinhoud PJ (1996) Cryopreservation of tobacco suspension cells by vitrification. PhD Thesis, Leiden University, Netherlands, 100 pp

Sakai A (1993) Cryogenic strategies for survival of plant cultured cells and meristems cooled to –196 °C. JICA GRP REF 6:5–26

Sakai A (1995) Cryopreservation of germplasm of woody plants. In: Bajaj YPS (ed) Biotechnology in agriculture and forestry, vol 32. Cryopreservation of plant germplasm I. Springer, Berlin Heidelberg New York, pp 53–69

Sakai A, Kobayashi S, Oiyama I (1990) Cryopreservation of nucellar cells of navel orange (*Citrus sinensis* Osb. Var. *brasiliensis* Tanaka) by vitrification. Plant Cell Rep 9:30–33

Scowcroft WR (1984) Genetic variability in tissue culture: impact on germplasm conservation and utilisation. IBPGR, Rome, Italy, 41 pp

Shimonishi K, Karube M, Ishikawa M (1993) Cryopreservation of taro (*Colocasia esculenta*) embryogenic callus by slow prefreezing. Jpn J Breed 43:187

Staritsky G, Dekkers AJ, Louwaars NP, Zandvoort EA (1986) In vitro conservation of aroid germplasm at reduced temperatures and under osmotic stress. In: Withers LA, Anderson PG (eds) Plant tissue culture and its agricultural applications. Butterworth, London, pp 227–283

Strauss MS, Michaud JD, Arditti J (1979) Seed storage and germination and seedling proliferation in taro, *Colocasia esculenta* (L.) Schott. Ann Bot 43:603–612

Strauss MS, Stephens GC, Gonzales CJ, Arditti J (1980) Genetic variability in taro, *Colocasia esculenta* (L.) Schott (Araceae). Ann Bot 45:429–437

Takagi H, Otoo E, Islam OM, Senboku T (1994) In vitro preservation of germplasm in root and tuber crops. I. Preliminary investigation of mid- and long-term preservation of yam (*Dioscorea* spp.) and taro (*Colocasia esculenta* (L.) Schott). Breed Sci 44 [Suppl 1]:273 (in Japanese)

Thinh NT (1997) Cryopreservation of germplasm of vegetatively propagated tropical monocots by vitrification. PhD Diss, Kobe University, Japan, 187 pp

Thinh NT, Takagi H, Sakai A (2000) Cryopreservation of in vitro-grown apical meristems of some vegetatively propagated tropical monocots by vitrification. In: Engelmann F, Takagi H (eds) Cryopreservation of tropical plant germplasm. Current research progress and application. IPGRI, Rome, pp 227–232. ISBN 9290434287

Uragami A, Sakai A, Nagai M (1990) Cryopreservation of dried axillary buds from plantlets of *Asparagus officinalis* L. grown in vitro. Plant Cell Rep 9:328–331

Velayudhan KC, Muralidharan VK, Amalraj VA, Thomas TA, Rana RS (1991) Studies on the morphology, distribution and classification of an indigenous collection of taro. J Root Crops 17:118–129

Wilson JE (1990) Agro-facts: taro breeding. IRETA Publications, Western Samoa, 51 pp

Wilson JE, Siemonsma JS (1996) Plant yielding non-seed carbohydrates. In: Flach M, Rumawas F (eds) Plant resources of South East Asia no 9. Backhuys Publishers, Leiden, Netherlands, pp 69–72

Winarno FG (1990) Food chemistry. Bogor Agricultural University, Bogor, Indonesia

Zandvoort EA (1987) In vitro germplasm conservation of tropical aroids. Acta Bot Neerland 36:150

II.7 Cryopreservation of *Hordeum* (Barley)

JUN-HUI WANG and CHUN-NONG HUANG

1 Introduction

1.1 Distribution and Important Species

Barley is the fourth most important cereal after wheat, rice and maize. Although its distribution is generally similar to wheat, barley can be grown in much drier and colder regions than wheat. It is distributed mainly over the middle latitudes of the earth especially of the northern hemisphere. Russia, Canada, United States, some European and Asian temperate countries are the world's leading producers of this crop.

The genus *Hordeum* is in the Triticeae tribe in the Poaceae and contains about 30 species. Only three subspecies (*distichon*, *hexastichon* and *intermedium*) of *H. vulgare* L. are cultivated. *H. spontaneum* and *H. argriocrithon* are their wild relatives. Among the other wild forms, *H. bulbosum* L., *H. murinum* L., *H. marinum* Huds., *H. brachyantherum* Nevski., and *H. brevisubulatum* (Trinius) Link are comparatively important in agriculture and genetics. For instance, *H. bulbosum* is employed to produce haploid barley plants. Wild species are annual or perennial, with chromosome number of 2n = 14, 28 or 42, and show many valuable traits in stress tolerance and disease resistance, which may be used to improve cultivated barley.

1.2 Routine Methods for Germplasm Conservation

Both cultivated and wild species have a vast number of varieties, which grow over broad environmental ranges. More than 1600 ecological or genetic accessions have been collected just for *H. spontaneum* (von Bothmer et al. 1992). The China National Academy of Agriculture together with some provincial subunits have collected 13,116 germplasm accessions of this plant (Gao and Ma 1996). However, a very limited number of cultivated varieties are adopted in modern agriculture, while others that may be of less importance at present are underutilized and may be completely lost. Resources of wild species are being eroded with expanding industrialization and environmental alteration. Therefore, germplasm conservation of this plant is increasingly important for

College of Life Science, Zhejiang University, 232 Wensan Road, Hangzhou 310012, China

Biotechnology in Agriculture and Forestry, Vol. 50
L.E. Towill and Y.P.S. Bajaj (Eds.) Cryopreservation of Plant Germplasm II
© Springer-Verlag Berlin Heidelberg 2002

its biodiversity. In recent years, some Triticeae species have become endangered despite the fact that their seeds are orthodox like barley. *Elymus foliosus* is no longer in existence, while *Elymus grandis* and *Psathyrostachys huashanica* are threatened with extinction (Lu 1995).

Barley seeds are desiccation-tolerant and can be stored by routine approaches. Barley seeds have been used to study longevity and equations have been developed to predict longevity under various conditions of temperature and moisture content (Ellis and Roberts 1980). Under conditions recommended for use in genebanks (ca. $-18\,°C$), longevities can be estimated to be many decades. However, there are few experimental data for such conditions. Practically speaking, in jars or pots with some desiccants, they can retain their germination capacity for 5–8 years under ambient temperatures, whereas in modern air-conditioned seed banks, the duration may reach about 30 years (Gao and Ma 1996). However, like other cereal seeds, barley seeds are very susceptible to mildew during room temperature storage if they are not dried sufficiently.

1.3 Necessity of Cryopreservation

Cryopreservation is an attractive technique for long-term conservation of plant germplasm. This technique has several unique advantages such as providing great longevities and enhanced safety. Reviews (Bajaj 1976, 1979; Withers 1978, 1987; Sakai 1986; Steponkus 1985; Engelmann 1991; Steponkus et al. 1992; Blakesley et al. 1996; Wang and Huang 1998b) and books (Kartha 1985; Bajaj 1995) detail preservation of plant materials at cryogenic temperatures (ca. -130 to $-196\,°C$).

Cryogenic storage not only benefits long-term storage of both desiccation-tolerant and desiccation-sensitive seeds, but also is important in plant biotechnology. It may aid germplasm preservation of vegetatively propagated plants, maintain the morphogenetic potential of cultured cells, and facilitate regeneration from young explants. Most transgenic cereal plants are obtained via transformation of immature embryos (Christou et al. 1991; Vasil et al. 1993; Ritala et al. 1994; Wan and Lemaux 1994), whose developmental stages restrict the efficiency of experiments.

Regeneration of green plants from barley suspension cells and protoplasts was less efficient compared with other major cereals and it was not until the 1990s that significant advances were attained (Jähne et al. 1991; Yan et al. 1991). However, the genotypic specificity and reproducibility of protocols developed in individual laboratories remain problems. Protoplast isolation and subsequent plant regeneration can be achieved from primary calluses that are derived from immature embryos or microspores (Kihara and Funatsuki 1995; Salmenkallio-Martilla and Kauppinen 1995; Stöldt et al. 1996). So far, establishment of protoplast-producing and embryogenic barley cultures is still a time-consuming job (Singh et al. 1997).

Cultured barley cells often lose totipotency drastically and also produce a high ratio of albino regenerants (Wang et al. 1992; Huang et al. 1993). It is believed that many biochemical (Stirn et al. 1995), genetic (Wang et al. 1992) and histological (Wang and Huang 1995) properties in relation to the embryogenic potential disappear rapidly during extended subculturing. In conclusion, cryogenic storage of barley cultures seems critically important.

2 Cryopreservation

2.1 Brief Review

Pioneering work in this respect emerged as early as a century ago. From 1897 to 1936, there were some accounts indicated that barley seeds could germinate unchanged after exposure to liquefied gases such as air, hydrogen and helium (see Stanwood 1985). Similar results were obtained using liquid nitrogen as refrigerant (Stanwood 1980; Withers 1982).

In 1982, suspension cultures (Withers 1982) and protoplasts (Takeuchi et al. 1982) of barley were successfully cryopreserved, but plant regeneration from revived cells was not examined or achieved. Five years later, Hahne and Lörz (1987) showed that embryogenic calluses retained their regeneration capacity after cryostorage and, 5 years after that, plant regeneration from frozen suspension cells (Fretz et al. 1992) and immature embryos (Huang et al. 1992) was reported. Fretz and Lörz (1995) further described the beneficial effects of abscisic acid and ascorbic acid for cryopreservation, and reported fertile plants developed from cryostored calluses of *Hordeum murimum*. We have also reported plant formation from suspension cells cryopreserved by vitrification (Huang et al. 1995b).

2.2 Methodology

2.2.1 Materials

Seeds and Their Derivatives. Dry seeds of *Hordeum vulgare* were obtained from seed stocks in our laboratory. To relieve dormancy, seeds were soaked in 50% H_2SO_4 for 1h and then washed in tap water for 30min. Germinating embryos occurred from these seeds on Murashige and Skoog (MS) basic medium (Murashige and Skoog 1962) after 24–36h (the externally visible part of the root apex was about 1mm long).

Young Explants. Immature embryos (1–1.5mm) were excised 10–14 days after pollination and cultured on semi-solid CC2.5 medium (CC basic medium [Portrykus et al. 1979] + 1g/l casein hydrolysate + 250mg/l inositol+0.5g/l proline + 2.5mg/l 2,4-D). Immature inflorescences (5–7mm) were sampled at the stamen and pistil differentiation stage from approximately 4-month-old plants (Wang et al. 1996) and cultured on MSB medium (MS basic medium +

0.15 g/l asparaginic acid + 0.55 g/l proline + 10 mg/l thiamine-HCl + 2.0 mg/l 2,4-D).

Embryogenic Cultures. Embryogenic calluses and suspension cells, both derived from culturing of barley mature embryos (Huang et al. 1993, 1995b), were propagated on semi-solid CC2.5 medium and liquid CC2.0 medium, respectively.

2.2.2 Preculture

Immature embryos were precultured in CC2.5 medium supplemented with 0.5 mol/l sorbitol for 1 day. Immature inflorescences were precultured in MSB medium containing 0.5 mol/l sucrose for 2 days. Embryogenic calluses and suspension cells were harvested at the exponential growth stage. They were first precultured in CC2.5 medium with 60 g/l sucrose for 3 days and then in CC2.5 medium with 0.4 mol/l sorbitol for 1 day, which was termed 'the 3 + 1 treatment'. Other preculture treatments were also tested.

2.2.3 Cryoprotectants

Cryoprotectant for young explants and suspension cultures in the two-step or the rapid freezing method was prepared by addition of 10% (w/v) dimethyl sulfoxide (DMSO) and 0.4 mol/l sorbitol to the corresponding culture medium. Cryoprotectant for germinating embryos was 5% DMSO in water. The vitrification solution was based on the PVS2 formula (Sakai et al. 1990), which consisted of 30% (w/v) glycerol, 15% (w/v) ethylene glycol, 15% (w/v) DMSO and 0.4 mol/l sucrose in culture medium.

    For cryostorage of young explants, each cryotube contained 0.4 ml cryoprotectant and ten samples. For cryostorage of embryogenic cultures, the total volume of contents was 0.75 ml and the packed-cell volume was about 40%.

2.2.4 Dehydration and Cooling

Two-Step Method. Cryoprotectant preincubation: samples were treated with the prechilled cryoprotectant in an ice bath for 15–30 min (for young explants and embryogenic cultures) or 1 day (for germinating embryos). Cooling: pretreated samples were cooled at 1 °C/min from 0 to −40 °C using a programmable freezer (WKL-95, Sihuan Scientific Machine Factory, China). Then, they were plunged immediately into liquid nitrogen.

Vitrification. Vitrification of cell suspensions used the following steps (Huang et al. 1995a). Loading: precultured cells were equilibrated in 25% PVS2 at 22 °C for 10 min. Dehydration: Loaded cells were transferred into cryotubes and exposed to 100% PVS2 in an ice bath for 7.5 min. Cooling: cryotubes were immersed directly in liquid nitrogen.

In the case of immature inflorescences, the loading step was omitted and the duration of dehydration in PVS2 was 5 min.

Rapid Freezing. Samples with or without cryoprotectant preincubation were quenched to −196 °C quickly.

2.2.5 Warming and Washing

After rapid warming of cryotubes in a 37 °C water bath, samples cryopreserved by the two-step or rapid freezing method were drained and then recultured in the dark. Samples cryopreserved by vitrification were first incubated in a medium containing 1.2 mol/l sorbitol for 25 min before reculture (Huang et al. 1995a).

2.2.6 Viability Assay and Regrowth

Survival of the cells was determined by TTC (triphenyltetrazolium chloride) reduction using the method developed by Towill and Mazur (1975) and slightly modified by Langis et al. (1989) for vitrification of *Brassica campestris* cell suspensions. Viability was expressed as % TTC reduction of the control per unit dry weight of cells. The FDA-PI (fluorescein diacetate-propidium iodide) double staining method was also employed to evaluate cell viability (Huang et al. 1986).

Treated immature embryos were regrown on induction medium (IM, the standard CC2.5 medium) or germination medium (GM, hormone-free CC medium). Proliferation of calluses or seedlings was judged by visual inspection.

Cryostored immature inflorescences were cultured on MSB medium. Secondary calluses that formed were transferred to differentiation medium (replacement of 2,4-D in the medium with 1.0 mg/l 6-benzylaminopurine) to induce plantlet regeneration.

Suspension cells cryopreserved by vitrification were plated on semi-solid 1.2 mol/l sorbitol-containing CC2.5 medium for 2 days and then on normal CC2.5 medium for 12–14 days. The differentiation condition of recovered cells was reported in our previous studies (Huang et al. 1995b).

2.3 Results

2.3.1 Cryopreservation of Barley Seeds

The procedure for cryopreservation of seeds, with a moisture content of 10–12% fresh weight basis, was relatively simple. Without any cryoprotectant, seeds retained their germination capacity and normal growth after direct exposure to liquid nitrogen. Cooling rate (direct freezing, or two-step method with different cooling rates), warming temperature (0, 25, or 37 °C) and storage duration (1, 15 or 50 days) had no effect on survival. However, seeds in which dormancy was broken always developed into plants with a yellowish appearance and with some calluses in the coleoptile parts. Nevertheless, normal growth was eventually recovered during the following growth.

The conditions for cryopreservation of germinating embryos became complicated as the water content increased. Survival was affected by some elements such as composition of cryoprotectant, duration and temperature of cryoprotectant treatment, cooling rates, and the transfer temperature. The control germination rate of *H. vulgare* L. cv. 1041 seeds was 76%, and the control survival of its germinating embryos was 100% because only live embryos were used in the experiments. After treatment in 5, 10 or 20% DMSO at 0 °C for 1 day, the survival of germinating embryos was 89, 81 and 45%, respectively. Thus, 5% DMSO was used in further experiments to reveal the suitable cooling rate. During direct freezing of 5% DMSO-treated samples, only 5% of the embryos survived. Using the two-step method at the transfer temperature of −80 °C, when the cooling rate was 0.5, 5.0 and 20 °C/min, the survival was 13, 68 and 77%, respectively.

2.3.2 Cryopreservation of Immature Embryos

Two methods (rapid freezing and two-step cooling) and two regrowth media (germination and induction) were tested in the cryopreservation of immature embryos. Recovery of samples treated by rapid freezing was better than that by the two-step method (Fig. 1). Precultured and then cryoprotectant-treated samples probably were sufficiently dehydrated before exposure to survive the rapid cooling in LN; the two-step method might have given excessive desiccation. Survivals on germination medium were less than those on induction medium. The reason was that, in the former case, revival of both the scutellum and plumular axes were required for the development of whole plants; while in the latter case, revival of either of these parts could ensure the initiation of callus.

Preculture was very important in improving the survival of immature embryos. Non-precultured samples showed very low regrowth on two media (Table 1). No difference was observed between the callus induction frequency of samples stored in LN for 3 months and that for 7 months. For cv. Sumai, for 3 months of storage, 16 of 19 embryos survived, and, for 7 months, 21 of 24 survived. Calli originating from revived embryos were able to develop into plants during differentiation.

2.3.3 Cryopreservation of Immature Inflorescences

Pregrowth and dehydration appeared critical for vitrification of young inflorescences. The preculture on 0.5 mol/l sucrose for 2 days was superior (Table 2), and the suitable dehydration period in PVS2 at 0 °C was 5 min (Fig. 2).

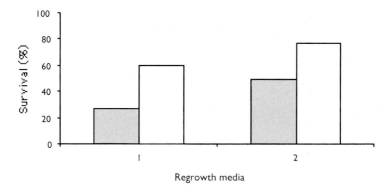

Fig. 1. Effect of cooling protocol and regrowth medium on survival of cryopreserved barley immature embryos cultured on germination medium (*left two columns, 1*) or induction medium (*right two columns, 2*). *Black columns* Cryopreserved by the two-step method; *open columns* cryopreserved by the rapid freezing method

Table 1. Effect of preculture and recovery medium on survival of barley immature embryos cryopreserved by the two-step method

		Cryopreservation		
		No. treated	No. surviving	Survival (%)
Precultured (1 d in 0.4 M	GM	18	7	39
sorbitol medium)	IM	8	4	50
Not-precultured	GM	22	1	4.5
	IM	25	1	4

GM, germination medium; IM, induction medium.

Table 2. Effect of preculture on survival of barley (cultivar Xu 9) immature inflorescences cryopreserved by vitrification

Preculture		Survival (%)	
Sucrose concentration (mol/L)	Duration of culture	Before cooling	After cooling
1.0	1	85.0	5.0
1.0	2	75.0	32.5
0.5	2	85.0	50.0

Samples were dehydrated in PVS2 at 0 °C for 5 min.

Three cultivars, 81G1, Gebeina and Xu 9, showed 100, 82.5 and 50% survival, respectively, after cryopreservation.

The difference between the two-step method and vitrification was remarkable. Survival of cv. Gebeina was 40% in the two-step method and 82.5% in vitrification. In the former method, growth only occurred after 7 days, whereas it was about 2–3 days in the latter.

No difference was observed in the time of secondary callus formation between cooled and control samples, i.e., for cultivar Xu 9, it was still 3 weeks

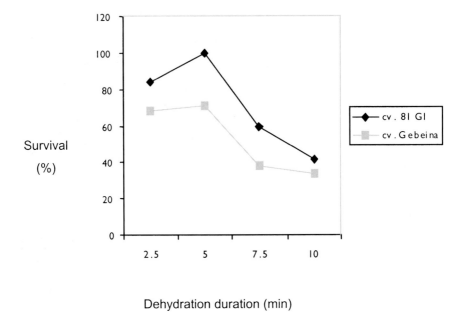

Dehydration duration (min)

Fig. 2. Effect of dehydration duration in PVS2 at 0 °C on the survival of barley immature inflorescences cryopreserved by vitrification

after induction, and for the other two cultivars, it was approximately after one to two passages of subculture. Green plants were regenerated from these calli (Fig. 3).

2.3.4 Cryopreservation of Cell Suspensions

Factors that improve survival of cell suspensions are well documented. We have previously investigated them in detail with rice cells using both a two-step (Yan et al. 1994) and a vitrification method (Huang et al. 1995a). Just as in rice suspension cells, preculture treatment and cryoprotectant preincubation ensured higher regrowth during two-step cryopreservation of barley cells (Table 3).

Preculture, loading and dehydration affected vitrification of barley cells. Influences of preculture protocols on TTC reduction, as a measurement of viability, are illustrated in Fig. 4; the 3+1 treatment gave the highest survival. As with rice, the appropriate loading duration (in 25% PVS2 at 22 °C) and dehydration duration (in PVS2 at 0 °C) for barley cells was 10 min and 7.5 min, respectively (data not shown). Revived cells formed regenerants (Fig. 5), but the frequency was rather low. Control cells also did not regenerate well because they were extensively subcultured.

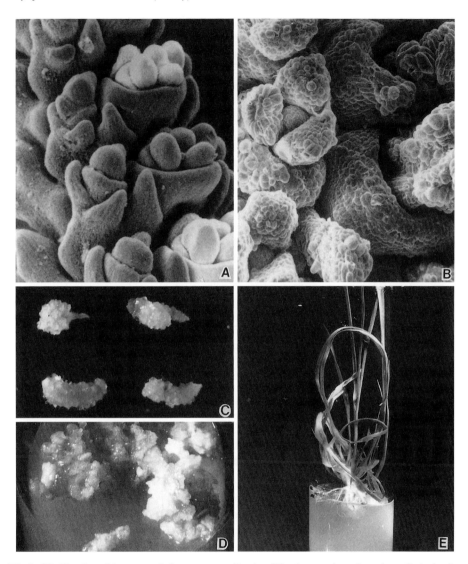

Fig. 3. Vitrification of immature inflorescences of barley (*Hordeum vulgare* L., subsp. *distinchon*) **A,B** Scanning electron micrographs of barley immature inflorescence, cv. 81G1; **A** an immature inflorence just after cryopreservation, 250×; **B** initiation of primary callus formation 4 days after cryopreservation, 125×; **C-E** cv. Xu 9; **C** primary callus after 10 days of reculture; **D** callus in subculture; **E** plantlet regeneration from cryopreserved immature inflorescences

2.3.5 Cryopreservation of Callus

About 40–60% of callus clusters survived when the two-step method was used. The starting materials were nodular and compact, and were derived from the culture of mature embryos. These are commonly referred to as embryogenic calli (Fig. 6A).

Table 3. Effect of preculture and cryoprotectant preincubation on the recovery percentage of barley suspension cells cryopreserved by the two-step method

Treatment before cryopreservation		Culture after thawing		
CC2.0 + 0.4 M sorbitol, 1 day	10% DMSO + 0.4 M sorbitol, 0 °C, 15 min	No. of clusters treated	No. of clusters show growth	Recovery (%)
–	–	15	0	0
+	–	30	14	47
+	+	55	45	82

+ subjected to this treatment; – Not subjected to this treatment.

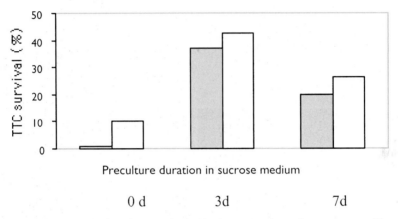

Fig. 4. Effect of preculture (days) on survival of barley suspension cells cryopreserved by vitrification. The concentration of sucrose was 60 g/l (in CC2.0 medium). *Black columns* cells were cryopreserved after the sucrose preculture; *open columns* after the sucrose preculture, cells were further precultured in 0.4 mol/l sorbitol-containing medium for 1 day before cryopreservation

Morphological studies of cells revealed interesting points. Young callus cells, containing abundant starch grains (amyloplasts), were less differentiated (Fig. 6B,C). After several subcultures, a large number of organized areas developed in the callus. These had cells with thickened walls and differentiated vascular elements (Wang and Huang 1995). Formation of primary roots was very prominent from these calli (Fig. 6D), and the calli lost the potential for shoot formation. Regeneration of plants from barley callus through embryogenesis was less efficient and became much more infrequent during sequential subcultures (Oka et al. 1995).

2.4 Discussion

2.4.1 Preculture

Regardless of cooling rate and warming temperature, barley dry seeds can be successfully cryopreserved without any pretreatments and cryoprotectants,

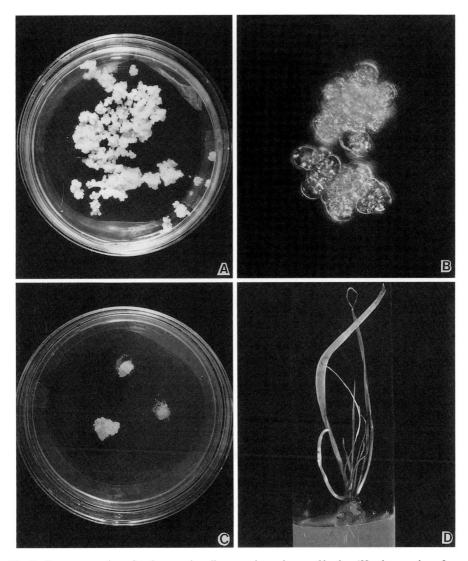

Fig. 5. Cryopreservation of embryogenic cell suspension cultures of barley *(Hordeum vulgare* L., subsp. *distinchon*, cv. ZhaoShu 3) by vitrification. **A** Embryogenic callus in friable appearance derived from culturing mature embryos; **B** differential-interference-contrast micrograph of cell suspensions; **C** recovery of cryopreserved cells; **D** plantlet regeneration from recovered cells

indicating that dry seeds are already sufficiently desiccated before cooling. Germinating embryos can be successfully cryopreserved by a two-step protocol without any precultures. Preculture on a suitable medium, however, was required for the cryopreservation of barley young explants and embryogenic cultures. The effects of preculture were studied using rice suspension cells (Yan et al. 1994; Wang et al. 1998). Here, preculture reduced the volume of vacuoles

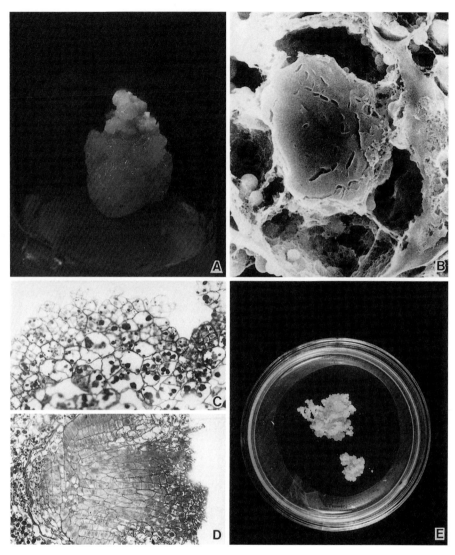

Fig. 6. Cryopreservation of embryogenic callus of barley (*Hordeum vulgare* L., subsp. *distinchon*, cv. ZhaoShu 3) by the two-step method. **A** Formation of nodular compact callus (commonly referred to as embryogenic callus) from primary callus (watery and soft); **B** scanning electron micrograph of an embryogenic cell by freezing cracking method, 7000×; **C,D** epoxy resin semi-thin section micrograph of callus, approximately 200×; **C** fresh callus; **D** subcultured callus with differentiation of a young root tip; **E** recovery of cryopreserved callus

and eliminated some osmotically sensitive mitochondria. Other than structural changes, reduction in total water content, increase in bound water ratio and accumulation of new proteins also occurred during this step. The adaptation both in structure and physiology improved the tolerance of cells to desiccation and freezing.

In the present study, the 3 + 1 treatment was best and is similar to, but simpler than, the 10 + 1 treatment that was employed in the vitrification of rice suspension cells (Huang et al. 1995a). Pregrowth in the presence of 60 g/l sucrose for 3–10 days contributed greatly to the freezing tolerance of rice cells, and pregrowth in 0.4 mol/l sorbitol for 1 day contributed greatly to the desiccation tolerance of rice cells. Survival was considerably improved when both preculture steps were used, i.e., the survival rate was higher than preculture in sucrose or in sorbitol alone. Probably, some interactions occurred between the two preculture steps (Huang et al. 1995a).

2.4.2 Two-Step Method and Vitrification Approach

There are some differences and similarities between the two-step method and vitrification. The former involves a freeze-induced cell dehydration during programmed cooling and intracellular vitrification during rapid quenching. In the latter, cells are dehydrated by an extremely concentrated vitrification solution, and both extracellular and intracellular solutions are probably vitrified with rapid cooling. The key to high survival rates in the two-step method is to carefully control the cooling procedure, whereas, in the vitrification method, the cryoprotectant exposure must be carefully controlled.

The latter method has several advantages, including higher survival levels, shorter lag periods and simplicity of manipulation. In this study, the lag period of growth for vitrified immature inflorescences was shorter than that for the two-step method.

2.4.3 Interdependence Among Cryopreservation Steps

The steps in a complete cryopreservation protocol complement each other. In the case of vitrification, if the loading duration is prolonged, the optimum dehydration length should be shortened. Likewise, in the two-step method, the cooling rate may be increased, if periods of incubation in the cryoprotectant are extended. Therefore, parameters of cryostorage steps permit some variation, providing that cells are still well protected and adequately dehydrated.

In the present work, rapid freezing was superior to two-step cooling in cryopreservation of immature embryos. However, an opposite result was reported for maize immature embryos (Laurie et al. 1996). This inconsistency may be attributed to the difference in preculturing or cryoprotectant pretreatment.

2.4.4 Dehydrins and Antifreeze Proteins

Approximately 70 complete amino acid sequences of dehydrins were published between 1988 and 1996 (Close 1997). These proteins as well as other LEAs (late embryogenesis abundant proteins) may protect plant seeds and vegetative tissues against water deficits or low temperature stress via chaperone-like properties, such as ion sequestering and membrane stabilization (Close 1996). Studies have begun to examine whether the desiccation sensitivity of recalcitrant seeds is related to an insufficient accumulation of LEAs (Kermode 1997).

Antifreeze proteins, first identified in polar fishes, also accumulate in freezing-tolerant overwintering cereals (Antikainen and Griffith 1997). These proteins can lower the freezing temperature noncolligatively and inhibit the growth of ice crystals. Fish antifreeze proteins were used for cryostudies of animal samples, but both positive and negative results were observed (Wang and Huang 1996). The function of these proteins during cryopreservation of rice cells was tested (Wang and Huang 1998a; Wang et al. 1999). At the proper concentration, antifreeze proteins may enhance viability through inhibition of ice recrystallization, whereas, at high concentration, they may decrease the survival rates by ice nucleation.

Barley has more than five types of dehydrins (Close et al. 1989) and knowledge about these may help improve the freezing and desiccation resistance. Tobacco suspension cells, being transformed by a lea3 gene, were vitrified successfully without any mannitol precultures (Reinhoud et al. 1997). Recently, using Western blotting, we have analyzed the dehydrins of precultured rice suspension cells and are currently carrying out similar work with barley cells.

3 Summary and Conclusions

Barley is an important cereal crop with numerous varieties and ecological types, and germplasm conservation is becoming increasingly important. Compared with other major cereals, establishment and maintenance of highly efficient and easily available plant regeneration systems are still difficult and time-consuming. Cryopreservation can arrest the genetic instability that occurs by continual culture of embryogenic lines. In this chapter, progress on cryostorage of barley seeds, germinating embryos, immature embryos, immature inflorescences, embryogenic suspension cells and calluses was described. In contrast to rice, cryopreservation of barley embryogenic cultures is not well described and the regeneration of fertile plants or transgenic plants from cryopreserved cultures has not yet been achieved.

Memorial and Acknowledgments. The first author and his colleagues commemorate Professor Chun-Nong Huang, a diligent and famous Chinese biologist, who directed the studies in the present paper. Tragically, on December 15th, 1997, he died from heart failure.

The studies were supported by three grants from the Natural Science Foundation of China (Project Nos. 39370086, 39570083, and 39900012) and a grant from Zhejiang Science and Technology Committee (Project No. 913276). Dr. Qin-Feng Yan, Dr. Hong-Wu Bian, Ms. Yi-Xiang Zhang, Dr. Feng Liu and Ms. Yong Zheng contributed greatly to the present paper.

References

Antikainen M, Griffith M (1997) Antifreeze protein accumulation in freezing-tolerant cereals. Physiol Plant 99:423–432

Bajaj YPS (1976) Gene preservation through freeze-storage of plant cell, tissue and organ culture. Acta Hortic 63:75–84

Bajaj YPS (1979) Technology and prospects of cryopreservation of germplasm. Euphytica 28: 267–285

Bajaj YPS (ed) (1995) Cryopreservation of plant germplasm I. Biotechnology in agriculture and forestry, vol 32. Springer, Berlin Heidelberg New York, 507 pp

Blakesley D, Pask N, Henshaw GG, Fay MF (1996) Biotechnology and the conservation of forest genetic resources: in vitro strategies and cryopreservation. Plant Growth Regul 20:11–16

Christou P, Ford TL, Kofron M (1991) Production of transgenic rice (*Oryza sativa* L.) plants from agronomically important indica and japonica varieties via electric discharge particle acceleration of exogenous DNA into immature zygotic embryos. Bio/Technology 9:957–962

Close TJ (1996) Dehydrins: emergence of a biochemical role of a family of plant dehydration proteins. Physiol Plant 97:795–803

Close TJ (1997) Dehydrins: a commonalty in the response of plant to dehydration and low temperature. Physiol Plant 100:291–296

Close TJ, Kortt AA, Chandler PM (1989) A cDNA-based comparison of dehydration-induced proteins (dehydrins) in barley and corn. Plant Mol Biol 13:95–108

Ellis RH, Roberts EH (1980) Improved equations for the predictions of seed longevity. Ann Bot 45:13–30

Engelmann F (1991) In vitro conservation of tropical plant germplasm – a review. Euphytica 57:227–243

Fretz A, Lörz H (1995) Cryopreservation of in vitro cultures of barley (*Hordeum vulgare* L. and *H. murinum* L.) and transgenic cells of wheat (*Triticum aestivum* L.). J Plant Physiol 146:489–496

Fretz A, Jähne A, Lörz H (1992) Cryopreservation of embryogenic suspension cultures of barley (*Hordeum vulgare* L.). Bot Acta 105:140–145

Gao D, Ma D (1996) Germplasm resource of barley. In: Lu L (ed) Barley science in China. Chinese Agriculture Press, Beijing, pp 153–173

Hahne G, Lörz H (1987) Cryopreservation of embryogenic callus cultures from barley (*Hordeum vulgare* L.). Plant Breed 99:330–332

Huang C-N, Cornej NJ, Bush DS, Jones RL (1986) Estimating viability of plant protoplasts using double and single staining. Protoplasma 135:80–87

Huang C-N, Yan Q-F, Wang J-H, Yu Z-Y, Du Y (1992) Studies on cryopreservation of barley (*Hordeum vulgare* L.) immature embryos. Bull Sci Technol 8:209–212

Huang C-N, Yan H, Yan Q, Zhu M, Yuan M, Xu A (1993) Establishment and characterization of embryogenic cell suspension cultures from immature and mature embryos of barley (*Hordeum vulgare* L.). Plant Cell Tissue Organ Cult 32:19–25

Huang C-N, Wang J-H, Yan Q-S, Zhang X-Q, Yan Q-F (1995a) Plant regeneration from rice (*Oryza sativa* L.) embryogenic suspension cells cryopreserved by vitrification. Plant Cell Rep 14:730–734

Huang C-N, Wang J-H, Yan Q-F, Yan Q-S, Zhang X-Q (1995b) Rapid establishment and cryopreservation of embryogenic cell suspension cultures of barley (*Hordeum vulgare* L.). Sci Agric Sin 28(3):21–27

Jähne A, Lazzeri PA, Lörz H (1991) Regeneration of fertile plants from protoplasts derived from embryogenic cell suspensions of barley (*Hordeum vulgare* L.). Plant Cell Rep 10:1–6

Kartha KK (ed) (1985) Cryopreservation of plant cells and organs. CRC Press, Boca Raton, 276 pp

Kermode AR (1997) Approaches to elucidate the basis of desiccation-tolerance in seeds. Seed Sci Res 7(2):75–96

Kihara M, Funatsuki H (1995) Fertile plant regeneration from barley (*Hordeum vulgare* L.) protoplasts isolated from primary calluses. Plant Sci 106:115–120

Langis R, Sohnabel B, Earle ED, Steponkus PL (1989) Cryopreservation of *Brassica compestrist* L. cell suspensions by vitrification. Cryo Lett 10:421–428

Laurie JD, Zhang G, McGann LE, Case DD (1996) Cryopreservation of maize immature embryos – embryo culture, cryopreservation, and effect of preculture conditions . In Vitro 32 (3, part II):107A–108A

Lu B (1995) Diversity and conservation of the Triticeae genetic resources. Chin Biodiv 3(2):63–68

Murashige T, Skoog F (1962) A revised medium for rapid growth and bioassays with tobacco tissue cultures. Physiol Plant 15:473–497

Oka S, Saito N, Kawaguchi H (1995) Histological observations on initiation and morphogenesis in immature and mature embryo derived callus of barley (*Hordeum vulgare* L.). Ann Bot 76:487–492

Portrykus I, Harms CT, Lörz H (1979) Callus formation from cell culture protoplasts of corn (*Zea mays* L.). Theor Appl Genet 54:209–214

Reinhoud PJ, Versteege I, Van Iren F, Kijne JW (1997) Induction of lea5 in tobacco suspension cells contributes to their tolerance to vitrification. Cryo Lett 18:69

Ritala A, Aspegren K, Kurten U, Salmenkallio-Marttila M, Mannonen L, Hannus R, Kauppinen V, Teeri TH, Enari TM (1994) Fertile transgenic barley by particle bombardment of immature embryos. Plant Mol Biol 24:317–325

Sakai A (1986) Cryopreservation of germplasm of woody plants. In: Bajaj YPS (ed) Biotechnology in agriculture and forestry, vol 1. Springer, Berlin Heidelberg New York, pp 113–129

Sakai A, Kobayashi S, Oiyama I (1990) Cryopreservation of nucellar cells of navel orange (*Citrus sinensis* var. brasiliensis Tanaka) by vitrification. Plant Cell Rep 9:30–33

Salmenkallio-Marttila M, Kaupinnen V (1995) Efficient regeneration of fertile plants from protoplasts isolated from microspore cultures of barley (*Hordeum vulgare* L.). Plant Cell Rep 14:253–256

Singh RR, Kemp JA, Kollmorgen JF, Qureshi JA, Fincher GB (1997) Fertile plant regeneration from cell suspension and protoplast cultures of barley (*Hordeum vulgare* cv. Schooner). Plant Cell Tissue Organ Cult 49:121–127

Stanwood PC (1980) Tolerance of crop seeds to cooling and storage in liquid nitrogen (–196 °C). J Seed Technol 5:26

Stanwood PC (1985) Cryopreservation of seed germplasm for genetic conservation. In: Kartha KK (ed) Cryopreservation of plant cells and organs. CRC Press, Boca Raton, pp 199–226

Steponkus PL (1985) Cryobiology of isolated protoplasts: application to plant cell cryopreservation. In: Kartha KK (ed) Cryopreservation of plant cells and organs. CRC Press, Boca Raton, pp 49–60

Steponkus PL, Langis R, Fujikawa S (1992) Cryopreservation of plant tissues by vitrification. In: Steponkus PL (ed) Advances in low-temperature biology, vol 1. JAI Press, London, pp 1–61

Stirn S, Mordhorst AP, Fuchs S, Lörz H (1995) Molecular and biochemical markers for embryogenic potential and regenerative capacity of barley (*Hordeum vulgare* L.) cell cultures. Plant Sci 106:195–206

Stöldt A, Wang X, Lörz H (1996) Primary callus as source of totipotent barley (*Hordeum vulgare* L.) protoplasts. Plant Cell Rep 16:137–141

Takeuchi M, Matsushima H, Sugawara Y (1982) Totipotency and viability of protoplasts after long-term freeze preservation. In: Fujiwara A (ed) Proceedings of the 5th international congress of plant tissue and cell culture. Maruzen, Tokyo, Japan, pp 797–798

Towill LE, Mazur P (1975) Studies on the reduction of 2,3,5-triphenyltetrazolium chloride as a viability assay for plant tissue cultures. Can J Bot 53:1097–1102

Vasil V, Srivastava V, Castillo AM, Fromm ME, Vasil IK (1993) Rapid production of transgenic wheat plants by direct bombardment of cultured immature embryos. Bio/Technology 11:1553–1558

von Bothmer R, Seherg O, Jacobsen N (1992) Genetic resources in the Triticeae. Hereditas 116:141–150

Wan Y, Lemaux PG (1994) Generation of large numbers of independently transformed fertile barley plants. Plant Physiol 104:37–48

Wang J-H, Huang C-N (1995) Histological studies on the dedifferentiation and redifferentiation pattern of barley (*Hordeum vulgare* L.) mature embryo cells. J Hangzhou University (Natural Science Edition) 22:102–106

Wang J-H, Huang C-N (1996) Application of antifreeze proteins in hypothermic preservation and cryopreservation. Chin J Cell Biol 18(3):107–111

Wang J-H, Huang C-N (1998a) Assessment of antifreeze proteins and water deficit proteins during cryopreservation of *Oryza sativa* and *Dendrobium candidum* cells. Abstracts of XVIII international congress of genetics, 10–15 Aug, Beijing, 186 pp

Wang J-H, Huang C-N (1998b) Progress on germplasm cryopreservation of woody plants. World For Res 11(5):6–11

Wang J-H, Yan Q-F, Huang C-N (1996) Plant regeneration from barley (*Hordeum vuglare* L.) immature inflorescences cryopreserved by vitrification. Acta Bot Sin 38:730–734

Wang J-H, Ge J-G, Liu F, Huang C-N (1998) Ultrastructural changes during cryopreservation of rice (*Oryza sativa* L.) embryogenic suspension cells by vitrification. Cryo Lett 19:49–54

Wang J-H, Bian H-W, Huang C-N, Ge J-G (1999) Studies on the application of antifreeze proteins in cryopreservation of rice embryogenic suspension cells. Acta Biol Exp Sin 32:271–276

Wang XH, Lazzeri PA, Lörz H (1992) Chromosomal variation in dividing protoplasts derived from cell suspension of barley (*Hordeum vulgare* L.). Theor Appl Genet 85:181–185

Withers LA (1978) Freeze-preservation of cultured cells and tissues. In: Thorpe T (ed) Frontiers of plant tissue culture. Calgary University, Calgary, pp 297–306

Withers LA (1982) The development of cryopreservation techniques for plant cell, tissue and organ culture. In: Fujiwara A (ed) Plant tissue culture. Tokyo, Japan, pp 793–794

Withers LA (1987) Long-term preservation of plant cells, tissues and organs. Oxf Surv Plant Mol Cell Biol 4:221–272

Yan Q-S, Zhang X-Q, Shi J-B, Li J-M (1991) Green plant regeneration from protoplasts of barley (*Hordeum vulgare* L.). Chin Sci Bull 36:932–935

Yan Q-F, Wang J-H, Huang C-N, Yan Q-S, Zhang X-Q (1994) Studies on cryopreservation of rice (*Oryza sativa* L.) suspension cultures. Acta Biol Exp Sin 27:399–409

II.8 Cryopreservation of *Humulus lupulus* L. (Hop)

M.A. REVILLA and D. MARTÍNEZ

1 Introduction

1.1 Plant Distribution and Importance

The cultivated hop (*Humulus lupulus*, Cannabinaceae) is a hardy herbaceous climbing plant that is indigenous in the northern hemisphere above 32° latitude. Germany, the USA, China and the Czech Republic are the main producer countries. The Spanish production represents around 5% of the total production of the European Union, having the third place in importance among the producing member countries behind Germany and the United Kingdom.

Hop cultivars are vegetatively propagated by layering vines or by rooting soft-wood cuttings during the growing season (Neve 1991). The aboveground parts of the plant die back to ground level each winter but the belowground rootstocks are perennial and can survive many years.

Hop bells are used almost exclusively for brewing, although they were first taken into cultivation for their herbal and medicinal properties. The commercial value of hops lies in the resins (especially α-acids and the essential oils) contained within the lupulin glands of the female cone-like inflorescences. These resins give beer its bitterness, aroma and flavor. Whereas male flowers develop only a small number of resin glands, they are abundant on the cones of the female plants (Neve 1991). Hop is dioecious and consequently plants are genetically very heterozygous. When plants are raised from seed the progeny are extremely variable and are of little value commercially. For this reason, hop gardens are invariably planted with material that has been propagated vegetatively from one of the established cultivars.

1.2 Methods for Storage and Need for Cryopreservation

The centre of origin of *H. lupulus* is considered to be China, from where it migrated east to America and west to Europe, resulting in two distinct populations. Such geographical distribution and evolution of the species is sup-

Departamento Biología de Organismos y Sistemas, Facultad de Biología, C/C. Rodrigo Uría s/n, Universidad de Oviedo, 33071 Oviedo, Spain

Biotechnology in Agriculture and Forestry, Vol. 50
L.E. Towill and Y.P.S. Bajaj (Eds.) Cryopreservation of Plant Germplasm II
© Springer-Verlag Berlin Heidelberg 2002

ported by differences in the Y chromosome between European and American hops (Neve 1991) and by the discovery of restriction fragment length polymorphism (RFLP) pattern characteristics of native American or wild and cultivated European and Asian genotypes (Pillay and Kenny 1996a). Very little is known about the origin of hop cultivation but the first selections were probably made from hop gardens created from wild hops (Neve 1991). In Europe, this resulted in the development of regional characteristic types that were named after the area in which they were grown (Saazer, Halertauer, Tettnanger) or, in the case of selections, after the growers who developed them (Fuggle, Golding). In the history of hop breeding wild hops have always played an important role as a genetic source (Salmon 1934; Haunold 1981).

At least 50 major hop cultivars are grown around the world. The differences between these cultivars are found on the basis of observations during the growing period and on the basis of morphological and agronomic characteristics (analysis of secondary metabolites – bitter resins and essential oils). In recent years, there have been reports on the use of molecular markers for evaluating genetic variability in hops (Brady et al. 1996; Pillay and Keny 1996a,b; Matousek et al. 1999). The characteristics of any variety vary considerably from season to season and from location to location. Wild hops are a source of valuable characteristics for the plant breeder, such as resistance to a new pest or disease and different oils or resin characteristics.

Hop germplasm is traditionally preserved in the form of field collections in areas of cultivation. However, plants conserved in natural conditions remain exposed to pests or pathogens that may act as vectors of virus transmission. Slower growth of the plant and flower cones, delay in flower onset and maturation, together with a lower production of α-acids are some of the drastic consequences of virus infection in hop plants (Probasco and Winslow 1986; Fric et al. 1997). Nuclear stocks of virus-free hop material have been proposed (Samyn and Welvaert 1983). Virus-free plants can be produced by in vitro culture of shoot meristems (Vine and Jones 1969; Adams 1975; Probasco and Winslow 1986; Heale et al. 1989). Nevertheless, management of large-scale in vitro collections poses numerous practical problems. Moreover, risks of contamination and of somaclonal variation increase with time. It is therefore essential to develop long-term conservation techniques to maintain and preserve germplasm collections representing a wide diversity of hop cultivars and their wild relatives for future plant breeding.

2 Cryopreservation of Shoot Tips

2.1 Plant Material

In Vitro Cutting. In vitro shoot cultures were initiated from rootstocks of virus-free hop plants (*Humulus lupulus* L.) taken in December. After surface sterilization (30 min in 4% sodium hypochlorite and three washes with sterile water) the rhizomes were cultured on trays with wet vermiculite, at 25 °C and 16-h photoperiod ($40\,\mu M\,m^{-2}\,s^{-1}$). After 3 months, 6–7 cm long shoots had been formed, which were cut into nodal segments (1–2 cm long), sterilized with the same procedure used for rhizomes, and cultured in 125-ml flasks with "cutting" medium (Adams 1975): salts of Murashige and Skoog (1962), vitamin mixture of Wetmore and Sorokin (1955), glucose (30 g/l), BAP (4.4 μM), IBA (0.5 μM), pH 5.2, and solidified with 0.7% agar (Roko, La Coruña, Spain). Culture conditions were as described for rhizomes. Transfers of the micro-shoots to fresh medium were made at 6-week intervals.

Cold acclimation of the micro-cuttings was performed at the end of the culture period, either under 8 h light at 12 °C and 16 h dark at 6 °C in a controlled environmental chamber or under dark conditions at 4 °C, both for 7–60 days. For experiments on cooling rates and differential scanning calorimetry, cultures were cold treated for 21 days under photoperiod conditions.

Shoot Tip Culture. Apical and axillary shoot tips (0.5–2.0 mm), comprising the meristematic dome with one or two pairs of foliar primordia, were excised under a stereomicroscope from 8–10 cm length shoots, 6 weeks after their last transfer. Both types of shoot tips were cultured on "apex" medium which differed from the "cutting" medium in the growth regulators added: BAP (4.4 μM) and GA_3 (0.28 μM). Culture conditions were the same as for in vitro cuttings.

2.2 Cryopreservation Procedures

To ensure survival of plant cells after plunging into liquid nitrogen (LN) and subsequent rewarming, it is necessary to prevent the formation of intracellular ice crystals by dehydrating the cells. Three procedures were examined to cryopreserve hop shoot tips. Survival was defined as the percentage of shoot tips recovering normal regrowth after cooling and warming.

Slow Cooling Procedure. The explants were precultured on a solid "apex" preculture medium supplemented only with DMSO (5%) or sucrose (0.3 M) or both together at the same concentrations, under the conditions used for micropropagation. After 2-day incubation, the shoot tips were transferred into cryotubes (2 ml), containing 0.5 ml of cryoprotective solution (only DMSO at 5, 10 or 20%, or DMSO plus sucrose, both at the same concentration, 5 or 10%) for 1 h at room temperature.

Samples were cooled at different rates (0.1, 0.25, 0.5, 0.75 and 10 °C/min) from +4 to −40 °C, using a programmable freezer, and directly immersed in LN. After rapid warming (2 min in a water bath at +40 °C) shoot tips were transferred to "apex" culture medium.

Vitrification Procedure. Shoot tips were precultured for 2 days on solid medium progressively enriched with sucrose (12 h in 0.3 M, 12 h in 0.5 M and

24 h in 0.75 M). After loading for 2 days with glycerol (2 M) and sucrose (0.4 M) in apex medium (30 shoot tips/20 ml medium), they were transferred to 20 ml of the vitrification solution (PVS2) (Sakai et al. 1990) in "apex" medium supplemented with sucrose (0.4 M) at 0 °C for different durations (0–3 h).

Cooling was performed by direct immersion in LN of 2-ml cryovials, each containing 10 shoot tips and 0.5 ml of PVS2. After rapid warming for 2 min in a bath water at +40 °C, the vitrification solution was replaced by apex culture medium supplemented with 1.2 M sucrose for 60 min (solution was replaced every 20 min). Finally, shoot tips were transferred to "apex" culture medium.

Encapsulation-Dehydration Procedure. Beads containing shoot tips were prepared according to the technique of Redenbaugh et al. (1986). Shoot tips were suspended in calcium-free "apex" medium supplemented with 3% Na-alginate solution and the mixture was dripped into apex medium containing 100 mM calcium chloride.

Beads containing one shoot tip each were first precultured in "apex" medium supplemented with sucrose either at different concentrations: 0.75 or 1.0 M for 1–7 days, or progressively increased from 0.3 to 0.5, 0.75 and 1.0 M (12 h in each concentration). Dehydration was carried out by placing encapsulated shoot tips in uncovered Petri dishes containing 30 g of dry silica gel and on top of a filter paper, in the air current of a laminar flow hood for up to 9 h.

Beads were then transferred to 2-ml cryovials (5 beads per vial) for cooling either rapidly by direct immersion in LN (200 °C/min) or slowly (0.1, 0.25, 0.5, 1.0, or 10.0 °C/min) from +20 to –40 °C followed by immersion in LN. After slow rewarming under the laminar air of a flow cabinet at room temperature, beads were transferred to Petri dishes containing solid apex medium. After 30 days developing shoots were cultured in cutting medium and afterwards were transferred to soil conditions.

2.3 Thermal Analysis

Differential scanning calorimetry (DSC) was performed using a Mettler TA 4000 system (TC 11 PC and DSC 30). Variations of enthalpy due to state changes of solutions or live materials can be measured by this technique during cooling as well as during rewarming. These events were estimated from temperature differences between two pans cooled or warmed at the same rate. One pan was empty and was used for reference; the second contained the sample. This technique could detect ice crystallization, melting, and glass transitions.

2.4 Results and Discussion

Very high survival was achieved after cryopreservation of hop shoot tips by the encapsulation-dehydration procedure. However, no survival was achieved using slow cooling or vitrification procedures. Survival of the control shoot tips, after 1 h of cryoprotection by the slow cooling or 1.5 h by the vitrification procedure, were very high (90–100%). However, no survival was obtained after cooling and warming for any of the cryoprotective mixes or precultures examined in the slow cooling or vitrification procedures.

Using encapsulation-dehydration, both control and cryopreserved shoot tips after 3 days of being subcultured on apex medium became green, at 7 days began to develop the first pair of leaves (Fig. 1A) and in 30–40 days resumed growth, producing new shoots by direct development of the apical dome, without callus formation (Fig. 1B). Rooting was produced in the same medium after another 30–40 days (Fig. 1C), and complete plants, 5–8 cm high, were transferred to soil conditions (Fig. 1D; Martínez et al. 1999). This procedure combines two cryoprotective treatments: preculture with sucrose and air-drying. Both of them have been used to discover the most favorable sucrose concentration and the optimal duration of dehydration and sucrose preculture, to induce cytoplasmic vitrification and avoid, therefore, the formation of intracellular ice crystals during subsequent rapid cooling in liquid nitrogen.

2.4.1 Effect of Sucrose Preculture

The effect of sucrose concentration in the preculture on tolerance of encapsulated shoot tips (non-cold acclimated) to dehydration and subsequent cooling in LN was studied by applying directly 0.75 or 1.0 M sucrose or by progressively increasing the sucrose concentration during preculture. The preculture duration was 2 days, and three different dehydration periods (3, 4 and 5 h) were examined (Fig. 2). Higher recoveries (70–80%) were obtained with 0.75 M sucrose and with progressive increases of sucrose together with 4 and 5 h of dehydration. However, direct preculture in 1.0 M sucrose using the same dehydration periods caused a considerable decrease in shoot recovery (40%).

The effect of the preculture period on shoot tip survival after cryopreservation was studied using exposure to two sucrose concentrations (0.75 and 1.0 M) for 1–7 days, and a dehydration period of 4 h (16% final water content; Fig. 3). Survival was about 30% for 1-day exposure, but notably increased to about 60 and 80% for 2–7 day preculture in 0.75 M sucrose. However, lower survivals (10–40%) were obtained when 1.0 M sucrose preculture was used. Preculture previous to the air dehydration was necessary for the survival of the frozen samples (Paulet et al. 1993; Vandenbussche et al. 1993; Fukai et al. 1994; Bachiri et al. 1995).

2.4.2 Effect of Dehydration

After 2-day preculture with 0.75 M sucrose, the water content of beads was 88% and decreased during dehydration, reaching 20, 16 and 14% after 3, 4 and 7 h of dehydration, respectively (Fig. 4). Encapsulated shoot tips with water contents between 20 and 13% gave very good shoot recovery (55–75%) after

Fig. 1. a Growth recovery from cryopreserved hop shoot tips (*Humulus lupulus* L. var. Nugget) trapped in alginate beads after 7 days of culture; **b** development of a new shoot after 30 days; **c** regeneration of complete plants after 60 days; **d** plants growing in the greenhouse. *Bar* 2 mm. (Martínez et al. 1999)

rapid cooling in LN. However, without dehydration as well as with 24 h of dehydration, no survival was obtained after exposure to LN (Fig. 4).

2.4.3 Effect of Cold Acclimation

In vitro donor cultures were cold acclimated under photoperiod or continuous dark conditions. Shoot recovery after cryopreservation was studied as a function of the duration of cold acclimation (0–60 days). The experimental

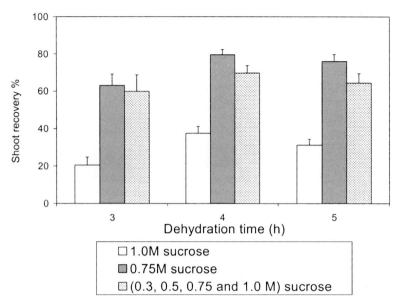

Fig. 2. Effect of dehydration period and sucrose concentration in the preculture (1.0; 0.75 M; and progressive increases in sucrose from 0.3, 0.5, 0.75 and 1.0 M for 2 days) on the percentage of shoot recovery from encapsulated shoot tips after cryopreservation. *Vertical bars* represent standard error. (Martínez et al. 1999)

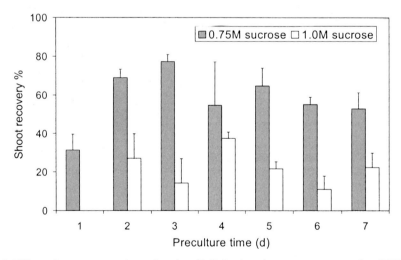

Fig. 3. Effect of sucrose preculture duration (1–7 days) and sucrose concentration (0.75 and 1.0 M) in the preculture on the percentage of shoot recovery from encapsulated shoot tips after cryopreservation. *Vertical bars* represent standard error. (Martínez et al. 1999)

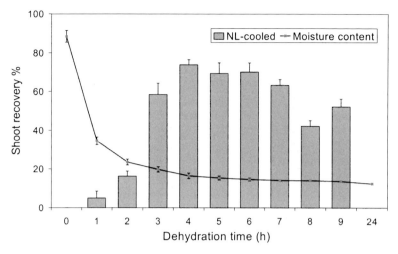

Fig. 4. Effect of dehydration period (0–24 h) on the percentage of shoot recovery from encapsulated shoot tips precultured for 2 days in progressively increased sucrose concentrations and cryopreserved. *Vertical bars* represent standard error. (Martínez et al. 1999)

conditions were those used previously (preculture with 0.75 M sucrose, a 4-h dehydration and direct immersion in LN). No differences in survival to cooling were obtained between the two cold-hardening treatments applied to the in vitro cultures. Survival after cryopreservation was about 70–80% for shoot tips from non-acclimated plant material. Acclimation periods comprised between 7 and 45 days increased the percentage of shoot recovery rate to 90–95% (Fig. 5).

Cold acclimation improved the resistance of the encapsulated plant material to cryopreservation through an increase in tolerance to both sucrose preculture and dehydration (Martínez and Revilla 1998). Cold storage of the donor cultures may increase tolerance to osmotic stress through changes in the levels of abscisic acid (Daie and Campbell 1981), lipids of plasma membrane (Sharom et al. 1994) or synthesis of specific proteins (Arora and Wisniewki 1994). Cold acclimation improved the efficiency of cryopreservation protocols for other plant species such as pear (Dereuddre et al. 1990), apple (Niino and Sakai 1992) and mulberry (Scottez et al. 1992).

2.4.4 Effect of the Subculture Number

In these experiments cuttings were taken from hop flowering plants in July. After surface sterilization, they were cultured in vitro as previously described. The effect of the number of subcultures of the cuttings on survival after cryopreservation was studied using shoot tips from cold acclimated (21 days under photoperiod conditions) and non-acclimated plant material (Martínez 1999). At least four subcultures were necessary to obtain the highest survival

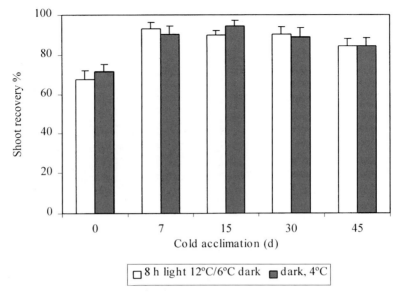

Fig. 5. Effect of cold acclimation period (0–45 days) under two different light and temperature conditions on the percentage of shoot recovery from encapsulated shoot tips precultured in 0.75 M sucrose for 2 days, dehydrated for 4 h and then immersed in LN. *Vertical bars* represent standard error

for shoot tips from non-acclimated plant material. However, with cold acclimated material, only one subculture was needed for high survival (Fig. 6).

2.4.5 Effect of Cooling Rate

Although high levels of survival were obtained after cryopreservation by direct immersion in LN (200 °C/min) of cryovials containing shoot tips trapped in alginate beads, it was necessary to study the effect of slow cooling rates on survival, as rapid cooling cannot be applied in a differential scanning calorimeter. DSC was used to examine the thermal transitions produced in samples during a cooling cycle for cryopreservation.

The effect of the cooling rate on the survival of the cryopreserved plant material was studied together with five dehydration periods of the samples (Fig. 7) and only marginal differences in percentages of shoot recovery were obtained. Survival was not related to cooling rate in pear (Scottez et al. 1992). By contrast, in *Vitis vinifera*, the greatest shoot survival percentages were obtained when slow cooling rates were applied (Plessis et al. 1993).

2.4.6 Effect of Genotype

Shoot tips were excised from cuttings after four subcultures and with or without cold acclimation (21 days under photoperiod conditions).

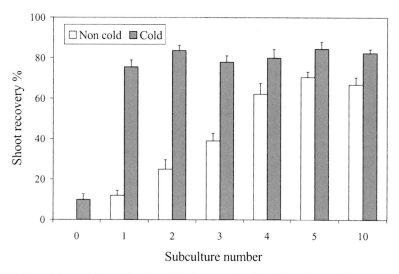

Fig. 6. Effect of the number of in vitro subcultures and cold acclimation on the percentage of shoot recovery after cryopreservation. Encapsulated shoot tips were precultured in 0.75 M sucrose for 2 days, dehydrated for 4 h, prior to immersion in LN. *Vertical bars* represent standard error

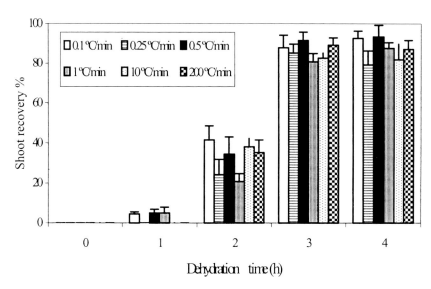

Fig. 7. Effect of cooling rate (°C/min) and dehydration period on the survival of encapsulated hop shoot tips excised from cold-acclimated in vitro cultures after cryopreservation under the same conditions as described in Fig. 6. *Vertical bars* represent standard deviation. (Martínez and Revilla 1998)

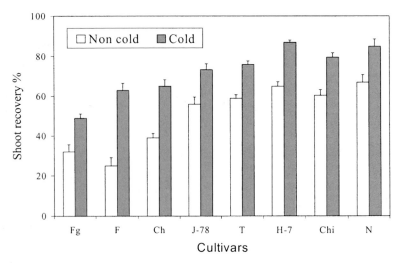

Fig. 8. Effect of a cold acclimation (21 days under 8 h light at 12 °C and 16 h dark at 12 °C) on survival after cryopreservation for eight hop cultivars: Fist Gold (*Fg*), Fuggle (*F*), Challenger (*Ch*), J-78, Taurus (*T*), H-7, Chinook (*Chi*) and Nugget (*N*). *Vertical bars* represent standard error. (Martínez and Revilla, unpubl.)

Cryopreservation was performed as described before: encapsulation, 0.75 M sucrose preculture for 2 days, 4-h dehydration, rapid cooling, slow warming and growth recovery in "apex" culture medium. The percentages of shoot recovery for eight hop varieties are shown in Fig. 8 (Martínez 1999). Cold acclimation of the in vitro shoots gave for any variety higher survivals to cryopreservation than non-acclimated plant material. The cultivars Nugget, Chinook and H-7 gave the highest survival rates. Differences in survival to cryopreservation have also been reported for different genotypes of other plant species (Niino et al. 1992; Paulet et al. 1993; Kohmura et al. 1994; Benson et al. 1996). In shoot tips of the genera *Coffea* it was necessary to apply a different preculture for each variety to obtain the highest survival (Mari et al. 1995).

2.4.7 Thermal Analysis

Differential scanning calorimetry (DSC) is a powerful tool for determining the relationship between water content and thermal events (Martínez et al. 1998). To determine the efficacy of dehydration times in producing stable and reproducible vitrification, DSC analysis was performed at several dehydration states (Fig. 9). Ice nucleation during cooling (N) and ice melting during warming (M) produced exotherm or endotherm peaks, respectively, at 0 and 1 h of dehydration during cooling or warming of the beads containing shoot tips. The formation of these peaks was related to the nil or very low survival rates obtained for the referred dehydration periods. The low level of enthalpy after 2-h dehydration (Fig. 9E,F) was correlated with large increases (ca. 50%)

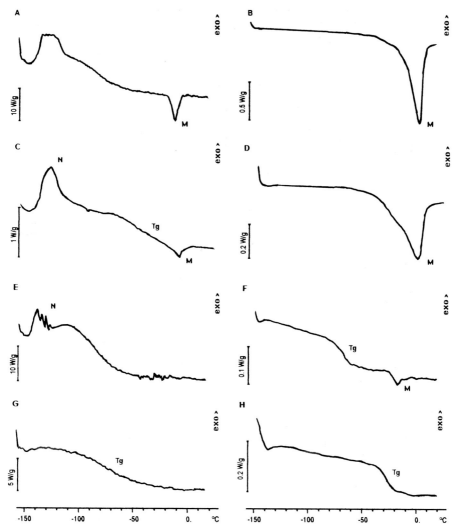

Fig. 9. Cooling (**A**, **C**, **E** and **G**) and warming (**B**, **D**, **F** and **H**) thermograms from encapsulated hop shoot tips. Beads were precultured in 0.75 M sucrose for 2 days and exposed to different dehydration periods in the air current of a laminar flow cabinet (0 h: A, B; 1 h: C, D; 2 h: E, F; 3 h: G, H). *N*: Nucleation. *M*: melting. *Tg*: glass transition. (Martínez and Revilla 1998)

in survival. However, vitrification (glass transition) was seen for 3- and 4-h dehydration periods (Fig. 9G,H) and was associated with disappearance of ice nucleation or melting peaks and with the recovery of high survivals. This event indicates that plant samples at this moisture content had been properly prepared for storage at ultra-low temperatures. Glass transition temperatures were about –65 and –55 °C, depending on water content. Recent experiments with the var. Nugget have shown similar survival for hop shoot tips stored below these transition temperatures, i.e., either under LN or in a freezer at –80 °C for 30 days (Martínez 1999).

The cooling and warming thermograms obtained for hop are in good accordance with those obtained for other plant materials, such as naked or encapsulated somatic embryos of carrot (Dereuddre et al. 1991), clumps of oil palm somatic embryos (Dumet et al. 1993), or encapsulated shoot tips of banana (Panis 1995). Steps of sucrose preculture and desiccation were essential to achieve glass transition during cooling and rewarming (Dumet et al. 1993; González-Arnao et al. 1996). A direct relationship between the formation of a glassy state and survival after the cryopreservation procedure has also been reported for shoot tips of other plant species (Dereuddre 1992; Benson et al. 1996) using vitrification or encapsulation-dehydration procedures.

3 Summary and Conclusions

The encapsulation-dehydration procedure described for hop allows recovery of very high percentages of shoots (90–100%) from encapsulated shoot tips after a short cold acclimation period (7 days) of the in vitro donor cultures, 2-day sucrose preculture, 4-h dehydration in the air of a flow cabinet, immersion in LN and slow warming. Survival was independent of cooling rates. DSC analyses have shown a direct relationship between the formation of a glassy state and survival after cryopreservation and existence of enthalpy variation (exothermic and endothermic peaks) and loss of survival. The absence of recrystallization processes during warming will allow the storage of hop germplasm at higher temperatures (freezers at –80 °C) avoiding the future use of liquid nitrogen reservoirs.

References

Adams AN (1975) Elimination of viruses from the hop (*Humulus lupulus*) by heat therapy and meristem culture. J Hortic Sci 50:151–160

Arora R, Wisniewki ME (1994) Cold acclimation in genetically related (sibling) deciduous and evergreen peach (*Prunus persica* L. Batsch). Plant Physiol 105:95–101

Bachiri Y, Gazeau C, Hansz J, Morisset C, Dereuddre J (1995) Successful cryopreservation of suspension cells by encapsulation-dehydration. Plant Cell Tissue Organ Cult 43:241–248

Benson EE, Reed BM, Brennan RM, Clacher KA, Ross DA (1996) Use of thermal analysis in the evaluation of cryopreservation protocols for *Ribes nigrum* L. germplasm. Cryo Lett 17: 347–362

Brady JL, Scott NS, Thomas MR (1996) DNA typing of hops (*Humulus lupulus*) through application of RAPD and microsatellite marker sequences converted to sequence tagged sites (STS). Euphytica 91:277–284

Daie J, Campbell WF (1981) Response of tomato plants to stressful temperatures. Plant Physiol 67:26–29

Dereuddre J (1992) Cryopreservation of in vitro cultures of plant cells and organs by vitrification and dehydration. In: Dattée Y, Dumas C, Gallais A (eds) Reproductive biology and plant breeding. Springer, Berlin Heidelberg New York, pp 291–300

Dereuddre J, Scottez C, Arnaud Y, Duron M (1990) Résistance d'apex axillaires de poirer (*Pyrus communis* L.) à une déshydratation puis à une congélation dans l'azote liquide: effet d'un traitement au froid des vitroplants. CR Acad Sci Paris 310 (Ser III):317–323

Dereuddre J, Hassen N, Blandin S, Kaminski M (1991) Resistance of alginate-coated somatic embryos of carrot (*Daucus carota* L.) to desiccation and freezing in liquid nitrogen. 2. Thermal analysis. Cryo Lett 12:135–148

Dumet D, Engelmann F, Chanbrillange N, Dereuddre J (1993) Importance of sucrose for the acquisition of tolerance to desiccation and cryopreservation of oil palm somatic embryos. Cryo Lett 14:243–250

Fric V, Krofta K, Svoboda P, Kopecky J (1997) The evaluation of the virus-free hop quality after five years growing in the Czech Republic. Rostlinná Vyroba 43:307–314

Fukai S, Togashi M, Goi M (1994) Cryopreservation of in vitro-grown *Dianthus* by encapsulation-dehydration. Tech Bull Fac Agric Kagawa Univ 46:101–107

González-Arnao MT, Moreira T, Urra C (1996) Importance of pregrowth with sucrose and vitrification for the cryopreservation of sugarcane apices using encapsulation-dehydration. Cryo Lett 117:141–148

Haunold A (1981) Hop production, breeding and variety development in various countries. J Am Soc Brew Chem 39:27–34

Heale JB, Legg T, Connell S (1989) *Humulus lupulus* L. (hop): in vitro culture; attempted production of bittering components and novel disease resistance. In: Bajaj YPS (ed) Biotechnology in agriculture and forestry, vol 7. Medicinal and aromatic plants II. Springer, Berlin Heidelberg New York, pp 264–285

Kohmura H, Ikeda Y, Sakai A (1994) Cryopreservation of apical meristems of Japanese shallot (*Allium wakegi* A.) by vitrification and subsequent high plant regeneration. Cryo Lett 15:289–298

Mari S, Engelmann F, Chabrillange N, Huet C, Micchaux-Ferrièe N (1995) Histo-citological study of apices of coffee (*Coffea racemosa* and *C. sessiliflora*) in vitro plantlets during their cryopreservation using the encapsulation-dehydration technique. Cryo Lett 16:289–298

Martínez D (1999) Crioconservación de ápices de tallos de lúpulo: análisis térmico. Tesis Doctoral, Universidad de Oviedo, Oviedo, Spain

Martínez D, Revilla MA (1998) Cold acclimation and thermal transitions in the cryopreservation of hop shoot tips. Cryo Lett 19:333–342

Martínez D, Revilla MA, Espina A, Jaimez E, García JR (1998) Survival cryopreservation of hop shoot tips monitored by differential scanning calorimetry. Thermochim Acta 317:91–94

Martínez D, Tamés RS, Revilla MA (1999) Cryopreservation of in vitro grown shoot-tips of hop (*Humulus lupulus* L.) using encapsulation/dehydration. Plant Cell Rep 19:59–63

Matousek J, Junker V, Vrba L, Schubert J, Patzak J, Steger G (1999) Molecular characterization and genome organization of 7SL RNA genes from hop (*Humulus lupulus* L.). Gene 239:173–183

Murashige T, Skoog FA (1962) A revised medium for rapid growth and bioassays with tobacco tissue cultures. Physiol Plant 15:473–497

Neve RA (1991) Hops. Chapman and Hall, London, pp 1–10

Niino T, Sakai A (1992) Cryopreservation of alginate-coated in vitro-grown shoot tips of apple, pear, and mulberry. Plant Sci 87:199–206

Niino T, Sakai A, Nojiri K (1992) Cryopreservation of in vitro-grown shoot tips of apple and pear by vitrification. Plant Cell Tissue Organ Cult 28:261–266

Panis B (1995) Cryopreservation of banana (*Musa* spp.) germplasm. Dissertatons de Agricultura, Katholieke Universitiet, Leuven, Belgium

Panis B, Totté N, Nimmen KV, Withers LA, Swennen R (1996) Cryopreservation of banana (*Musa* spp.) meristem cultures after preculture on sucrose. Plant Sci 121:95–106

Paulet F, Engelmann F, Glaszmann J (1993) Cryopreservation of apices of in vitro plantlets of sugarcane (*Saccharum* sp. hybrids) using encapsulation/dehydration. Plant Cell Rep 12:525–529

Pillay M, Kenny ST (1996a) Structure and inheritance of ribosomal DNA variants in cultivated and wild hop, *Humulus lupulus* L. Theor Appl Genet 93:333–340

Pillay M, Kenny ST (1996b) Random amplified polymorphic DNA (RAPD) markers in hop, *Humulus lupulus*: level of genetic variability and segregation in F_1 progeny. Theor Appl Genet 92:334–339

Plessis P, Leddet C, Collas A, Dereuddre J (1993) Cryopreservation of *Vitis vinifera* L. cv. Chardonnay shoot tips by encapsulation-dehydration: effects of pretreatment, cooling and postculture conditions. Cryo Lett 14:309–320

Probasco G, Winslow S (1986) The use of shoot-tip culture to eliminate viruses from hop varieties grown in the United States. MBAA Tech Q 23:26–31

Redenbaugh K, Paasch BD, Nichol JW, Kessler ME, Viss PR, Walker KA (1986) Somatic seeds: encapsulation of asexual plant embryos. Biotech 4:797–801

Sakai A, Kobayashi,S, Oiyama I (1990) Cryopreservation of nucellar cells of navel orange (*Citrus sinensis* Osb. var. *brasiliensis* Tanaka) by vitrification. Plant Cell Rep 9:30–33

Salmon ES (1934) Two new hops: Brewer's Favourite and Brewer's Gold. Agric Coll Wye 34:93–105

Samyn G, Welvaert W (1983) Producing a "nuclear stock" of virus-free hop plants. Med Fac Land Bouww Rijsunic Gent 48: 877–881

Scottez C, Chevreau E, Godard N, Arnaud Y, Duron M, Dereuddre J (1992) Cryopreservation of cold acclimated shoot tips of pear in vitro cultures after encapsulation-dehydration. Cryobiology 29:691–700

Sharom M, Willemot C, Thompson J (1994) Chilling injury induces changes in membranes of tomato fruit. Plant Physiol 105:305–308

Vandenbussche B, Demeulemeester MAC, De Proft MP (1993) Cryopreservation of alginate-coated in vitro grown shoot-tips of chicory (*Cichorium intybus* L.) using rapid freezing. Cryo Lett 14:259–266

Vine SJ, Jones OP (1969) The culture of shoot tips of hop (*Humulus lupulus* L.) to eliminate viruses. J Hortic Sci 44:281–284

Wetmore RH, Sorokin S (1955) On the differentiation of xylem. J Arnold Arboretum 36:305–317

II.9 Cryopreservation of *Mentha* (Mint)

LEIGH E. TOWILL

1 Introduction

1.1 Distribution and Important Species of *Mentha*

The genus *Mentha* is in the Lamiaceae and is divided taxonomically into five sections (Chambers and Hummer 1994a). The 19 species are all herbaceous perennials, except for the annual *M. micrantha* (Table 1), and have origins in Europe, Asia, Australia, Japan and Morocco. A discussion of the taxonomy of mint and some of the difficulties involved therein can be found in Harley and Brighton (1977) and Tucker et al. (1980). Chromosome numbers range from 18 to 96 among species. A base number of 12 is common, but within the section Pulegium base numbers of 10 and 12 are described (Murray et al. 1971; Chambers and Hummer 1994a). Hybridization among species within the section *Mentha* has been widely recognized and several hybrids are named, the most important one being peppermint (*Mentha × piperita*) arising from crosses between *M. aquatica* and *M. spicata*.

Various mints are grown for fresh or dried use as food or herbal products, but the major economic value is as a field crop for the production of extractable oils by steam distillation.

The major species of commerce are peppermint and spearmint (*Mentha spicata*), although numerous others are used around the world. Uses of extracted oil are in the food, cosmetic, tobacco, chemical and pharmaceutical industries. The constituents are complex and differ among species. Peppermint produces predominantly the monoterpenes, menthone and menthol, whereas spearmint produces carvone. Monoterpenes are phytotoxic and are produced in oil glands on leaves and, to a lesser extent, on stems. The quality of the oil in terms of constituent content and proportions is very important for any use. Thus, plantings must be from defined and consistent stock.

USDA-ARS National Seed Storage Laboratory, 1111 S. Mason St., Fort Collins, Colorado 80521, USA

Biotechnology in Agriculture and Forestry, Vol. 50
L.E. Towill and Y.P.S. Bajaj (Eds.) Cryopreservation of Plant Germplasm II
© Springer-Verlag Berlin Heidelberg 2002

Table 1. The genus *Mentha*

Species	Common Name	Origin	Chromosome Number, 2n
Section *Audibertia*			
Mentha requienii Benth.	Corsican mint	Corsica, Sardinia	18
Section *Eriodontes*			
Mentha australis R.Br.	Australian mint	Australia	72
Mentha cunninghamii Benth.	Maori mint	New Zealand	72
Mentha diemenica Spreng.	Slender mint		
var. *diemenica*		Australia	120
var. *serphllifolia* (Benth.) J.H. Willis		Australia	
Mentha grandiflora Benth.		Australia	
Mentha japonica (Miq.) Makino	Hime-akka	Japan	50
Mentha laxiflora Benth.	Forest mint	Australia	
Mentha satureoides R.Br.	Creeping mint	Australia	
Section *Mentha*			
Mentha aquatica L.	Water mint	Europe	96
Mentha arvensis L.	Corn or Field mint		
var. *arvensis*		Europe	72
var. *canadensis* L.		North America, Asia	96
Mentha longifolia L. (Hudson)	Horse mint	Europe, Middle East	24, 48
Mentha spicata L.	Spearmint	unknown	36, 48
Mentha suaveolens Ehrh.	Apple mint	Europe	24
Section *Preslia*			
Mentha cervina L.		Europe	36
Section *Pulegium*			
Mentha gatlefossei Maire		Morocco	40, 48
Mentha micrantha (Benth.) Schost.		S.E. Russia,	
Mentha pulegium L.	Pennyroyal	Europe	20, 30, 40

Modified from Chambers and Hummer, 1994b.

1.2 Need for Conservation

As with any economically important crop, the conservation of diversity is crucial for developing improved lines. Sterility and incompatibility in *Mentha* restrict conventional breeding progress, but there is considerable interest in genetically engineering the crop to alter and improve oil content and to enhance disease resistance. Although breeding and engineering efforts in mint are minor relative to agronomic crops, advances have been made and the future for developing improved lines is bright. Most field mint is vegetatively propagated, and hence is subject to clonal selection procedures. Conservation is needed to preserve the status of lines to insure uniform, consistent quality of the economic product.

There is little information on germplasm maintained within repositories. The Agricultural Research Service/US Department of Agriculture maintains both seed and clonal selections of about 400 accessions of mint species at the National Clonal Germplasm Repository (NCGR) in Corvallis, Oregon

(Chambers and Hummer 1992, 1994b). Current holdings can be accessed on the Internet at http://www.ars-grin.gov/PacWest/Corvallis/ncgr/. Clones are maintained within pots in a greenhouse; in vitro plants are maintained as a backup.

1.3 Propagules for Preservation of *Mentha* Germplasm

Seeds. The collection of seeds from native populations can provide diversity for improvement, although incompatibilities and sterility in some materials limit usefulness. There is limited information available on mint seed storage characteristics, but the seeds are desiccation-tolerant and should show considerable longevity if stored at low temperatures. Viability of *Mentha arvensis* seeds declined rapidly over a year if they were not stored with a desiccant (Ikeda et al. 1960). As exists in seeds from some other members of the Lamiaceae, hard seeds and dormancy may confound estimates of viability. Seeds from *Mentha aquatica* require cold stratification for 30 days to remove dormancy (Grime et al. 1981). Germination was highest with a 20°C day/15°C night regime. The NCGR-Corvallis method chills surface-sterilized seed for 30 days at 5°C and then examines germination over a 30-day period (Hummer 1998).

Pollen. There are no published reports to my knowledge on the preservation of mint pollen. Genera within the Lamiaceae possess either bicellular or tricellular pollen (Brewbaker 1967). In general, bicellular pollen, if small, is desiccation-tolerant and shows considerable longevity if it is dehydrated and stored at low temperatures; tricellular pollens often have some desiccation-sensitivity that limits storage capabilities (Towill 1985). The optimum moisture content for the greatest longevity depends on the storage temperature, with a slightly increasing moisture content being beneficial at lower temperatures (Buitink et al. 1998). Thus, it is difficult to predict storage characteristics for mint pollen. Although feasible, the collection of adequate amounts of pollen from mint flowers is tedious and probably would limit the usefulness of this approach for germplasm preservation per se.

Vegetative Plants. The lines used in commerce are heterozygous and preservation of the clone is the only method to retain its unique characteristics. Mint plants are fairly prolific and can be readily cut and divided. Lines can be maintained in pots in greenhouses with precautions to avoid stolons from establishing in adjacent pots. Propagation is by rooting of stolon segments and above-ground stems. Rooting single node sections provides rapid multiplication (El-Keltawi and Croteau 1986).

In Vitro Culture. In vitro lines of most mint species can be established by meristem tip or shoot tip culture, and propagation is easily accomplished by axillary buds or nodal sections (Rech and Pires 1986). Single shoot elongation occurs on a medium without growth regulators and multiple shoot formation occurs on media with benzyladenine (BA) or kinetin. The initial establishment of the axenic line may be difficult since many mint lines harbor non-specific, endophytic bacteria (Buckley et al. 1995; Reed et al. 1995). Some plants contained more than one endophyte. Antibiotic treatments could eliminate these bacteria but the antibiotic(s) efficacy differed depending on the contaminant(s) within the line.

A number of research and propagation laboratories maintain in vitro cultures of mint under usual growth conditions (short-term storage) for distribution. The generation of disease-free stocks using meristem culture and in vitro propagation has allowed development of a nuclear program to provide growers with tested and defined clones. Little longevity data has been published. Storage at 25 °C in Parafilm-wrapped test tubes retained viability in cultures from most species for about 6–8 months (Towill, unpubl.). Medium-term conservation using slow-growth procedures has been reported for some mint species (Reed 1999); samples held in heat-sealed plastic storage bags were viable after storage for 30 months at +4 °C.

Lack of disease resistance, for example peppermint's susceptibility to *Verticillium dahliae*, is a concern for sustained yield and increasing longevity in perennial mint plantings. Improvement by conventional crossing is difficult and suitable sources of resistance are rare, and, hence, transgenic approaches are being examined. Several tissues and organs have been examined for callus production and shoot regeneration (Van Eck and Kitto 1990). Shoot organogenesis from leaf disks is feasible (Van Eck and Kitto 1992; Caissard et al. 1996; Faure et al. 1998) and co-cultivation of disks with *Agrobacterium tumefaciens* has produced transgenic plants (Spencer et al. 1993; Berry et al. 1996; Niu et al. 1998). Production of somatic hybrids by protoplast fusions also has the potential for crop improvement (Sato et al. 1993; Krasnyanski et al. 1998), but regenerants from *M.* × *piperita* and *M.* × *citrata* have shown variation for oil production (Chaput et al. 1996).

Callus and suspension cultures from mint are easily established, but give low oil yield. They do, however, possess some ability to convert exogenously supplied monoterpenes to desired products (Park and Kim 1998). Other research has focused on producing proliferative shoot cultures that may yield oil or be able to perform metabolic conversions (Rhodes et al. 1991). Although these systems are not yet productive, a need will exist to cryopreserve these cultures once suitable lines are created.

2 Cryopreservation

Cryopreservation has as its goal the preservation of stock for extended durations. Use of preexisting apical or axillary shoot tips is preferred for germplasm preservation to minimize unwanted variation. Some progress has been made using regeneration from leaf disks and it might be projected that preservation of these shoot tip clusters may prove useful. There are no reports on somatic embryo proliferation in *Mentha* species so, at present, this does not appear to be a useful route for preservation. Likewise, preservation of *Mentha* spp. cell suspensions or callus has not been reported.

Several reviews describe the state of the art in cryopreservation (Steponkus et al. 1992; Sakai 1993; Bajaj 1995; Benson 1999). In developing a useful protocol a number of factors must be evaluated, and these factors can be grouped into a plant culture phase, a cryogenic phase and a recovery phase. Cryopreservation studies for *Mentha* to date have utilized shoot tips from in vitro plants (Towill 1988, 1990; Hirai and Sakai 1999; Sakai et al. 2000). The latter two studies used encapsulation methods for preservation and will be discussed briefly later. The former two examined both two-step cooling and solution-based vitrification protocols with emphasis on the cryogenic phase for interactions of factors. I report here some observations from our studies.

In vitro plants were used as a source of shoot tips for all studies described below. Stock cultures of the different species were obtained from NCGR-Corvallis, OR. In vitro stocks were propagated by nodal section culture on a Murashige and Skoog (1962) mineral medium with 3% sucrose, solidified with 0.7% agar. Axillary buds were directly excised from in vitro plants about 6–12 weeks old. In some cases, single node sections were excised and cultured on the above medium for 2–3 days to promote expansion of the axillary bud, similar to the system described for potato by Bouafia et al. (1996). Regrowth of treated shoot tips was assessed on a MS medium with 0.7% agar, 3% sucrose, $0.1\,mg\,ml^{-1}$ indolebutyric acid and $0.5\,mg\,ml^{-1}$ BA. Most shoot tips directly developed into shoots on this medium, although some species did develop some callus from the basal, cut surface.

2.1 Cryopreservation by Two-Step Cooling Methods

The first studies examined use of conventional cryopreservation methods, including cryoprotectant types and concentrations, temperatures and rates of cryoprotectant addition and dilution, need for external nucleation, cooling rates, transfer temperature to liquid nitrogen and warming rates. The following briefly summarizes these studies with implications for routine cryopreservation. Survival after cryogenic exposure was obtained using a 1- or 2-day preculture duration in growth medium followed by application of 12% dimethyl sulfoxide (DMSO) in growth medium (Towill 1988). Preculture in 4% DMSO with 0.5–0.75 M sucrose gave more consistent viability and regrowth with cryopreserved samples, although some survival occurred with no preculture (Towill, unpubl.). After equilibration and ice nucleation at −5 °C, samples were cooled at 0.25 °C/min to −35 °C and then immersed in liquid nitrogen (LN). Samples were warmed in a +40 °C water bath. Survival of samples cooled to about −35 °C and then warmed was about the same or slightly lower than samples held at −5 °C, but survival declined considerably with cooling to lower temperatures. There was no significant effect of rate of addition and dilution of the cryoprotectant. At cooling rates greater than about 0.5 °C/min, survival decreased. Holding at −35 °C for 1 h prior to LN

Table 2. Two-step cooling of *Mentha* sp.: 10% DMSO, 10% glucose, 10% PEG-8000[a]

Line	Percentage Survival[b]	
	Control	LN
M. dumetorum	100 (100)	87 (100)
M. longifolia	100 (100)	75 (78)
M. maximiliana	100 (100)	77 (80)
M. niliaca	100 (100)	60 (100)
M. piperita subsp. *citrata*	100 (100)	71 (88)
M. piperita	100 (100)	78 (86)
M. pulequium	100 (100)	14 (67)
M. spicata	100 (100)	46 (54)
M. villosa	75 (100)	60 (100)

[a] Samples exposed to cryoprotectant solution for 1 h at room temperature and cooled at 0.25 °C/min to −35 °C, held for 1 h and then immersed in LN; samples warmed in +40 °C water bath.
[b] Percentage of shoot tips showing growth (percentage of growing shoot tips producing a shoot).

immersion was beneficial, but very low viability occurred in samples held for 24 h at −35 °C. Rapid warming from LN was beneficial. The cryoprotectant mixture of 10% DMSO, 10% glucose and 10% polyethylene glycol, MW 8000 (PEG) (Finkle and Ulrich 1979) gave very high levels of viability with several species of *Mentha*, although some did not respond as well (Table 2).

An obvious question was whether the mint plants regenerated from treated shoot tips developed normally. To test this we generated greenhouse plants of *Mentha aquatica* × *M. spicata* from about 300 cryoprotectant-treated shoot tips and from about 700 LN-exposed shoot tips. Plants were examined for growth habit, leaf shape and color, hair production on leaves, runner production, inflorescent type and flower shape and color. Although this was a qualitative assessment with greenhouse plants, no abnormal types were observed. Most mint species did not form extensive callus on the recovery medium described and it was quite apparent that the shoots regenerated came from either the apical meristem or meristems within the axils of the leaf primordia.

The major disadvantage with two-step procedures for mint was experimental variation. Part of this variation may be due to inconsistent nucleation. Nucleation of the sample solution was important for high levels of survival. Glass vials cooled in methanol baths could be easily nucleated but nucleation was more difficult in pulsed LN vapor coolers. Survival was always higher using glass vials compared to plastic vials. This may relate to ease of nucleation or rate of cooling in the second step from −35 °C to LN.

2.2 Cryopreservation by Vitrification

The main advantage offered by vitrification is simplicity of cooling. Cooling rate equipment is not needed and the potential exists for performing cryopre-

Table 3. Variation in survival and shoot regeneration of shoot tips from *Mentha aquatica* × *M. spicata* after vitrification using two solutions

Vitrif. solution: Experiment #	Percentage Survival[a]			
	1 M DMSO; 35% EG + 10% PEG[b]		1 M EG; 20% PG + 16% EG + 10% PEG[c]	
	Exposed	Exposed, LN	Exposed	Exposed, LN
1	100 (75)	56 (37)	100 (92)	92 (91)
2	86 (61)	31 (17)	100 (100)	46 (83)
3	92 (77)	74 (76)	83 (90)	90 (84)
4	100 (56)	42 (20)	100 (100)	78 (89)
5	91 (70)	75 (50)	100 (100)	87 (85)
6	96 (70)	48 (36)	100 (100)	85 (82)
7	73 (62)	59 (46)		
8	75 (89)	67 (67)		
Average	89 (70)	56 (44)	97 (97)	80 (86)
SE	9 (13)	13 (18)	6 (3)	16 (3)

[a] Percentage of shoot tips showing growth (percentage of growing shoot tips developing into shoots).
[b] Shoot tips loaded with 1 M DMSO, then exposed to 1 M DMSO + 35% EG + 10% PEG.
[c] Shoot tips loaded with 1 M EG, then exposed to 20% PG + 16% EG + 10% PEG.

servation with a minimum of facilities. A readily available source of LN and suitable storage tanks are still needed. Samples must be retrieved from storage without inadvertent warming of other stored samples. Retrieval of most shoot tips is by culture methodology and a culture facility is also needed.

Vitrification was examined as an alternative method for cryopreservation because data obtained from two-step cooling experiments often showed considerable variability. Studies concentrated on the cryogenic phase of the process. The first studies examined different vitrification solutions for toxicity and effectiveness in obtaining survival after LN exposure. Some of these solutions were patterned after those used in animal systems. It is well recognized that cooling rate under less than optimal solution conditions influences vitrification (Sutton 1991). Thus rapid cooling on either paper strips or small envelopes of aluminum foil was used in these studies.

Several vitrification solutions were identified, two of which were examined more extensively (Table 3). A loading phase was used in most studies. Shoot tips were incubated in either DMSO or ethylene glycol (EG) and then transferred to a vitrification solution composed of the loading solution plus either EG, propylene glycol (PG) or EG + PG. Several interactions were apparent. A mixture of EG, PG and PEG was less toxic and gave somewhat higher levels of viability and of shoot growth than one with DMSO, EG and PEG. Results across experiments were fairly consistent. Duration in the vitrification solution was important. The data from Table 4 obtained with two other vitrification solution variants illustrate the observation that vitrification-solution exposed samples retained high levels of viability, but effectiveness was reduced with extended exposure. The use of the PG + EG + PEG vitrification solution was applicable to shoot tips from several mint species (method A in Table 5).

Table 4. Effect of time in concentrated vitrification solution on survival of shoot tips from *Mentha aquatica* × *M. spicata*

Vitrification solution[a]	Time in solution	% Survival[b], exposed	% Survival[b], LN
40% PG + 7.8%	1 hr	100 (100)	87 (81)
DMSO + 10% PEG	3 hr	100 (100)	33 (62)
50% PG + 7.8%	1 hr	100 (100)	100 (74)
DMSO + 10% PEG	3 hr	83 (100)	46 (77)

[a] Loaded in 7.8% DMSO for 1 h at room temperature before exposure to vitrification solution at 0 °C.
[b] Percentage of shoot tips showing growth, (percentage of growing shoot tips producing sustained shoot growth).

Table 5. Survival of excised shoot tips from mint species after exposure to vitrification solution and LN immersion[a]

Species/Line	Method[b]	Survival, not cooled	Survival, LN-exposed
Mentha sp.			
M. aquatica × *M. spicata*	A	97	80
	B	92	83
M. aquatica	B	58	33
M. arvensis	B	50	39
M. canadensis	A	75	19
	B	83	62
M. × *dumatorum*	B	73	54
M. × *gracillis*	B	69	45
M. maximilliana	A	90	96
	B	92	4
M. niliaca	A	100	83
	B	90	100
M. × *piperata*	B	100	81
M. spicata	A	100	96
	B	100	76
M. × *verticillata*	B	92	83

[a] Values presented are the percentage of shoot tips that survived as assessed by regrowth.
[b] Method A. Load in 1 M EG for 60 min; vitrify in 20% PG + 16% EG + 10% PEG.
Method B: Load in 20% PVS2; expose for 5 min to 60% PVS2, 5 min to 80% PVS2 and 5 min to 100% PVS2; vitrify in foil envelopes.

With the subsequent publication of the recipe for the vitrification solution, PVS2, by Sakai et al. (1991), studies were initiated to determine its effect with mint. First tests showed good levels of survival without preculture and subsequent tests used shoot tips directly after excision. The loading phase for vitrification utilized 20% PVS2 for 60 min at room temperature. Five-minute exposures to 60, 80 and 100% PVS2 were successively applied prior to cooling. Results were comparable to those obtained with the PG + EG + PEG vitrification solution described above (Table 5). These studies were carried out using plants of a similar age and performed over a short time period. The PVS2 method gave better survival for some species, yet for others the opposite was true. Extensive replications for each species were not performed, but our experience suggests that the methods are very comparable. Several other tests

Table 6. Storage of vitrified *Mentha* spp. shoot tips at −180 °C[a]

Storage Time	% Survival (% forming shoots)		
	312[b]	573[c]	130[d]
PVS exposed	100 (95)	83 (89)	100 (100)
LN, 1 hr	70 (94)	42 (70)	56 (78)
LN, 1 wk	81 (85)	13 (67)	38 (100)
LN, 1–2 mo	91 (90)	28 (83)	52 (83)
LN, 3–4 mo	68 (93)		43 (80)
LN, 7–9 mo	65 (87)	37 (67)	48 (90)
LN, 12–13 mo	80 (100)	54 (100)	

[a] Samples vitrified in PVS2 solution within semen straws; shoot tips exposed to 20% PVS2 for 60 min, 60% PVS2 for 5 min, 80% PVS2 for 5 min and 100% PVS2 for 5 min. Samples annealed at −160 °C for 15 min before immersing in LN. Samples stored in vapor phase over LN.
[b] Line 312: *Mentha aquatica* × *Mentha spicata*.
[c] Line 573: *Mentha* × *rotundifolia*.
[d] Line 130: *Mentha* × *maximiliana*.

were performed. Exposure to the 100% PVS2 solution for up to 45 min did not decrease viability in exposed samples, but was not more effective in cryopreserving samples than the graded exposure method (5 min at each concentration).

Application of a loading phase was beneficial in most studies, although survival was obtained without loading. Application of 2 M glycerol + 0.4 M sucrose provides an effective alternative loading solution for some systems (Matsumoto et al. 1994, 1995), and has been described for mint (Hirai and Sakai 1999). We subsequently have confirmed that loading for 30 min with 2 M glycerol plus either 0.4 M or 0.6 M sucrose with subsequent exposure to PVS2 gave high levels of viability after LN exposure.

Most vitrification studies described here used rapid cooling of paper or foil strips or envelopes to maximize survival. The use of foil envelopes with subsequent placement in LN-cooled vials is readily performed and has been used by the DSMZ German Collection of Microorganisms and Cell Cultures, Braunschweig, for routine cryopreservation of potato (*Solanum tuberosum*) shoot tips (Schafer-Menuhr et al. 1997). Semen straws were used for many studies with PVS2 and also are a usable storage system. The storage of vitrified shoot tips from three mint lines in semen straws was examined over a year (Table 6). No decrease in viability was observed and shoot formation was high in all samples. Cooling of 1-ml samples in plastic cryovials by immersion in LN has given more variable results in our studies. The use of partially solidified nitrogen (ca. −210 °C) as a coolant, allowing a faster cooling rate, has not improved survival of samples within cryovials.

In some systems, cracking of the vitrification solution can lead to physical damage of the specimen (Kasai et al. 1996; Pegg et al. 1997). Samples were placed in semen straws and cooled by exposure to −160 °C for 15 min prior to immersion in LN. Annealing at −160 °C reduced the frequency of vitrification

Table 7. Effect of warming method on survival of PVS2-treated and LN-cooled shoot tips from *Mentha aquatica* × *M. spicata*

Treatment			% Survival[a]
PVS2 exposure			100 (−)
PVS2 and LN exposure[b]:			
Cooling	**Warming**	**Cracking**	
annealed	10 sec air, 25 °C water	no	77 (9)
annealed	30 sec air, 25 °C water	no	57 (16)
annealed	10 sec air, 40 °C water	no	69 (12)
annealed	air thaw only	no	54 (12)
annealed	water directly	yes	58 (12)
annealed	very slow warm	no	8 (5)
direct	very slow warm	yes	0 (−)

[a] Percentage regrowth of treated shoot tips (standard error).
[b] Exposure to LN is by annealing at −150 °C for 15 min and then immersion in LN; direct is immersion from 25 °C into LN; very slow warming is exposure to −70 °C for 15 min, then to −30 °C for 15 min and then to 0 °C; cracking refers to visible cracks present in the vitrification solution in the semen straw.

Table 8. Effect of cracking method on survival of PVS2-treated and cooled shoot tips from *Mentha longifolia* (L.) L

Treatment	Survival[a]
Untreated	100 (−)
PVS2 exposure	100 (−)
PVS2 and LN exposure[b]	
1. annealed samples (not cracked)	55 (11)
2. cirect LN exposure (cracked)	69 (24)
3. annealed and shattered (cracked)	65 (12)

[a] Percentage regrowth of treated shoot tips (standard error).
[b] Annealed samples were held at ca −150 °C for 15 min prior to immersing in LN; direct exposure is by plunging the straw into LN; annealed samples were shattered by tapping with cold forceps; all were warmed by holding in air for 10 sec followed by 1 min in 25 °C water.

solution cracking during subsequent cooling. High levels of survival and shoot formation were found in most species tested. By varying the cooling and warming regimes we could control cracking and examine its effect on survival. Viability was not reduced by cracking occurring in either the cooling or warming regime, nor by cracking artificially induced by tapping with cooled forceps (Tables 7 and 8). No visible damage to shoot tips was apparent.

The data in Table 7 also shows that viability is not influenced by alterations in warming rate that might occur through minor variations in handling. Very slow rates allowed sufficient time for devitrification events to occur within the cells, leading to lethal consequences.

Recent studies have shown that lines of *M. spicata* can be cryopreserved using a combination of alginate encapsulation and desiccation by exposure to PVS2 (Hirai and Sakai 1999) or by exposure to silica gel (Sakai et al. 2000).

Cold acclimation of the in vitro sections was used to improve survival. Thus, mint can be cryopreserved using several methods that ultimately depend on vitrification of the cellular contents during cooling.

3 Summary

Both slow-cooling and vitrification methods gave survival in several mint species. Higher levels of shoot formation would be desirable in some species and numerous possibilities exist for altering the procedure. Preliminary tests with alginate-encapsulation, dehydration tests, as used for shoot tips from several species, have also been successful (Towill, unpubl.). All studies described above emphasized the cryogenic phase of cryopreservation and more studies on both the culture phase and recovery phase are warranted. For example, the application of a cold period to in vitro plants from species that acclimate has improved the survival for cryo-treated shoot tips (Reed 1988). Many mint species do cold acclimate so this represents a strategy to improve overall levels of survival. Cold acclimation for 1 month did enhance cryo-reservability using two-step cooling (Towill, unpubl.) and vitrification (Hirai and Sakai 1999). Preculture of excised shoot tips in sucrose solutions has been very useful for a range of species and optimization of concentrations and duration of exposure is a logical route to improve survival.

Thus, most mint species seem amenable to cryopreservation and sufficient information exists to apply the methodology to collections. Which method is actually applied to a practical cryopreservation process depends on issues of effectiveness and cost. Vitrification under the conditions presented shows less variation, but improvements described above may make the two-step cooling process as dependable. Both two-step cooling and vitrification have about the same labor requirements in shoot tip harvesting, culture and manipulations.

Acknowledgements. This study was supported in part by a grant from the Mint Industry Research Council.

References

Bajaj YPS (1995) Cryopreservation of cells, tissues, organs and recalcitrant seeds. In: Bajaj YPS (ed) Biotechnology in agriculture and forestry, vol 32. Cryopreservation of plant germplasm I. Springer, Berlin Heidelberg New York, pp 3–28

Benson EE (1999) Cryopreservation. In: Benson EE (ed) Plant conservation biotechnology. Taylor and Francis, London, pp 83–95

Berry C, Van Eck JM, Kitto SL, Smigocki A (1996) Agrobacterium-mediate transformation of commercial mints. Plant Cell Tissue Organ Cult 44:171–181

Bouafia S, Jelti N, Blanc A, Bonnel E, Dereuddre J (1996) Cryopreservation of potato shoot tips by encapsulation-dehydration. Potato Res 39:69–78

Brewbaker J (1967) The distribution and phylogenetic significance of binucleate and trinucleate pollen grains in the angiosperms. Am J Bot 54:1069–1083

Buckley PM, DeWilde TN, Reed BM (1995) Characterization and identification of bacteria isolated from micropropagated mint plants. In Vitro Cell Dev Biol 31:58–64

Buitink J, Walters C, Hoekstra F, Crane J (1998) Storage behavior of *Typha latifolia* pollen at low water contents: interpretation on the basis of water activity and glass concepts. Physiol Plant 103:145–153

Caissard JC, Faure O, Jullien F, Colson M, Perrin A (1996) Direct regeneration in vitro and transient GUS expression in *Mentha × piperita*. Plant Cell Rep 16:67–70

Chambers HL, Hummer KE (1992) Clonal repository houses valuable mint collection in Corvallis, Oregon. Diversity 8:31–32

Chambers HL, Hummer KE (1994a) Chromosome counts in the *Mentha* collection at the USDA-ARS National Clonal Germplasm Repository. Taxon 43:423–432

Chambers HL, Hummer KE (1994b) Mints in the U.S.D.A. germplasm repository. Herb Spice Med Plant Dig 12:1–5

Chaput MH, San H, de Hys L, Grenier E, David H, David A (1996) How plant regeneration from *Mentha* x *piperita* L. and *Mentha × citrata* Ehrh. leaf protoplasts affects their monoterpene composition in field conditions. J Plant Physiol 149:481–488

El-Keltawi NE, Croteau R (1986) Single-node cuttings as a new method of mint propagation. Sci Hortic 29:101–105

Faure O, Diemer F, Moja S, Jullien F (1998) Mannitol and thidiazuron improve in vitro shoot regeneration from spearmint and peppermint leaf disks. Plant Cell Tissue Organ Cult 52:209–212

Finkle BJ, Ulrich JM (1979) Effects of cryoprotectants in combination on the survival of frozen sugarcane cell. Plant Physiol 63:598–604

Grime JP, Mason G, Curtis AV, Rodman J, Band SR, Mowforth MAG, Neal AM, Shaw S (1981) A comparative study of germination characteristics in a local flora. J Ecol 69:1017–1059

Harley RM, Brighton CA (1977) Chromosome numbers in the genus *Mentha* L. J Linn Soc Bot 74:71–96

Hirai D, Sakai A (1999) Cryopreservation of in vitro grown shoot-tip meristems of mint (*Mentha spicata* L.) by encapsulation vitrification. Plant Cell Rep 19:150–155

Hummer K (1998) Seed germination: *Mentha*. NCGR-Corvallis operations manual. Station report, Corvallis, OR, pp 7–9

Ikeda N, Udo S, Saisho I, Minakata S (1960) Studies on the storage of mint seed. Okayama Univ Fac Agric Sci Rep 16:1–5

Kasai M, Zhu SE, Pedro PB, Makamura K, Sakurai T, Edashige K (1996) Fracture damage of embryos and its prevention during vitrification and warming. Cryobiology 33:459–464

Krasnyanski S, Ball TM, Sink KC (1998) Somatic hybridization in mint: identification and characterization of *Mentha piperita* (+) hybrid plants. Theor Appl Genet 96:683–687

Matsumoto T, Sakai A, Yamada K (1994) Cryopreservation of in vitro grown apical meristems of wasabi (*Wasabia japonica*) by vitrification and subsequent high plant regeneration. Plant Cell Rep 13:442–446

Matsumoto T, Sakai A, Yamada K (1995) Cryopreservation of in vitro grown apical meristems of lily by vitrification. Plant Cell Tissue Organ Cult 41:237–241

Murashige T, Skoog F (1962) A revised medium for rapid growth and bioassays with tobacco tissue cultures. Physiol Plant 15:473–497

Murray MJ, Marble PM, Lincoln DE (1971) Inter-subgeneric hybrids in the genus *Mentha*. J Heredity 62:363–366

Niu X, Lin K, Hasegawa PM, Bressan RA (1998) Transgenic peppermint (*Mentha × piperita* L.) plants obtained by cocultivation with *Agrobacterium tumefaciens*. Plant Cell Rep 17:165–171

Park SH, Kim SU (1998) Modified monoterpenes from biotransformation of (–)-isopiperitenone by suspension culture of *Mentha piperita*. J Nat Prod 61:354–357

Pegg DE, Wusteman MC, Boylan S (1997) Fractures in cryopreserved elastic arteries. Cryobiology 34:183–192

Rech EL, Pires MJP (1986) Tissue culture propagation of *Mentha* spp. by the use of axillary buds. Plant Cell Rep 5:17–18

Reed BM (1988) Cold acclimation as a method to improve survival of cryopreserved *Rubus* meristems. Cryo Lett 9:166–171

Reed BM (1999) In vitro storage conditions for mint germplasm. HortScience 34:350–352

Reed BM, Buckley PM, DeWilde TN (1995) Detection and eradication of endophytic bacteria from micropropagated mint plants. In Vitro Cell Dev Biol 31:53–57

Rhodes MJC, Spencer A, Hamill JD (1991) Plant cell culture in the production of flavour compounds. Biochem Soc Trans 19:702–705

Sakai A (1993) Cryopreservation strategies for survival of plant cultured cells and meristems cooled to −196 °C. In: Cryopreservation of plant genetic resources. Japan International Cooperation Agency, Japan, pp 4–26

Sakai A, Kobayashi S, Oiyama I (1991) Survival by vitrification of nucellar cell of navel orange (*Citrus sinensis* var. *brasiliensis* Tanaka) cooled to −196 °C. J Plant Physiol 137:465–470

Sakai A, Matsumoto T, Hirai D, Niino T (2000) Newly developed encapsulation-dehydration protocol for plant cryopreservation. Cryo Lett 21:53–62

Sato H, Enomoto S, Oka S, Hosomi K, Ito Y (1993) Plant regeneration from protoplasts of peppermint (*Mentha × piperita* L.). Plant Cell Rep 12:546–550

Schafer-Menuhr A, Schumacher HM, Mix-Wagner G (1997) Long-term storage of old potato varieties by cryopreservation of shoot-tips in liquid nitrogen. Plant Genet Res Newslett 111:19–24

Spencer A, Hamill JD, Rhodes MJC (1993) In vitro biosynthesis of monoterpenes by *Agrobacterium* transformed shoot cultures of two *Mentha* species. Phytochemistry 32:911–919

Steponkus PL, Langis R, Fujikawa S (1992) Cryopreservation of plant tissues by vitrification. In: Steponkus PL (ed) Advances in low temperature biology, vol 1. JAI Press, London, pp 1–61

Sutton RL (1991) Critical cooling rates to avoid ice crystallization in solutions of cryoprotective agents. J Chem Soc Faraday Trans 87:101–105

Tal B, Rokem JS, Goldberg I (1983) Factors affecting growth and product formation in plant cells grown in continuous culture. Plant Cell Rep 2:219–222

Towill LE (1985) Low temperature and freeze/vacuum-drying preservation of pollen. In: Kartha KK (ed) Cryopreservation of plant cells and organs. CRC Press, Boca Raton, pp 171–198

Towill LE (1988) Survival of shoot tips from mint species after short-term exposure to cryogenic conditions. HortScience 23:839–841

Towill LE (1990) Cryopreservation of isolated mint shoot tips by vitrification. Plant Cell Rep 9:178–180

Tucker AO, Harley RM, Fairbrothers DE (1980) The Linnaean types of *Mentha* (Lamiaceae). Taxon 29:233–255

Van Eck JM, Kitto SL (1990) Callus initiation and regeneration in *Mentha*. HortScience 25: 804–806

Van Eck JM, Kitto SL (1992) Regeneration of peppermint and orange mint from leaf disks. Plant Cell Tissue Organ Cult 30:41–49

II.10 Cryopreservation of *Panax* (Ginseng)

Kayo Yoshimatsu and Koichiro Shimomura

1 Introduction

The genus *Panax* (family Araliaceae) comprises about half a dozen species. The old Greek term "Panax" implies all healing or a panacea (Kains 1958) and almost all *Panax* spp. have been used in folk medicine. One of the most famous species, and now uncommon in natural habitats, is *P. ginseng* C. A. Meyer. Roots from this species have held an honoured place in Chinese medicine for over 4000 years.

1.1 Plant Distribution and Important Species

Species of *Panax* grow widely in the Northern Hemisphere, from the Himalayas through China and Japan to North America. *Panax* spp. consist of *P. ginseng* (Korea, China and Russia), *P. japonicus* (Japan and China), *P. quinquefolium* (North America), *P. pseudoginseng* var. *notoginseng*, *P. pseudoginseng* var. *elegantior*, *P. pseudoginseng* var. *wangiatior*, *P. pseudoginseng* var. *bipinnatifidus*, *P. stipleanathus* (China), *P. pseudoginseng* var. *himalaicus* (Himalaya), and *P. vietnamensis* (Vietnam). The former four species are currently used in commerce with *P. ginseng* having the biggest market (Shoyama et al. 1995).

    P. ginseng had a natural range in Northeast China, Korea and the Far-East part of Siberia (Shibata 1977). The plant is now cultivated in China, Korea, Siberia and Japan to supply the market for pharmaceuticals. The most expensive drug (ginseng root) is derived from Korean root. The plant, perennial, erect, approximately 50 cm tall with a crown of dark green verticillate leaves and small green flowers giving rise to clusters of bright red berries, is cultivated under thatched covers and harvested when 6 years old. Sun-drying of the roots, after removal of the rootlets and the outer layers or quickly passing through hot water, produces white ginseng. Red ginseng is obtained by first steaming the root, followed by artificial drying and then sun-drying. Ginseng has been used for the treatment of anaemia, diabetes, gastritis, sexual impotence and many conditions arising from onset of old age (Evans 1989).

Tsukuba Medicinal Plant Research Station, National Institute of Health Sciences, 1 Hachimandai, Tsukuba, Ibaraki, 305-0843 Japan

Biotechnology in Agriculture and Forestry, Vol. 50
L.E. Towill and Y.P.S. Bajaj (Eds.) Cryopreservation of Plant Germplasm II
© Springer-Verlag Berlin Heidelberg 2002

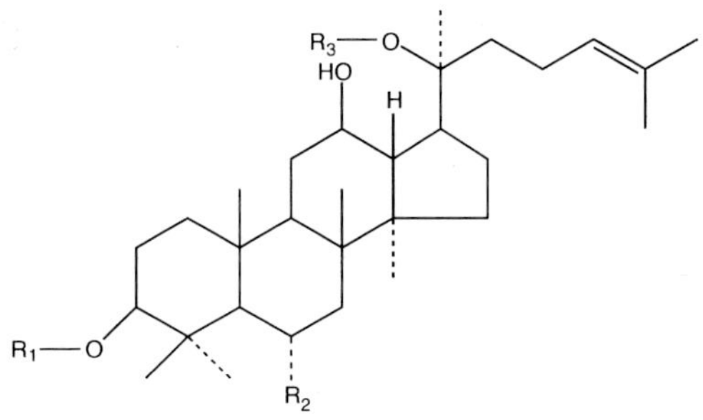

	R1	R2	R3
ginsenoside Rb1	Glc(β1→2)Glc–	H	Glc(β1→6)Glc–
ginsenoside Rc	Glc(β1→2)Glc–	H	Ara(f)(α1→6)Glc–
ginsenoside Rd	Glc(β1→2)Glc–	H	Glc–
ginsenoside Re	H	Rha(α1→2)Glc–*O*–	Glc–
ginsenoside Rg1	H	Glc–O–	Glc–

Fig. 1. Ginsenosides of *Panax ginseng* quantified by high-performance liquid chromatography (Yoshimatsu et al. 1996)

Ginseng contains many dammarane and oleanane saponins of which the biological activity has been studied widely (Fig. 1; Shibata 1977).

P. japonicus C. A. Meyer is indigenous in Japan (Shibata 1977). Its rhizome has been utilized for several hundred years in Japan as a substitute for *P. ginseng* root in Chinese medicine, especially for gastroenteric disorders, antitussive, expectorants and antipyretics (Saito et al. 1977). The drug is also used as a hair-growing tonic in Japan (Shoyama et al. 1995).

P. quinquefolium L. grows in rich woods throughout eastern and central North America, especially along the mountains from Quebec and Ontario, south to Georgia. It was used by the North American Indians (Grieve 1995). Enormous demand for *P. ginseng* roots in China led to an extensive search for a substitute resulting in the discovery in 1716 of *P. quinquefolium* near Montreal, Canada (Bailey 1958). This root was favorably received by the Chinese, and soon became an important article of export from the eastern USA and Canada via Hong Kong (Evans 1989). It is mainly cultivated in north-central Wisconsin.

P. pseudoginseng var. *notoginseng* (Burk.) Hoo et Tseng originated in South-West China, 'Tun-nan and Kwan-si' is an important Chinese crude drug used as a remedy for bruises and a hemostatic for hemorrhage (Shibata 1977). It is cultivated in Kumming province in China.

1.2 Various Methods for the Storage of Germplasm of Ginseng

Field-cultivation of the plants, that is conservation as plants in a field collection, is the main method for storage of ginseng germplasm. For field-cultivation, ginseng plants require recognition of their peculiarities. Since the plants cannot be propagated vegetatively, seedlings are usually employed.

However, the seeds collected from ripe fruits should be stored and treated properly for germination. *P. ginseng* seeds must be stratified with fine sand to encourage embryo growth artificially because the embryo is immature when seeds are gathered (Choi 1988). If the seeds are dried too much, they become dormant and breaking dormancy is not usually easy. Thus, storage of germplasm of ginseng by seeds (seed bank) is usually not applicable.

1.3 Need for Conservation/Cryopreservation

P. ginseng has long been recognizes all over the world for its important clinical value. Its range has been reduced due to habitat alteration or destruction resulting from human activities and to excessive over-collection and uprooting of plants. Naturally occurring *P. ginseng* is now rare. It has long been extinct in the Zhongyuan region including southern Shanxi, southern Hebei, Henan and western Shandong. *P. ginseng* has been identified for preservation in China and the Changbai Shan Nature Reserve has afforded it a degree of protection. It has been widely cultivated in Northeast China, and introduced to Hebei, Shanxi, Shaanxi, Hubei, Guangxi, Sichuan, Yunnan, etc. (Yang 1992).

P. ginseng plants start to flower from the age of 3 years and to set fruit when 5–6 years old. For cultivation, seeds are stratified at low temperature before sowing; cracked seeds are sown immediately after collection. Two- to 3-year-old seedlings are transplanted to the field, protected from cold in winter and from drought in spring. Young plants require special attention because of their delicacy. Shading, weeding and controlling diseases are necessary at the seedling stage (Yang 1992).

Because of the troublesome cultivation of the plants and risks of unfavorable change of the ecosystem, e.g., unusual weather, disaster and diseases, conservation other than a field collection is needed. Extensive studies on in vitro culture of *P. ginseng* have been conducted (references cited in Shoyama et al. 1995). In vitro culture can preserve germplasm in a limited space under a disease-free condition. However, in vitro cultures are always exposed to the danger of microbial contamination or unexpected malefaction of the culture facility. Thus, a combination of in vitro culture and cryopreservation may enhance germplasm conservation. Cryopreservation, where the metabolic function of the living cells is arrested, has proven reliable for long-term preservation (Bajaj 1995).

2 Cryopreservation

2.1 Review of the Cryopreservation Work on *P. ginseng*

Several studies for the cryopreservation of in vitro *P. ginseng* cultures are listed in Table 1. Cell suspensions of *P. ginseng* were first cryopreserved by Butenko et al. (1984). They investigated the effects of cold hardening and sucrose con-

Table 1. Cryopreservation studies using in vitro cultures of *P. ginseng*

Material	Method	max. post thaw viability	Reference
Cell suspensions	Slow freezing	51.1%	Butenko et al. (1984)
Cell suspensions	Slow freezing	40%	Seitz and Reinhard (1987)
Cell suspensions	Slow freezing	not described	Mannonen et al. (1990)
Hairy roots	Vitrification	60%	Yoshimatsu and Shimomura (1996)

centration on the survival of *P. ginseng* cells after storage in liquid nitrogen (LN). Cells grown in modified Murashige and Skoog (MS; Murashige and Skoog 1962) media were precultured with 7–25% sucrose at 2–10 °C for 5–20 days. After the cells had settled, the ice-cold cryoprotectant solution (DMSO, glycerol, sucrose or their combination) was added gradually over 1 h. The ampoules containing the cells were slowly cooled as follows; –5 to –7 °C followed by ice crystallization initiation by submerging the ampulla holder into LN for 5 s with subsequent holding at –5, and –7 °C for 20 min; to –30 °C at a rate of –0.5 °C/min; from –30 to –70 °C at –9 °C/min, and then the ampoules were immersed in LN. After storage for a few months, the ampoules were warmed rapidly by shaking in water at 20 or 40 °C. When the cryoprotectant contained DMSO, the cells were washed twice with nutrient medium. Cells were cultured on agar medium. The best survival (51.1%) was obtained when the cells were cold-hardened at 2 °C for 18 days in the presence of 20% sucrose and the use of sucrose alone (20%) as the cryoprotectant.

Seitz and Reinhard (1987) tried to optimize the protocol with regard to preculture conditions and cryoprotectants using ginsenoside-producing cell suspension cultures of *P. ginseng*. Actively growing cells were transferred to preculture media and suspended twice. Preculture media were prepared by adding the appropriate additive to standard medium. The combination of the preculture additive and culture condition was as follows; 6% mannitol/25 °C/3 days; 0.2 or 1.0 M sorbitol/25 °C/16 h; 20% sucrose/25 °C or from 25 to 4 °C stepwise/18 days (Butenko et al. 1984). After cryoprotectant (0.5 M DMSO + 0.5 M glycerol + 1.0 M sucrose, 1.0 M sorbitol, 1.0 M sorbitol + 5% DMSO, 20% sucrose or 10% sucrose + 10% glycerol) treatment for 1 h, the samples were cooled at a rate of –1 °C/min, then kept for 40 min at –35 °C, and subsequently immersed directly in LN. In their experiments, the best post-thaw viability (40%), which was immediately determined after thawing, and good recovery growth were obtained when the cells were precultured with 20% sucrose with gradual reduction of the culture temperature from 25 to 4 °C over 18 days and use of 20% sucrose alone as cryoprotectant. They also investigated the post-thaw characteristics, growth and ginsenoside productivity. The growth patterns, measured by fresh and dry weight, cell number, and packed-cell volume, and ginsenoside yield and pattern of frozen-thawed and control cultures were identical.

Mannonen et al. (1990) preserved *P. ginseng* cell cultures either in LN or under mineral oil for 6 months and compared their growth behavior and ability to produce ginsenosides after a recovery period with cultures maintained by frequent subcultivation during the same period. Their cryogenic protocol was as follows; cells precultured with 6% mannitol for 3 days, treated with a mixture of 0.5 M DMSO/0.5 M glycerol/1 M sucrose, cooled at a rate of −1 °C/min to −35 °C, held at −35 °C for 30 min and then immersed in LN. For storage under mineral oil, 0.5 g of callus at the exponential growth phase (2 weeks old) was transferred to normal solidified medium in a wide-mouthed tube and covered with a 2.5-cm overlay of sterile liquid paraffin. Tubes were kept at 10 °C for 6 months, after which the calli were transferred to normal growth medium for recovery. They demonstrated that neither growth kinetics nor the degree of vacuolization that occurred during growth was affected by either storage protocol. However, some changes in secondary metabolism were found with preservation under mineral oil whereas none occurred with the cryogenic method.

Recently, we applied a vitrification method to cryopreserve *P. ginseng* hairy roots and compared its efficacy with cold storage at 4 °C (Yoshimatsu et al. 1996). The results obtained from this study indicated that vitrification is promising for long-term preservation of hairy roots. Indeed, hairy roots of several different plant species were cropreserved successfully by the vitrification method (Yoshimatsu et al. 2000).

2.2 Methodology for Cryopreservation of *P. ginseng* Hairy Roots by Vitrification

Vitrification involves treatment of the cultures with large concentrations of cryoprotectants and use of plunge cooling. Because it is a simple protocol (Fig. 2), the method has been applied to a wide range of plant meristems (see Sakai 1993); however, there were no reports on cryopreservation of hairy roots by vitrification before our report (Yoshimatsu et al. 1996).

If the cells are sufficiently dehydrated and intracellular contents concentrated and then cooled to a very low temperature, cells will be vitrified

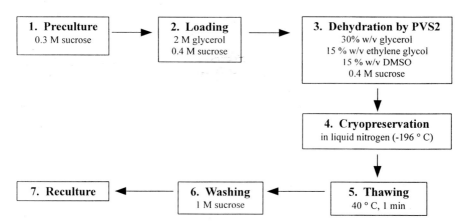

Fig. 2. Procedure for cryopreservation of *P. ginseng* hairy roots

(solidified without crystallization) and avoid lethal intracellular freezing. During the protocol, it is necessary to prevent injury by chemical toxicity or excessive osmotic stresses. In our method, based on the two-step vitrification procedure developed by Matsumoto et al. (1994; Fig. 2), the hairy roots are preliminarily treated with a less toxic loading solution, 2 M glycerol and 0.4 M sucrose, and subsequently dehydrated with PVS2. Without loading, the direct exposure of roots to a highly concentrated vitrification solution may be harmful. The cryotube containing the roots is then directly immersed in LN.

2.2.1 Induction of Hairy Roots

Petioles were cut from new leaves of 4-year-old *P. ginseng* plants cultivated in a field at Tsukuba Medicinal Plant Research Station, NIHS, Japan. The petioles were surface-sterilized with 2% sodium hypochlorite (Yoshimatsu and Shimomura 1992). The petiole segments (ca. 1 cm in length) were vertically placed onto 1/2 MS solid medium and directly infected with *Agrobacterium rhizogenes* ATCC 15834 at 25 °C under 16 h/daylight ($5 \mu E \, m^{-2} s^{-1}$). The roots that appeared at the infected sites were excised and placed onto a half strength MS solid medium containing 500 mg/l Claforan (Hoechst Japan Ltd.). This procedure was repeated until the bacteria had been eliminated. Since the axenic roots thus obtained did not proliferate, they were transferred onto the solid medium containing 1 mg/l IBA. The roots began to proliferate vigorously even on the phytohormone-free (HF) medium and could be maintained as hairy roots after their growth had been stimulated by addition of IBA. The transformation of hairy roots was confirmed by an opine assay (agropine and mannopine, Petit et al. 1983) and PCR analysis (Jaziri et al. 1994).

2.2.2 Culture of Hairy Roots

The hairy roots were cultured either on 1/2 MS solid medium (25 ml/9 cm i.d. petri dish) or in various basal liquid media [1/2 MS, Gamborg B5 (B5, Gamborg et al. 1968), Woody Plant (WP, Lloyd and McCown 1980), Root Culture (RC, Thomas and Davey 1982) and root culture medium containing 100 mg/l myoinositol (RCI)]; ca. 100 mg fresh weight inoculum; 50 ml/100 ml flask, rotated at 100 rpm) in the dark, or under 16 h/daylight at 15, 20 or 25 °C (Fig. 3A). The media were adjusted to pH 5.7 before autoclaving at 121 °C for 15 min and solidified with 0.2% Gelrite. All media contained 30 g/l sucrose, except the preculture medium for cryopreservation, which contained 0.3 M sucrose.

2.2.3 Cryopreservation of Hairy Roots

Root tips (ca. 1 mm) of the hairy roots grown either on 1/2 MS solid (Tables 2–4) or in liquid (Table 5) medium were precultured on 1/2 MS solid medium containing 0.3 M sucrose for 1–3 days at 25 °C in the dark. Root tips were transferred into 2-ml plastic cryotubes (10–15 root tips per tube) and treated with cryoprotective solution (2 ml; 2 M glycerol and 0.4 M sucrose in MS basal medium) for 10 min. The solution was replaced with PVS2 solution (Sakai et al. 1990); 2 ml; 30% w/v glycerol, 15% w/v ethylene glycol, 15% w/v DMSO and 0.4 M sucrose in MS basal medium). After a few minutes at 25 °C (see Tables 2–5), 1 ml of PVS2 solution was removed and the tubes were directly immersed in LN. For the experiment involving the addition of phytohormone to the preculture medium, IAA (5 mg/l), 2,4-D (0.1 or 0.5 mg/l) or TIBA (0.1 or 0.5 mg/l) was added to the preculture medium (1/2 MS solid medium containing 0.3 M sucrose) and the root tips were treated using the procedure mentioned above.

2.2.4 Viability and Recovery Assessment of Cryopreserved Root Tips

The vitrified hairy root tips held in LN for at least 4 days were rapidly warmed at 40 °C in a water bath and washed with MS medium containing 1.0 M sucrose (2 ml) for 10 min. For recovery, the root tips were cultured in the dark on HF 1/2 MS solid medium containing 3% sucrose at 25 °C; root regeneration and elongation during reculture were evaluated. Viability was examined by fluorescein diacetate (FDA) staining (Benson and Hamill 1991) immediately after thawing and washing, and subsequent microscopic observation of fluorescence in the tissue. FDA stock solution was prepared (0.1 mg FDA/20 ml acetone; kept at −20 °C) and diluted 50-fold with distilled water before use. Whole root tips were incubated in an FDA solution (final concentration: 0.1 mg/l) at 25 °C for 30 min, and those that exhibited bright fluorescence similar to the untreated control were judged viable (Fig. 3B). To define the time periods of preculture and PVS2 treatment, the root tips were treated using the same procedure except for immersion in LN, and their viability was tested by FDA staining (Table 2). To determine the recovery growth (elongation and new root proliferation, n = 2–9), independently regenerated hairy roots (ca. 1 cm) were subcultured on fresh HF 1/2 MS solid medium after 7 weeks of regrowth (Fig. 4). To evaluate ginsenoside production, independently regenerated hairy roots cultured on HF 1/2 MS solid medium for 12 weeks were further transferred into HF 1/2 MS liquid medium and cultured at 25 °C in the dark for 6 weeks (ca. 100 mg fresh weight inoculum; 50 ml/100-ml flask; rotated at 100 rpm; $n = 2$) (Figs. 3E and 5).

2.2.5 Extraction and Analysis of Ginsenosides

Extraction and determination of ginsenosides were performed following the procedure reported by Kanazawa et al. (1989). Lyophilized powder (ca. 50 mg) was refluxed with 7 ml methanol for 1 h. This was repeated three times. The combined methanolic extract was evaporated to dryness and dissolved in 2 ml distilled water. This solution (0.5 ml) was applied onto a Sep-Pak C18 cartridge

(Millipore) equilibrated with 5 ml methanol and 5 ml distilled water. After washing with 5 ml distilled water and 5 ml 30% aqueous methanol solution, the ginsenoside fraction was eluted with 5 ml methanol. The eluate was concentrated to 2 ml on a water bath (below 50 °C) and then an appropriate volume was injected into a high performance liquid chromatograph equipped with a Wakosil II 5C18AR column (4.6 mm i.d. × 250 mm, Wako Chemicals, Japan), mobile phase; acetonitrile/water (3:7 for ginsenoside Rb1, Rc and Rd; 2:8 for ginsenoside Re and Rg1), flow rate; 1 ml/min, room temperature, UV detection at 203 nm.

2.2.6 PCR Analysis

DNA extraction and PCR analysis were carried out according to the procedures of Jaziri et al. (1994). The primers for Tʟ-DNA were 5′-ATG-GAATTAGCCGGACTAAACG-3′ and 5′-ATGGATCCCAAATTGCTAT-TCC-3′, which are complementary to the 5′ coding sequence of *rol A* and the 3′ coding sequence of *rol B*, respectively. Those for Tʀ-DNA were 5′-CGGAAATTGTGGCTCGTTGTGGAC-3′ and 5′-AATCGTTCAGAGAG CGTCCGAAGTT-3′, complementary to the 5′ and 3′ flanking sequences of the *ags* gene, respectively (Slightom et al. 1986). For the positive control of Tʟ and Tʀ-DNAs, the DNAs extracted from the relevant cosmids pLJ 1 and pLJ 85 (Jouanin 1984) were used as templates, respectively.

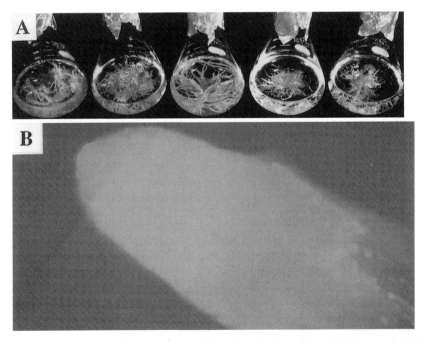

Fig. 3. A *Panax ginseng* hairy roots cultured in various basal media at 25 °C in the dark for 8 weeks. *Left* to *right* 1/2 MS, WP, B5, RC and RCI medium, respectively. **B** FDA staining of cryopreserved hairy root tip. (Parts **C–F** see next page)

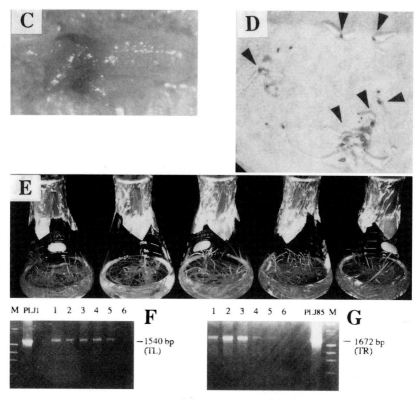

Fig. 3. C Emergence of new root meristem from cryopreserved root tip. **D** Regeneration of hairy roots from cryopreserved root tips precultured with 2,4-D (*arrowhead*). The hairy roots were cultured on 1/2 MS solid medium at 25 °C in the dark for 4 weeks. **E** Untreated hairy roots and those regenerated from cryopreserved root tips (*left* to *right* untreated control, precultured without phytohormone, precultured with IAA, TIBA or 2,4-D before cryopreservation, respectively). **F** Amplification of TL-DNA fragment. *Lane M* λ/Hind III digest-φX174/Hae III digest (TOYOBO). *Lanes 1–4* Hairy roots regenerated from cryopreserved root tips (from 1 to 4: precultured with 2,4-D, TIBA, IAA or without phytohormone, respectively); *lane 5* untreated hairy root; *lane 6* non-transformed root. **G** Amplification of TR-DNA fragment. Explanation of lanes is the same as **F**. (Yoshimatsu et al. 1996)

2.3 Results and Discussion

2.3.1 Culture of Hairy Roots

To determine the optimum culture conditions, roots were cultured in various HF basal liquid media (1/2 MS, B5, WP, RC and RCI) in the dark, or under 16 h/daylight at 15, 20 or 25 °C. Although slow growth of *P. ginseng* hairy roots (3-fold increase after 3 weeks of culture) in HF liquid medium was reported earlier (Yoshikawa and Furuya 1987), the roots in the study grew faster in HF liquid medium. Best growth was observed in 1/2 MS liquid medium at 25 °C in the dark (Fig. 3A, ca. 40-fold increase after 8 weeks of culture); roots cultured in this condition produced almost the same level of ginsenoside Rb1 as 4-year-old ginseng roots and a higher level of ginsenoside Rg1. Therefore, 1/2 MS medium was used for further experiments.

2.3.2 Cryopreservation of Hairy Roots

Table 2. Effect of duration of preculture and PVS2 treatment on the viability percentage of *P. ginseng* hairy roots, as determined by FDA staining (after Yoshimatsu et al. 1996)

PVS2 treatment (min)	Preculture (days)		
	1	2	3
5	80.0	40.0	75
7	44.4	50.0	**50.0**
10	**44.4**	**60.0**	50.0
12	57.1	44.4	14.3

Viability%: No. of root tips with fluorescence/No. of segments tested (n = 6 to 10) × 100.
Bold letters: samples with a relatively higher percentage of stained cells and/or stronger fluorescence.

Table 3. Effect of duration of preculture and PVS2 treatment on the viability percentage estimated by FDA staining and recovery (root regeneration and elongation after 3 weeks of reculture) of cryopreserved *P. ginseng* hairy roots (After Yoshimatsu et al. 1996)

Preculture (days)	PVS2 treatment (min)	Viability	Recovery
1	11	0	11.1
2	11	0	30.0
3	9	33.3	50.0

Callus cultures of *P. ginseng* were successfully cryopreserved, and retained their biosynthetic potential (Butenko et al. 1984; Seitz and Reinhard 1987; Mannonen et al. 1990). Root tips (Bajaj 1987) and hairy root cultures (Benson and Hamill 1991) are, however, believed to be more stable in culture and yield of secondary metabolites.

We first estimated the effect of duration of preculture and PVS2 treatment on the viability of root tips without immersion in LN (Table 2). Over 50% of root tips was viable in three treatments. These three combinations of preculture (day) and PVS2 treatment (min) were: 1 day/10 min, 2 days/10 min and 3 days/7 min (Table 2). Root tips treated by these three regimes showed a relatively high percentage of stained cells and/or a stronger fluorescence. The combination of preculture and PVS2 treatment were 1 day/10 min, 2 days/10 min and 3 days/7 min (Table 2). Segments were treated with these three combinations with a slight prolongation of PVS2 treatment (11 or 9 min), immersed in LN and held for 4 days. After rapid warming and washing, viability was immediately determined by FDA staining and recovery was estimated from root regeneration and elongation after 3 weeks of reculture (Table 3). Although the data differed between viability and recovery, the best results (33% viability and 50% recovery) were obtained with 3-day preculture and 9-min PVS2 treatment. For the viability test, we did not count the partially or weakly stained segments as surviving in order to avoid an overestimation of survival. It is possible that partially or weakly stained root tips retained the capability of regeneration, thus leading to a higher percentage recovery.

Table 4. Effect of phytohormone addition to the preculture medium on the recovery percentage (root regeneration and elongation) of cryopreserved *P. ginseng* hairy roots (After Yoshimatsu et al. 1996)

Preculture medium	Duration in liquid nitrogen	
	4 days	15 weeks
0.3 M sucrose	20.0	0
0.3 M sucrose + IAA 5 mg/l	27.0	28.6 ± 16.5
0.3 M sucrose + 2,4-D 0.1 mg/l	60.0	54.2 ± 11.0
0.3 M sucrose + TIBA 0.5 mg/l	20.0	24.6 ± 9.2

Preculture: 25 °C, dark, 3 days. PVS2 treatment, 25 °C, 8 min. Culture after cryopreservation: HF 1/2 MS solid medium containing 3% sucrose, 25 °C, dark for 50 days for 4 days duration in liquid nitrogen (left) and 29 days for 15 weeks duration in liquid nitrogen (right)

Root meristems of cryopreserved root tips exerted bright fluorescence after FDA staining (Fig. 3B). After 1 week of reculture at 25 °C in the dark, the surface of the root cap cracked and subsequently new root meristems emerged from the root tip (Fig. 3C). These observations suggest that the actively proliferating part of the hairy root was successfully preserved in LN and could regenerate new roots after warming. To improve recovery of treated root tips, we investigated the effect of auxin addition to the preculture medium on the survival of cryopreserved hairy roots; auxin generally enhances cell division (Table 4). TIBA (auxin polar transport inhibitor) was also examined because it stimulated an increase in endogenous auxin levels in cultured root segments of *Cephaelis ipecacuanha* (Yoshimatsu and Shimomura 1994). The segments precultured were dehydrated with PVS2 for 8 min and cryopreserved for 4 days or 15 weeks. Among the phytohormones tested, the best recovery was observed on the segments precultured with 0.1 mg/l 2,4-D (Fig. 3D). In this condition, there were recovery rates of 60% after 4 days of cryopreservation and 54% after 15 weeks of cryopreservation. This enhancement of recovery by addition of 2,4-D was significant ($p < 0.01$, ANOVA, $n = 3$) compared with HF treatment (15 weeks data). The recovery rate fluctuated when the segments were precultured without phytohormone. *Beta vulgaris* hairy root tips in the early growth stage were more amenable to cryopreservation than those in the stationary growth stage (Benson and Hamill 1991). There-fore, the fluctuation of recovery in HF treatment was thought to be due to inherent physiological differences among hairy root tips and pretreatment with auxin or TIBA might overcome this difficulty by promoting a desirable physiological state before cryopreservation.

2.3.3 Properties of Cryopreserved Hairy Roots

It is particularly important that cryopreserved plant cells remain capable of producing cells or tissue identical with the non-treated ones. The growth of hairy roots regenerated from cryopreserved root tips was compared with that of untreated ones (Fig. 4). Although the regenerated roots proliferated fewer new lateral roots than the control, they elongated similarly to the control. This suggests that root proliferation is more susceptible to the direct effect arising from the cryogenic procedure than root elongation. However, they grew well and no differences were observed as compared to the control when further transferred into 1/2 MS liquid medium and cultured at 25 °C in the dark (Fig. 3E).

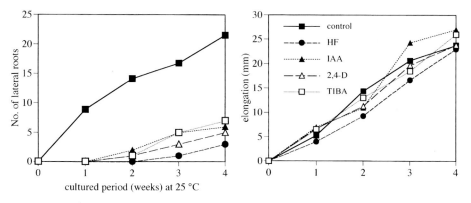

Fig. 4. Recovery growth (*above* elongation of mother root; *below* new root proliferation) of *P. ginseng* hairy roots regenerated from cryopreserved root tips (*n* = 2–9) and untreated control (without cryopreservation). The root tips were precultured with IAA (5 mg/l), 2,4-D (0.1 mg/l), TIBA (0.5 mg/l) or without phytohormone (HF) before cryopreservation. (Yoshimatsu et al. 1996)

Untreated hairy root cultures produced much more ginsenoside Rb1 than field-grown ginseng, which produced ginsenoside Re as a main saponin. This might be one of the differences between field-grown roots and cultured roots (Yoshimatsu and Shimomura 1992). In the case of *P. ginseng* cell cultures, the same ginsenoside patterns (ginsenoside Rb1 as a main saponin) were observed (Seitz and Reinhard 1987). The regenerated hairy roots from cryopreserved root tips produced almost the same levels of ginsenosides, showing the same

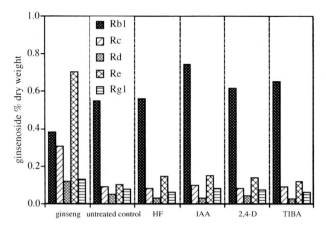

Fig. 5. Ginsenoside production in *P. ginseng* hairy root cultures regenerated from cryopreserved root tips and 4-year-old root of *P. ginseng* cultivated in the field from seedling. The culture conditions of hairy roots are the same as Fig. 3E (Yoshimatsu et al. 1996). *Rb1* Ginsenoside Rb1; *Rc* ginsenoside Rc; *Rd* ginsenoside Rd; *Re* ginsenoside Re; *Rg1* ginsenoside Rg1 (see Fig. 1)

patterns as those in the untreated control (Fig. 5). This result indicates that cryopreservation of root tips did not influence the biochemical capabilities of hairy roots.

The value of hairy root cryopreservation depends on the molecular stability of the T-DNA. Therefore, we examined the existence of T-DNAs by PCR analysis as a preliminary investigation for genetic stability, and confirmed the existence of T-DNAs (T_L- and T_R-DNAs) in the hairy roots regenerated from cryopreserved root tips (Fig. 3F,G).

Initially, hairy roots grown on solid medium were used for cryopreservation because of their reduced water content. The previous results were obtained with these roots, but hairy root cultures grew much more rapidly and vigorously in liquid culture than on solid culture. Therefore, cryopreservation of *P. ginseng* hairy roots cultured in the liquid medium was studied (Table 5).

Table 5. Effect of phytohormone addition to the preculture medium on the recovery percentage of cryopreserved *P. ginseng* hairy roots which had been cultured in the liquid medium

Preculture medium	Recovery percentage
0.3 M sucrose	83.3 ± 19.1
0.3 M sucrose + 2,4-D 0.1 mg/l	70.7 ± 17.8
0.3 M sucrose + 2,4-D 0.5 mg/l	85.9 ± 12.2
0.3 M sucrose + TIBA 0.1 mg/l	17.4 ± 6.5
0.3 M sucrose + TIBA 0.5 mg/l	43.7 ± 29.9

Preculture: 25 °C, dark, 1 day. PVS2 treatment: 25 °C, 10 min. Culture after cryopreservation: HF 1/2 MS solid medium containing 3% sucrose, 25 °C, dark, 30 days.

Root tips (ca. 1 mm) of hairy roots in HF 1/2 MS liquid medium were excised and precultured on 0.3 M sucrose medium without phytohormone or with either 2,4-D (0.1 or 0.5 mg/l) or TIBA (0.1 or 0.5 mg/l) for 1 day. After preculture the root tips (10 segments/tube) were transferred into a cryotube (2 ml), loaded at 25 °C for 10 min, dehydrated with PVS2 at 25 °C for 10 min and then cryopreserved. Recovery rates of over 70% were obtained either with or without 2,4-D (0.1 and 0.5 mg/l).

3 Summary and Conclusions

Ginseng is very expensive because of its long-term conventional cultivation (5–7 years, Choi 1988). As an alternative, the production of ginsenosides by hairy root cultures has been studied (Yoshikawa and Furuya 1987); however, hairy roots required phytohormones for satisfactory growth. We also established *P. ginseng* hairy root culture by infecting *Agrobacterium rhizogenes* ATCC 15834 onto petiole segments of field-grown plants and found that our hairy roots grew vigorously and produced far more ginsenosides than the hairy roots previously reported. For the induction of hairy roots, it is indispensable that T-DNA of Ri plasmid in *Agrobacterium rhizogenes* be integrated into plant genomic DNA and successfully expressed (Saito et al. 1992). The site of integration of T-DNA into plant nuclear DNA is apparently random. In general, a number of transformed clones are induced by infection and each has a different biosynthetic capability. They are valuable for differential screening. Therefore, stable long-term preservation of transformed cultures is advantageous because maintenance of these cultures requires extensive labor, space and facilities.

Recently, the vitrification procedure for cryopreservation was applied to a wide range of plant shoot tips (Sakai 1993). This method may be profitable for routine preservation of hairy root clones because it eliminates the need for controlled slow freezing and permits cells and shoot tips to be cryopreserved by direct transfer into LN. Although cryopreservation of *P. ginseng* cell cultures has been reported (Butenko et al. 1984; Seitz and Reinhard 1987; Mannonen et al. 1990), there have been no reports on cryopreservation of *P. ginseng* hairy roots.

By the optimization of the vitrification protocol including material culture, preculture and PVS2 treatment, recovery rates of over 70% were obtained. Although several studies revealed that the preculture conditions, such as relatively high sucrose concentration or cold hardening, greatly influence freezing or dehydration tolerance (Butenko et al. 1984; Seitz and Reinhard 1987; Sakai 1993), enhanced sucrose concentration to 0.5 M and cold hardening at 4 °C decreased the recovery rate for hairy roots. For successful cryopreservation, use of actively growing hairy roots in liquid medium was effective. Hairy roots regenerated from cryopreserved root tips grew well and maintained their capability to produce ginsenosides and conserve T-DNAs. These results are promising for further research on ginsenoside production of *P. ginseng* because transformed cultures can be maintained cryogenically.

Acknowledgements. The authors thank Dr. Akira Sakai for his valuable advice on this research. This study was supported in part by the Ministry of Health and Welfare, Health Sciences Research Grants, Special Research.

References

Kains MG (1958) Ginseng. In: Bailey LH (ed) The standard encyclopedia of horticulture, vol 1. Macmillan, New York, pp 1338–1339

Bajaj YPS (1987) Cryopreservation of potato germplasm. In: Bajaj YPS (ed) Biotechnology in agriculture and forestry, vol 3. Potato. Springer, Berlin Heidelberg New York, pp 472–486

Bajaj YPS (1995) Cryopreservation of plant cell, tissue, and organ culture for the conservation of germplasm and biodiversity. In: Bajaj YPS (ed) Biotechnology in agriculture and forestry, vol 32. Cryopreservation of plant germplasm I. Springer, Berlin Heidelberg New York, pp 3–28

Benson EE, Hamill JD (1991) Cryopreservation and post freeze molecular and biosynthetic stability in transformed roots of *Beta vulgaris* and *Nicotiana rustica.* Plant Cell Tissue Organ Cult 24:163–171

Butenko RG, Popov AS, Volkova LA, Chernyak ND, Nosov AM (1984) Recovery of cell cultures and their biosynthetic capacity after storage of *Dioscorea deltoidea* and *Panax ginseng* cells in liquid nitrogen. Plant Cell Lett 33:285–292

Choi KT (1988) *Panax ginseng* C. A. Meyer: micropropagation and the in vitro production of saponins. In: Bajaj YPS (ed) Biotechnology in agriculture and forestry, vol 4. Medicinal and aromatic plants I. Springer, Berlin Heidelberg New York, pp 484–500

Evans WC (1989) Ginseng. In: Evans WC (ed) Pharmacognosy, 13th edn. Baillière Tindall, London, pp 490–491

Gamborg OL, Miller RA, Ojima K (1968) Nutrient requirements of suspension culture of soybean root cells. Exp Cell Res 50:151–158

Grieve M (1995) Ginseng. In: Botanical.com, a modern herbal, Electric Newt (http://www.botanical.com/botanical/mgmh/g/ginseng15.html).

Jaziri M, Homès J, Shimomura K (1994) An unusual root tip formation in hairy root culture of *Hyoscyamus mutics.* Plant Cell Rep 13:349–352

Jouanin L (1984) Restriction map of an agropine-type Ri plasmid and its homologies with Ti plasmids. Plasmid 12:91–102

Kanazawa H, Nagata Y, Matsushima Y, Tomoda M, Takai N (1989) High-performance liquid chromatographic analysis of ginsenosides in pharmaceutical preparations. Shoyakugaku Zasshi 43:121–128

Lloyd G, McCown B (1980) Commercially-feasible micropropagation of mountain laurel, *Kalmia latifolia,* by use of shoot-tip culture. In: Int Plant Propag Soc Comb Proc for 1980, 30:21–427

Mannonen L, Toivonen L, Kauppinen V (1990) Effects of long-term preservation on growth and productivity of *Panax ginseng* and *Catharanthus roseus* cell cultures. Plant Cell Rep 9:173–177

Matsumoto T, Sakai A, Yamada K (1994) Cryopreservation of in vitro-grown apical meristems of wasabi (*Wasabia japonica*) by vitrification and subsequent high plant regeneration. Plant Cell Rep 13:442–446

Murashige T, Skoog F (1962) A revised medium for rapid growth and bioassays with tobacco tissue culture. Physiol Plant 15:473–497

Petit A, David C, Dahl GA, Ellis JG, Guyon P, Casse-Delbart F, Tempé J (1983) Further extension of the opine concept: plasmids in *Agrobacterium rhizogenes* cooperate for opine degradation. Mol Gen Gent 190:204–214

Saito H, Lee YM, Takagi K, Shibata S, Shoji J, Kondo N (1977) Pharmacological studies of Panacis Japonici Rhizoma. I. Chem Pharm Bull 25:1017–1025

Saito K, Yamazaki M, Murakoshi I (1992) Transgenic medicinal plants: *Agrobacterium*-mediated foreign gene transfer and production of secondary metabolites. J Nat Prod 55:149–162

Sakai A (1993) Cryogenic strategies for survival of plant cultured cells and meristems cooled to –196 °C. In: Sakai A (ed) Cryopreservation of plant genetic resources, technical assistance activities for genetic resources projects, ref no 6. Japan International Cooperation Agency, Food and Agriculture Research and Development Association (FARDA), Tokyo, pp 5–26

Sakai A, Kobayashi S, Oiyama I (1990) Cryopreservation of nucellar cells of navel orange (*Citrus sinensis* Osb. var. *brasiliensis* Tanaka) by vitrification. Plant Cell Rep 9:30–33

Seitz U, Reinhard E (1987) Growth and ginsenoside patterns of cryopreserved *Panax ginseng* cell cultures. J Plant Physiol 131:215–223

Shibata S (1977) Saponins with biological and pharmacological activity. In: Wagner H, Wolff P (eds) New natural products and plant drugs with pharmacological, biological or therapeutical activity. Springer, Berlin Heidelberg New York, pp 177–196

Shoyama Y, Matsushita H, Zhu XX, Kishira H (1995) Somatic embryogenesis in ginseng (*Panax* species). In: Bajaj YPS (ed) Biotechnology in agriculture and forestry, vol 31. Somatic embryogenesis and synthetic seed II. Springer, Berlin Heidelberg New York, pp 343–356

Slightom JL, Durand-Tardif M, Jouanin L, Tepfer D (1986) Nucleotide sequence analysis of TL-DNA of *Agrobacterium rhizogenes* agropine type plasmid. J Biol Chem 261:108–121

Thomas MR, Davey MR (1982) Plant tissue culture media. 5. Root culture medium. In: Montagu MV (ed) EMBO course: the use of Ti plasmid as cloning vector for genetic engineering in plants. Lab Genetics, Rijks Universiteit, Gent, Belgium, 4–23 Aug, 1982, p 109

Yang Y (1992) *Panax ginseng* C. A. Mey. In: Fu L, Jin J (eds) China plant red data book – rare and endangered plants, vol 1. Science Press, Beijing, pp 178–179

Yoshikawa T, Furuya T (1987) Saponin production by cultures of *Panax ginseng* transformed with *Agrobacterium rhizogenes*. Plant Cell Rep 6:449–453

Yoshimatsu K, Shimomura K (1992) *Cephaelis ipecacuanha* A. Richard (Brazilian ipecac): micropropagation and the production of emetine and cephaeline. In: Bajaj YPS (ed) Biotechnology in agriculture and forestry, vol 21. Medicinal and aromatic plants IV. Springer, Berlin Heidelberg New York, pp 87–103

Yoshimatsu K, Shimomura K (1994) Plant regeneration on cultured root segments of *Cephaelis ipecacuanha* A. Richard. Plant Cell Rep 14:98–101

Yoshimatsu K, Yamaguchi H, Shimomura K (1996) Traits of *Panax ginseng* hairy roots after cold storage and cryopreservation. Plant Cell Rep 15:555–560

Yoshimatsu K, Touno K, Shimomura K (2000) Cryopreservation of medicinal plant resources: retention of biosynthetic capabilities in transformed cultures. In: Engelmann F, Takagi H (eds) Cryopreservation of tropical plant germplasm, current research progress and application. International Plant Genetic Resources Institute, Rome, pp 77–88

II.11 Cryopreservation of In Vitro Grown Apical Shoot Tips of *Wasabia japonica* (Wasabi) by Different Procedures

Toshikazu Matsumoto

1 Introduction

Wasabi, Japanese horseradish (family Cruciferae), is an important crop for Japanese foods such as sushi, sashimi and soba. The roots contain a pungent ingredient (sinigrin). Myrosinase is responsible for the development of the flavor and pungency of wasabi roots (Ohtsuru and Kawatani 1979). Wasabi also contains other useful compounds, such as peroxidase, that is widely used as a label enzyme in clinical diagnosis and immunoassay analysis (Taniguchi et al. 1988). The plant occurs in remote and inaccessible cool mountain springs and streams in Japan. The optimal growth temperature is about 15 °C, and the growth is inhibited at high temperatures.

Since seeds are sensitive to desiccation, the cultivars and strains of wasabi are maintained as living material in the field of some repositories in Japan. A conventional method of cultivation known as a field gene bank has been used for the maintenance of wasabi germplasm. The conventional method has some disadvantages, such as a high cost for long-term maintenance, and the risk of genetic loss due to unpredictable weather conditions and diseases. To overcome these problems, in vitro preservation of wasabi germplasm at 20 °C (Yamada and Haruki 1992) and at −5 °C (Matsumoto and Nako 1999) were developed as methods for short- or medium-term storage. The use of these methods provides advantages such as a high capacity to store germplasm in a small space, and elimination of the risk of unfavorable natural conditions. In recent years, cryopreservation has become a very important tool for long-term storage of germplasm and experimental materials with unique attributes using minimal space and maintenance requirements without causing genetic alterations (Sakai 1997). The development of a simple and reliable method for cryopreservation would allow for a much more widespread use of cryopreserved cultured cells, shoot tips and somatic embryos. In addition, the number of species or cultivars that need to be cryopreserved has increased sharply over the last few years. Thus, cryopreservation may be an alternative approach for long-term storage of germplasm and valuable cell lines using a minimum of space and maintenance.

Recently, simplified cryogenic procedures such as vitrification (Langis et al. 1990; Sakai et al. 1990), encapsulation/dehydration (Fabre and Dereuddre

Shimane Agricultural Experiment Station, Ashiwata 2440, Izumo, Shimane, 693-0035, Japan

Biotechnology in Agriculture and Forestry, Vol. 50
L.E. Towill and Y.P.S. Bajaj (Eds.) Cryopreservation of Plant Germplasm II
© Springer-Verlag Berlin Heidelberg 2002

1990) and encapsulation/vitrification (Matsumoto et al. 1995a) have been developed. In this study, the survival of wasabi shoot tips cooled to −196 °C by different cryogenic procedures were compared under well-optimized conditions, and the most suitable method for wasabi shoot tips was determined.

2 Cryopreservation

2.1 Plant Materials

In vitro grown plantlets of wasabi (*Wasabia japonica* Matsumura, cv. Shimane No.3) were mainly used in the present study. Stock cultures of wasabi plants were maintained on modified Murashige and Skoog (MS; Murashige and Skoog 1962) basal medium (half-strength of ammonium nitrate and potassium nitrate, termed 1/2 MS medium) containing 0.1 mg l^{-1} benzyladenine (BA), 3% (w/v) sucrose and 0.2% (w/v) gellan gum at pH 5.8 (Yamada and Haruki 1992). Stock cultures were subcultured every 35 to 40 days. These tissue-cultured plantlets were grown on 5 ml medium in test tubes (18 mm diameter) under white fluorescent light (52 μm mol s^{-1} m^{-2}), 16-h photoperiod at 20 °C. Apical shoot tips about 1.0 mm in length were dissected from about 30 mm long 30- to 40-day-old plantlets. Shoot tips were precultured on a solidified 1/2 MS medium containing various concentrations of sucrose in Petri dishes (9 cm in diameter) at 20 °C under continuous light. Precultured shoot tips were treated with various cryoprotectants before air-drying or exposure to PVS2 (vitrification solution).

2.2 Vitrification Method

Ten precultured shoot tips were placed in a 2-ml cryotube and then treated with cryoprotective solutions for 20 min at 25 °C. After removing the cryoprotective solution, 2.0 ml of PVS2 were added and gently mixed. After removing the PVS2, this step was repeated once and held at 25 or 0 °C for various lengths of time. The shoot tips were finally suspended in 1.0 ml of PVS2. Cryotubes were plunged into liquid nitrogen (LN) and held there for at least 1 h. PVS2 contains 30% glycerol, 15% ethylene glycol (EG) and 15% dimethyl sulfoxide (DMSO) in 0.4 M sucrose solution at pH 5.8 (Sakai et al. 1990). After rapid warming in a water bath at 40 °C, PVS2 was drained from the cryotubes and replaced with 1.2 M sucrose solution and held for 20 min at 25 °C. Shoot tips were transferred onto sterilized filter paper discs over solidified 1/2 MS medium containing 0.1 mg l^{-1} BA, 3% sucrose and 0.2% gellan. After 1 day, the shoot tips were transferred onto a fresh filter paper disc in a Petri dish containing the same medium.

Shoot tips were precultured with 0.3 M sucrose and glycerol for 1 day, and exposed to PVS2 before immersion in LN. Sucrose, glucose, fructose

and glycerol were determined from shoot tips using liquid-liquid chromatography (LC).

2.3 Encapsulation/Dehydration Method

Shoot tips precultured with 0.3 M sucrose for 1 day were encapsulated in alginate gel beads according to the technique of Bapat et al. (1987) with some
alterations. Shoot tips were suspended in calcium-free MS medium supplemented with 3% (w/v) Na-alginate solution and 0.4 M sucrose. The mixture
was dispensed from a sterile disposable plastic syringe (1 ml) into 100 ml of
culture medium that contained 100 mM calcium chloride plus 0.4 M sucrose
and held for 30 min at 25 °C. Beads of about 5 mm in diameter each contained
one meristem. Encapsulated shoot tips were treated in MS medium supplemented with 0.8 M sucrose with or without different cryoprotectants for 16 h
at 25 °C and then dehydrated in Petri dishes (9 cm in diameter) containing
50 g dry silica gel held at 25 °C for up to 10 h. After dehydration, about 10 dried
beads were placed in a 2-ml cryotube and then directly immersed into
LN. Samples were warmed by placing the cryotubes in a water bath at 40 °C.
Water contents of beads and shoot tips were expressed on a fresh weight basis
(FW). The dry weights were determined after drying for about 100 h at 80 °C.
After rapid warming, beads were transferred onto a solidified medium (see
below).

2.4 Encapsulation/Vitrification Method

Shoot tips precultured with 0.3 M sucrose for 1 day were encapsulated in alginate beads (about 3 mm in diameter) containing a mixture of 2 M glycerol plus
0.4 M sucrose. The encapsulated shoot tips were dehydrated with 50 ml of
PVS2 in a 100-ml glass beaker at 100 rpm on a rotary shaker at 0 °C for various
lengths of time. Ten encapsulated dehydrated shoot tips were suspended in
0.7 ml of PVS2 in 2-ml cryotubes and were then plunged into LN. After rapid
warming in a 40 °C water bath, PVS2 was drained from the tubes and 2 ml of
1.2 M sucrose solution were added and the mixture was held for 30 min at
25 °C. The beads were then transferred onto a solidified medium.

2.5 Viability and Plant Growth

Cryopreserved shoot tips were cultured on 1/2 MS medium containing
0.1 mg l^{-1} BA, 3% sucrose and 0.2% gellan gum at pH 5.8 under the standard
conditions described above. Observations were made weekly on growth.
Shoot formation was recorded as the percentage of the total number of shoot
tips that formed normal shoots 21 days after plating. Ten apical shoot tips
were tested for each of four replicates for each treatment.

3 Results

3.1 Vitrification Method

In a preliminary experiment, shoot tips were precultured on solidified 1/2 MS medium supplemented with different concentrations (0.3–0.7 M) of sucrose at 20 °C for 1 day and then dehydrated with PVS2 for 30–40 min at 0 °C before being immersed into LN. Little or no survival was observed in the vitrified and warmed apical shoot tips cooled to −196 °C (data not shown). To enhance survival, shoot tips precultured with 0.7 M sucrose for 1 day were treated with various cryoprotective solutions for 20 min at 25 °C before dehydration with PVS2. The cryoprotective treatment (Nishizawa et al. 1992, 1993) was very effective in improving the survival of vitrified shoot tips cooled to −196 °C. The highest rate of shoot formation was observed in the shoot tips treated with a mixture of 2 M glycerol plus 0.4 M sucrose (Matsumoto et al. 1994). Based on this result, this mixture was adopted as the cryoprotective solution for wasabi apical shoot tips. The optimum sucrose concentration for preculture was determined. Shoot formation was greatest in the vitrified apical shoot tips precultured with 0.3 M sucrose (Fig. 1).

The effects of preculturing and cryoprotective treatment on the survival of vitrified shoot tips are summarized in Table 1. The apical shoot tips treated with a mixture of 2 M glycerol plus 0.4 M sucrose following preculture with 0.3 M sucrose for 1 day produced the highest shoot formation after cooling to −196 °C.

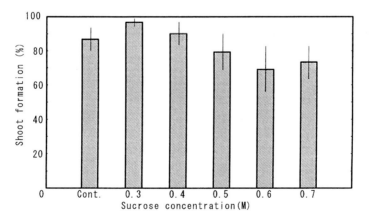

Fig. 1. Effect of preculture in different concentrations of sucrose on the shoot formation of wasabi shoot tips cooled to −196 °C by vitrification. (Matsumoto 1997). Shoot tips were precultured on solidified 1/2 MS medium supplemented with various concentrations of sucrose for 1 day at 20 °C, then treated with a mixture of 2 M glycerol plus 0.4 M sucrose for 20 min at 25 °C before dehydration with PVS2 for 50 min at 0 °C. Sufficiently dehydrated shoot tips were plunged into LN. Each treatment contained four replications each of approximately 10 shoot tips. The *vertical bars* represent standard error

Table 1. Effects of preculturing and cryoprotective treatment on the shoot formation of wasabi shoot tips cooled to −196 °C by vitrification

Period of preculture (day)	Cryoprotective treatment	Shoot formation (% ± S.E.)
−	−	10.0 ± 1.4
−	+	73.3 ± 2.4
1	−	61.2 ± 2.7
1	+	100
3	−	25.1 ± 2.7
3	+	40.0 ± 2.7

Meristems following preculturing and cryoprotective treatment were dehydrated with PVS2 for 10 min at 25 °C and immersed in LN. Preculturing: 0.3 M sucrose at 20 °C; Cryoprotective treatment: a mixture of 2 M glycerol plus 0.4 M sucrose for 20 min at 25 °C. Each treatment contained four replications each of approximately 10 shoot tips.

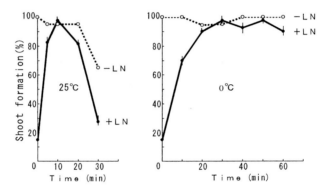

Fig. 2. Effect of time of exposure to PVS2 at 25 or 0 °C on the shoot formation of apical shoot tips cooled to −196 °C by vitrification (Matsumoto et al. 1994). Excised shoot tips were precultured with 0.3 M sucrose for 1 day and then treated with a mixture of 2 M glycerol and 0.4 M sucrose for 20 min at 25 °C. Shoot tips following preculture and cryoprotective treatment were dehydrated with PVS2 at 25 or 0 °C for various lengths of time before being immersed in LN (l). Each treatment contained four replications each of approximately 10 shoot tips. *Bar* represents standard error. Control (m): same as treated with PVS2 without cooling to −196 °C

To determine the optimum time of exposure to PVS2 at 25 or 0 °C, precultured, cryoprotected shoot tips were dehydrated with PVS2 for various lengths of time prior to a immersion in LN. Exposure to PVS2 for different lengths of time resulted in a variable extent of shoot formation (Fig. 2). The highest shoot formation was obtained with shoot tips treated with PVS2 for 10 min at 25 °C or for 30–50 min at 0 °C, respectively. Apical shoot tips treated with PVS2 for up to 20 min at 25 °C or for up to about 60 min at 0 °C without cooling in LN (control) retained high levels of shoot formation (about 90%).

Successfully vitrified and warmed shoot tips remained green continuously after plating, resumed growth in about 3 days and developed shoots within 2 weeks without intermediary callus formation. Fluorescence microscopic examination of longitudinal sections through the meristematic dome of vitrified shoot tips after 3 days of reculture revealed that, in most of the shoot tips,

Fig. 3.A Shoots 20 days after reculture (*bar* = 5 mm) and **B** plantlet 60 days after reculture (*bar* = 20 mm) developed from wasabi shoot tips cooled to –196 °C by vitrification. (Matsumoto et al. 1994)

Table 2. Effect of sucrose concentration in combination with 0.5 M glycerol on the preculturing of wasabi shoot tips cooled to –196 °C by vitrification

Sucrose conc. (M)	Shoot formation (% ± S.E.) Glycerol concentration (M)	
	0	0.5
0	3.3 ± 3.3	10.0 ± 5.8
0.1	20.0 ± 11.5	26.7 ± 8.8
0.3	50.0 ± 5.8	82.5 ± 4.8
0.5	33.3 ± 3.3	46.7 ± 12.0

Meristems were precultured on solidified 1/2 MS medium supplemented with various concentration of sucrose and glycerol for 1 day at 20 °C, then dehydrated with PVS2 for 10 min at 25 °C without cryoprotective treatment. Sufficiently dehydrated shoot tips were plunged into LN. Shoot formation (%): percent of shoot tips that produced normal shoots 21 days after reculture. Each treatment contained four replications each of approximately 10 shoot tips.

the domes appeared to be viable, based on FDA staining (Widholm 1972) (data not shown). Figure 3a shows shoots formed from the vitrified and cooled apical shoot tips after 20 days of reculture. A rooted plantlet developed from cryopreserved shoot tips by vitrification is shown in Fig. 3b. Almost all of the shoots formed roots on hormone-free solidified 1/2 MS medium and were successfully acclimated. No morphological abnormalities were observed in the plants developed from cryopreserved apical shoot tips of wasabi.

Shoot formation using the same vitrification procedure was compared for three other cultivars of wasabi. High levels of shoot formation were obtained for each cultivar tested (data not shown).

The optimum glycerol concentration in combination with 0.3 M sucrose for vitrification without cryoprotective treatment of a mixture of 2 M glycerol plus 0.4 M sucrose was determined. As shown in Table 2, the shoot tips precul-

Table 3. Effects of preculturing and cryoprotective treatment on the shoot formation of wasabi shoot tips cooled to −196 °C by vitrification

Preculture	Cryoprotective treatment	Shoot formation (% ± S.E.)
0.3 M suc	−	61.2 ± 2.7
0.3 M suc + 0.5 M gly	−	85.0 ± 2.9
0.3 M suc	+	97.5 ± 0.4

Cryoprotective treatment: Precultured shoot tips were treated with a mixture of 2 M glycerol plus 0.4 M sucrose for 20 min at 25 °C before exposure to PVS2 for 10 min at 25 °C. Each treatment contained four replications each of approximately 10 shoot tips.

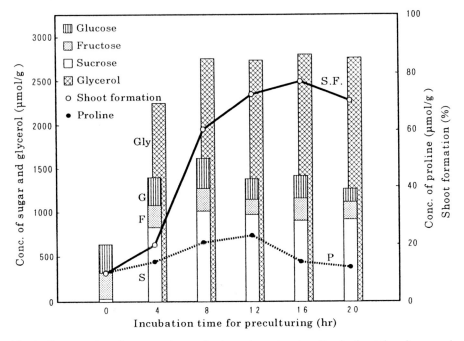

Fig. 4. Concentration of sucrose, glucose, fructose, glycerol and proline in shoot tips after preculture for different durations in 0.3 M sucrose plus 0.5 M glycerol (Matsumoto et al. 1998a). *SF* Shoot formation (%) of vitrified wasabi shoot tips with duration of preculture; *P* proline; *S* sucrose; *G* glucose; *F* fructose; *Gly* glycerol

tured with 0.3 M sucrose plus 0.5 M glycerol gave the greatest percentage of recovery. However, this percentage was about 10% lower than that obtained with the cryoprotective treatment of a mixture of 2 M glycerol plus 0.4 M sucrose (Table 3).

Sucrose, glucose, fructose and glycerol contents of shoot tips were determined after preculture with 0.3 M sucrose and 0.5 M glycerol. As shown in Fig. 4, contents of sugar (sucrose, glucose and fructose) and glycerol increased remarkably in the shoot tips after 4 h of preculture and reached a maximum after 8 h of preculture. It is very interesting that, during the preculture, proline gradually increased with time up to 12 h of preculture. An increase in proline was also observed in the preculture with 0.3 M sucrose alone (Matsumoto et al. 1998a). The extent of shoot formation of vitrified shoot tips significantly increased during 4 to 8 h of preculture and reached a maximum at a 16-h preculture. Thus, for wasabi shoot tips, a 16-h preculture is necessary to gain maximal shoot formation. The time lag between development of tolerance to vitrification and accumulation of sugar and glycerol during preculture may be due to the time required to penetrate and distribute into the meristematic cells after being taken up in the precultured apical shoot tips.

3.2 Encapsulation/Dehydration Method

To optimize the factors controlling the percentage of shoot formation of wasabi shoot tips by the encapsulation/dehydration technique, preculturing conditions, prior to encapsulation, were first examined. The highest percentage of shoot formation was obtained in the shoot tips precultured with 0.3 M sucrose (data not shown).

In the encapsulation/dehydration technique, resistance to dehydration and cooling was induced by preculturing encapsulated shoot tips in sucrose-enriched medium for 16–24 h. Thus, the effect of treatment with different concentrations of sucrose before dehydration was tested with shoot tips initially encapsulated into beads containing 0.4 M sucrose. An exposure to 0.8 M sucrose produced the greatest shoot formation after cooling in LN following dehydration (Table 4). The rates of shoot formation of dehydrated shoot tips at various water contents before and after immersion in LN are shown in Fig. 5. Shoot formation increased with decreasing water content and reached the highest rate (approx. 65%) at 18–20% water contents (Matsumoto et al. 1995a). Subsequently, the shoot formation decreased due to desiccation injury.

To increase further the dehydration tolerance, encapsulated shoot tips were treated with a mixture of 0.8 M sucrose plus different concentrations of glycerol for 16 h before air-drying. As shown in Table 5, a mixture of 0.8 M sucrose plus 1 M glycerol produced considerably higher levels of shoot for-

Table 4. Effect of sucrose treatment on shoot formation of encapsulated wasabi shoot tips cooled to −196 °C

Conc. of sucrose	Shoot formation (% ± S.E.)
Non-treated	0
0.6 M	17.5 ± 3.5
0.8 M	68.3 ± 10.4
1.0 M	25.0 ± 7.1

Precultured shoot tips were encapsulated with alginate gel beads with 0.4 M sucrose, then treated with various concentrations of sucrose for 16 hr at 20 °C before dehydration and cooling to −196 °C. Shoot formation (%): percent of shoot tips that produced normal shoots 21 days after reculture. Each treatment contained four replications each of approximately 10 shoot tips.

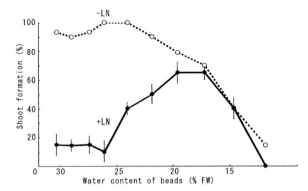

Fig. 5. Effect of water content on shoot formation of alginate-coated shoot tips of wasabi cooled to −196 °C (Matsumoto et al. 1995a). Shoot tips precultured on solidified 1/2 MS medium supplemented with 0.3 M sucrose for 1 day at 20 °C were encapsulated into alginate beads and then treated with 0.8 M sucrose solution for 16 h at 20 °C before air-drying. Each treatment contained four replications each of approximately 10 shoot tips. *Vertical bars* represent standard error. −LN means treated control without cooling to −196 °C

Table 5. Effect of glycerol concentration with 0.8 M sucrose on the shoot formation of wasabi shoot tips cooled to −196 °C by encapsulation/dehydration

Sucrose (M)	Glycerol (M)	Shoot formation (% ± S.E.)
0.8	0	64.9 ± 7.9
0.8	0.5	65.2 ± 2.9
0.8	1.0	79.1 ± 4.8
0.8	1.5	74.1 ± 9.8
0	1.0	0

Precultured shoot tips were encapsulated with alginate gel beads with 0.4 M sucrose, then treated with 0.8 M sucrose plus various concentrations of glycerol for 16 hr at 20 °C before dehydration for 7 hr and cooling to −196 °C. Shoot formation (%): percent of shoot tips that produced normal shoots 21 days after reculture. Each treatment contained four replications each of approximately 10 shoot tips.

mation (approx. 79% at 21–22% water content) than the shoot tips treated with 0.8 M sucrose alone, when treated for 16 h. Further treatment for up to 24 h produced lower levels of shoot formation. However, a mixture of 0.8 M sucrose in combination with EG or DMSO did not improve the rate of shoot formation due to toxic effects during dehydration (Matsumoto 1997).

3.3 Encapsulation/Vitrification Method

Shoot tips precultured with 0.3 M sucrose for 1 day were encapsulated in alginate gel beads with various cryoprotective solutions for 30 min at 25 °C before dehydration by PVS2. A mixture of 2 M glycerol plus 0.4 M sucrose produced

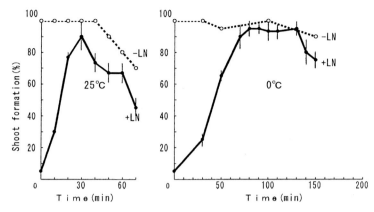

Fig. 6. Effect of time of exposure to PVS2 at 25 or 0 °C on the shoot formation of wasabi shoot tips cooled to –196 °C by encapsulation/vitrification (Matsumoto et al. 1995a). Excised apical shoot tips were precultured with 0.3 M sucrose for 1 day and then encapsulated with a mixture of 2 M glycerol and 0.4 M sucrose for 30 min at 25 °C. Then encapsulated shoot tips were dehydrated with PVS2 at 25 or 0 °C for various lengths of time before being immersed in LN. Each treatment contained four replications each of approximately 10 shoot tips. Vertical bars represent standard error. Control (m): same as treated with PVS2 without cooling to –196 °C

the greatest shoot formation (Matsumoto et al. 1995a). Effect of exposure to PVS2 at 25 or 0 °C on the shoot formation of encapsulated vitrified shoot tips cooled to –196 °C was investigated. The greatest shoot formation (95%–100%) was obtained in the shoot tips treated with PVS2 for 30 min at 25 °C or 80–100 min at 0 °C, respectively (Fig. 6). Finally, the effect of vitrification solution on shoot formation of encapsulated vitrified shoot tips cooled to –196 °C was tested. The greatest percentages of shoot formation were obtained in encapsulated shoot tips dehydrated with PVS2 and PVS3 (Nishizawa et al. 1993). However, EG-based vitrification solutions yielded lower rates of shoot formation than that of glycerol-based solutions. In the case of encapsulated and vitrified shoot tips (5 mm in diameter), this shoot formation after cooling to –196 °C was about 65%, which was lower than that of 3 mm beads due to crystallization of beads (data not shown). Crystallization was observed in cryopreserved beads during thawing in a water bath at 40 °C. Successfully encapsulated, vitrified and warmed shoot tips remained green continuously 3 days after plating, and developed shoots within 1 week without intermediary callus formation. Almost all the shoots formed roots on hormone-free solidified 1/2 MS medium and were successfully acclimated. No morphological abnormalities were observed in the plants developed from cryopreserved apical shoot tips by encapsulation/vitrification.

3.4 Comparison Among Cryogenic Procedures

Shoot formation of wasabi shoot tips cooled to –196 °C was compared with three different cryogenic procedures. Vitrification and encapsulation/vitrification produced much higher levels of shoot formation than that of encapsula-

Table 6. Shoot formation, shoot length and time used for dehydration of wasabi shoot tips cooled to −196 °C by different cryogenic protocols

Cryogenic protocol	Shoot formation (% ± S.E.)	Shoot length (mm)	Time used for dehydration (min)
Vitrification[a]	97.5 ± 1.0	10.6 ± 4.0	10 at 25 °C
Encapsulation/dehydration[b]	67.1 ± 8.9	6.3 ± 3.6	420 at 25 °C
Revised Encapsulation/ dehydration[b]*	79.1 ± 4.8	9.0 ± 2.6	420 at 25 °C
Encapsulation/vitrification[a]	96.7 ± 2.9	12.2 ± 3.6	100 at 0 °C

[a] Precultured meristems were treated with a mixture of 2 M glycerol plus 0.4 M sucrose at 25° C for 20 min, and then dehydrated with PVS2 before cooling in LN.
[b] Precultured meristems were encapsulated with alginate gel beads, and then treated with 0.8 M sucrose or plus 1 M glycerol* at 20 °C for 16 hr before dehydration and cooling to −196 °C. Shoot formation (%): percent of meristems that produced normal shoots 21 days after reculture.

tion/dehydration (Table 6). In addition, in the vitrified shoot tips with or without encapsulation, shoot formation and shoot growth after reculture were faster than those of the encapsulated dried shoot tips. The time used for dehydration was greatly decreased in the encapsulation/vitrification compared with that of encapsulation/dehydration.

4 Discussion

For successful cryopreservation, it is essential to avoid lethal intracellular freezing, which occurs without cryopreservation during rapid cooling using LN (Sakai and Yoshida 1967). Thus, in any cryogenic procedure, cells and shoot tips have to be sufficiently dehydrated to avoid intracellular freezing and to be vitrified upon rapid cooling in LN. Vitrification refers to the physical process by which a highly concentrated cryoprotective solution supercools to very low temperatures and eventually solidifies into a metastable glass without undergoing crystallization at the glass transition temperature (Fahy et al. 1984). In the vitrification method, the direct exposure of cells and shoot tips to PVS2 causes harmful effects due to osmotic stress and chemical toxicity. Thus, a successful cryopreservation by vitrification requires careful control of the dehydration procedure with PVS2. In addition, it is necessary to increase the dehydration tolerance of cells and shoot tips to be cryopreserved by preconditioning, such as preculture and cryoprotective treatment, before dehydration. The injurious effects caused by direct exposure to PVS2 can be eliminated or reduced by optimizing the exposure time, adding a gradual amount of PVS2, or a gradual dehydration process followed by dehydrating specimens at 0 °C. For vitrification, a cryoprotective treatment of a mixture of 2 M glycerol plus 0.4 M sucrose appeared promising as a cryoprotectant for enhancing dehydration tolerance of shoot tips of wasabi (Matsumoto et al. 1994), lily (Matsumoto et al. 1995b), statice (Matsumoto et al. 1998b), grape (Matsumoto and Sakai 2000), subtropical *Diospyros* (Matsumoto et al. 2000), hairy roots of horseradish (Phunchindawan 1997) and *Panax gingeng* (Yoshimatsu et al. 1996) and shoot primordia of statice (Matsumoto et al. 1997). However, we do not know the mechanism by which the treatment of glycerol plus sucrose enhances dehydration tolerance to PVS2. It may be supposed that the increase of sugar and glycerol along with proline in the cells during preculture (equivalent to about 0.8 M sucrose solution based on concentrations that initially induce plasmolysis) mitigates not only excess osmotic stress by PVS2, but also positively enhances the dehydration tolerance by increasing stability of membranes and modifying cell structures under conditions of severe dehydration (Crowe et al. 1987, 1988; Steponkus et al. 1992). Understanding cryo-tolerance is very important for the further development of cryobiology and its application. The natural cryoprotective mechanisms present in many seeds and plants may provide a new source of pretreatment chemicals (Close et al. 1989; Skriver and Mundy 1990; Luo and Reed 1997).

In the present study, the injurious effects mentioned above were effectively overcome by preculturing excised shoot tips on sucrose-enriched medium for 1 day (a significant increase in cell sap concentration of up to about 0.6 M), followed by a cryoprotective treatment with a mixture of 2 M glycerol plus 0.4 M sucrose for 20 min at 25 °C (note: the cells of shoot tips plasmolysed considerably, but glycerol and sucrose did not penetrate into the cytosol for 20 min as observed through a cytosolic volume change) before dehydration with PVS2 (the plasmolysis proceeds intensively). During preculture on sucrose-enriched medium for approximately 1 day, sugar and proline were greatly increased in the shoot tips (Matsumoto et al. 1998a), which, in turn, may have enhanced the stability of membranes under conditions of severe dehydration (Crowe et al. 1987). Additionally, Reinhoud et al. (1995) succeeded in the cryopreservation of cultured tobacco cultured cells by vitrification. It was clearly demonstrated that the tolerance of tobacco cells precultured with 0.3 M mannitol and exposed to PVS2 for 1 day was the combined result of the cell's responsiveness to mild osmotic stress caused by preculturing; in particular, the production of ABA, the accumulation of mannitol during preculture, proline and certain proteins including late embryogenesis abundant (LEA).

For the many herbaceous plants tested, overnight preculture of excised shoot tips with 0.3 M sucrose appeared to be inconsequential for producing a high level of recovery growth by vitrification. It was further demonstrated that cryoprotective solution was very effective in increasing tolerance to freeze-dehydration down to −30 °C and to dehydration using PVS2 (Sakai et al. 1991; Nishizawa et al. 1992). The protective effect of a mixture of 2 M glycerol plus 0.4 M sucrose in the cell's peri-protoplasmic space may be due to mitigation of a large osmotic stress by severe dehydration with PVS2 in addition to some mechanism of action that minimizes the injurious membrane changes from severe dehydration (Crowe et al. 1988; Steponkus et al. 1992). Plasmolysis might reduce the generation of mechanical stress on plasma membranes, which might be produced by deformation of the cell wall during extracellular freezing (Jitsuyama et al. 1997).

In our experiments, precultured and cryoprotected shoot tips were dehydrated with PVS2 at 25 °C for 10 min or at 0 °C for 50 min (Fig. 1), and survived subsequent rapid cooling and rewarming during the vitrification procedure with only a slight additional decrease in survival. Thus, it can be postulated that the shoot tips acquired dehydration tolerance to the PVS2 application and that optimizing the exposure time improves tolerance to cryopreservation by vitrification. These results suggest that under well-optimized conditions of cryogenic procedures, acquisition of PVS2 tolerance is sufficient for specimens to survive cryopreservation by vitrification.

In the encapsulation/dehydration technique, osmotic extraction of water by sucrose exposure results in progressive dehydration; additional loss of water is obtained by evaporation and subsequent increase of sucrose concentration in the beads. The sucrose molarity increased markedly during the drying process and reached or exceeded the saturation point of the sucrose solution resulting in glass transition during cooling to −196 °C (Dereuddre et al. 1991). In this technique, induction of dehydration tolerance and deep freezing is induced by preculturing encapsulated shoot tips with 0.8 M sucrose and eliminates the use of other cryoprotectants such as DMSO, EG and glycerol (Dereuddre et al. 1990; Fabre and Dereuddre 1990). We have confirmed that a mixture of glycerol in combination with 0.8 M sucrose produces much higher levels of shoot formation in the encapsulated dried shoot tips of lily (Matsumoto et al. 1995b), wasabi (Matsumoto 1997) and statice (Matsumoto et al. 1998b) than those treated with 0.8 M sucrose alone. Thus, glycerol contributes to minimizing injurious changes, probably membrane, resulting from severe dehydration. However, EG and DMSO in combination with sucrose were toxic during the dehydration process. Only glycerol in combination with 0.8 M sucrose did not show any toxic effects after severe dehydration by airdrying; rather it increased considerably the dehydration tolerance. The encapsulation/dehydration technique was successfully applied to a wide range of materials (Niino and Sakai 1992; Suzuki et al. 1994). However, lower levels of survival and later recovery growth were observed when compared with shoot tips cryopreserved by vitrification (Matsumoto and Sakai 1995). In the former technique, encapsulated shoot tips were treated with 0.8 M sucrose for 16 h to induce dehydration tolerance before air-drying (Fabre and Dureddre 1990). Overnight treatment with 0.8 M sucrose produced a much lower level of recovery growth (approx. 65%) than those of vitrified shoot tips of wasabi (approx. 95%) with or without encapsulation. Thus, treatment with 0.8 M sucrose alone appears to be insufficient to produce a higher level of recovery growth. In the revised encapsulation/dehydration technique, the recovery growth was significantly improved (from 65–80%) provided it included treatment with a mixture of 0.8 M sucrose plus 1 M glycerol.

The encapsulation/dehydration technique is easy to handle and simplifies the dehydration process; however, it is laborious and time consuming when compared with the vitrification method. In the vitrification method, it is difficult to treat carefully a large number of shoot tips at the same time. Therefore, we developed an encapsulation/vitrification method (Matsumoto and

Sakai 1995). With this method is easy to handle and treat a large number of shoot tips at the same time. Furthermore, this encapsulation/vitrification method may be more suitable for small specimens, such as hairy roots and callus. This method was also applied for shoot tips of lily (Matsumoto et al. 1996), statice (Matsumoto et al. 1998b), strawberry (Hirai et al. 1998) and shoot primordia of horseradish (Phunchindawan 1997). The recovery growth of encapsulation/vitrification is at the same level as that of vitrification, and much earlier than when using encapsulated air-dried shoot tips of wasabi. The same results were observed when comparing shoot formation after immersion in LN among vitrification, encapsulation/dehydration and encapsulation/vitrification for shoot tips of lily (Matsumoto et al. 1996) and statice (Matsumoto et al. 1998b).

The vitrification method significantly decreased the time used for dehydration and simplified the cryogenic procedures. More recently, the vitrification method was successfully applied to about 20 tropical monocotyledonous plants (Thinh 1997). Thus, the vitrification method seems promising for the cryopreservation of shoot tips and somatic embryos.

5 Summary and Conclusions

To establish an effective and reliable protocol for cryopreservation of in vitro grown shoot tips of wasabi, different cryogenic protocols were tested. They were: vitrification, encapsulation/dehydration (E/D) and encapsulation/vitrification (E/V). These methods were compared for the recovery growth after cryopreservation. The rate of shoot formation was highest in vitrification and E/V. The lowest was in E/D. However, when encapsulated shoot tips were treated with a mixture of 0.8 M sucrose and 1 M glycerol before dehydration, the shoot formation was increased about 15% over the former method. These different cryogenic procedures were also applied for other cultivars of wasabi. Vitrified shoot tips with and without encapsulation produced shoots much earlier than the encapsulated dried shoot tips.

It can thus be concluded that the most promising cryogenic procedure of wasabi shoot tips appears to be vitrification in terms of both its high recovery growth and the simplicity for the procedure.

References

Bapat VA, Mhatre M, Rao PS (1987) Propagation of *Morus indica* L. (Mulberry) by encapsulated shoot buds. Plant Cell Rep 6:393–395

Close TJ, Kortt AA, Chandler PM (1989) A cDNA-based comparison of dehydration-induced proteins (dehydrins) in barley and corn. Plant Mol Biol 13:95–108

Crowe JH, Crowe JF, Carpenter LM, Wistrom CA (1987) Stabilization of dry phospholipid bilayers and proteins by sugars. Biochem J 242:1–10

Crowe JH, Crowe JF, Carpenter LM, Rudolph AS, Wistrom CA, Spargo BJ, Anchordoguy TJ (1988) Interaction of sugars with membranes. Biochem Biophys Acta 947:367–384

Dereuddre J, Scottez C, Arnaud Y, Duron M (1990) Resistance of alginate-coated axillary shoot tips of pea tree (*Pyrus communis* L. cv. Beurre Hardy) in vitro plantlets to dehydration and subsequent freezing in liquid nitrogen: effects of previous cold hardening. CR Acad Sci Paris 310 (Ser III):317–323

Dereuddre J, Blandis S, Hassen N (1991) Resistance of alginate-coated somatic embryos of carrot (*Daucus carota* L.) to desiccation and freezing in liquid nitrogen. 1. Effects of preculture. Cryo Lett 12:125–134

Fabre J, Dereuddre J (1990) Encapsulation-dehydration: a new approach to cryopreservation of *Solanum* shoot tips. Cryo Lett 11:413–426

Fahy GM, MacFarlane DR, Angell CA, Meryman HT (1984) Vitrification as an approach to cryopreservation. Cryobiology 21:407–426

Hirai D, Shirai K, Shirai S, Sakai A (1998) Cryopreservation of in vitro-grown shoot tips of strawberry by encapsulation/vitrification. Euphytica 101:109–116

Jitsuyama Y, Suzuki T, Harada T, Fujikawa S (1997) Ultrastructural study on mechanism of increased freezing tolerance due to extracellular glucose in cabbage cells. Cryo Lett 18:33–44

Langis R, Schnabel-Preikstas BJ, Earle FD, Steponkus PL (1990) Cryopreservation of carnation shoot tips by vitrification. Cryobiology 27th Annual Meeting Abstract, pp 657–658

Luo J, Reed B (1997) Abscisic acid-responsive protein, bovine serum albumin, and proline pretreatments improve recovery of in vitro currant shoot-tip shoot tips and callus cryopreserved by vitrification. Cryobiology 34:240–250

Matsumoto T (1997) Cryopreservation of in vitro-grown apical meristems of wasabi (*Wasabia japonica*). PhD Thesis, Department of Agronomy, Kobe University, Kobe

Matsumoto T, Nako Y (1999) Effect of dimethyl sulfoxide on in vitro storage of wasabi meristems at low temperature. Plant Biotech 16:243–245

Matsumoto T, Sakai A (1995) An approach to enhance dehydration tolerance of alginate-coated dried meristems cooled to −196 °C. Cryo Lett 16:299–306

Matsumoto T, Sakai A (2000) Cryopreservation of grape in vitro-cultured axillary shoot tips by three-step vitrification. In: Engelmann F, Takagi H (eds) Cryopreservation of tropical germplasm. Current research progress and application. IPGRI, Rome, pp 266–267

Matsumoto T, Sakai A, Yamada K (1994) Cryopreservation of in vitro-grown apical meristems of wasabi (*Wasabia japonica*) by vitrification and subsequent high plant regeneration. Plant Cell Rep 13:442–446

Matsumoto T, Sakai A, Takahashi C, Yamada K (1995a) Cryopreservation of in vitro-grown apical meristems of wasabi (*Wasabia japonica*) by encapsulation-vitrification method. Cryo Lett 16: 189–206

Matsumoto T, Sakai A, Yamada K (1995b) Cryopreservation of in vitro-grown apical meristems of lily (*Lilium japonicum*) by vitrification. Plant Cell Tissue Org Cul 41:231–241

Matsumoto T, Sakai A, Takahashi C, Yamada K (1996) Cryopreservation in vitro-grown apical meristems of lily (*Lilium* L.) by encapsulation-vitrification method. Plant Tissue Cult Lett 13: 29–34

Matsumoto T, Nako Y, Takahashi C, Sakai A (1997) Induction of in vitro cultured masses of shoot primordia of hybrid statice and its cryopreservation by vitrification. HortScience 32:309–311

Matsumoto T, Sakai A, Nako Y (1998a) A novel preculturing for enhancing the survival of in vitro-grown meristems of wasabi (*Wasabia japonica*) cooled to −196 °C by vitrification. Cryo Lett 19:27–36

Matsumoto T, Takahashi C, Sakai A, Nako Y (1998b) Cryopreservation of in vitro-grown apical meristems of hybrid statice by three different procedures. Sci Hort 76:105–114

Matsumoto T, Mochda K, Itamura H (2000) Cryopreservation of dormant shoot tips of *Diospyros* by vitrification. Abstracts of the International Conference on Science and technology for managing plant genetic diversity in the 21st century, p 30

Murashige T, Skoog F (1962) A revised medium for rapid growth and bioassays with tobacco tissue cultures. Physiol Plant 15:473–497

Niino T, Sakai A (1992) Cryopreservation of alginate-coated in vitro shoot tips of apple, pear and mulberry. Plant Sci 87:199–206

Nishizawa S, Sakai A, Amano Y, Matsuzawa T (1992) Cryopreservation of asparagus (*Asparagus officinalis* L. Osb.) embryogenic cells and subsequent plant regeneration by a simple freezing method. Cryo Lett 13:379–388

Nishizawa S, Sakai A, Amano Y, Matsuzawa T (1993) Cryopreservation of asparagus (*Asparagus officinalis* L.) embryogenic suspension cells and subsequent plant regeneration by vitrification. Plant Sci 91:67–73

Ohtsuru M, Kawatani H (1979) Studies on the myrosinase from *Wasabia japonica*: purification and some properties of wasabi myrosinase. Agric Biol Chem 43:2249–2255

Phunchindawan M (1997) Cryopreservation of useful plant resources: Application of encapsulation-dehydration method to preservation of hairy root cultures and microalgae. PhD Thesis, Osaka University, Faculty of Pharmaceutical Sci, Osaka

Reinhoud PJ, Schrijnemakers EWM, Iren F, Kijne JW (1995) Vitrification and a heat-shock treatment improve cryopreservation of tobacco cell suspension compared to two-step freezing. Plant Cell Tissue Org Cult 42:261–267

Sakai A (1997) Potentially valuable cryogenic procedures for cryopreservation of cultured plant shoot tips. In: Razdan MK, Cocking EC (eds) Conservation of plant genetic resources in vitro. Science Publishers, New Hampshire, pp 53–66

Sakai A, Yoshida S (1967) Survival of plant tissue at super-low temperature. Effects of cooling and rewarming rates on survival. Plant Physiol 42:1695–1701

Sakai A, Kobayashi S, Oiyama I (1990) Cryopreservation of nucellar cells of navel orange (*Citrus sinensis* Osb. var. *brasiliensis* Tanaka) by vitrification. Plant Cell Rep 9:30–33

Sakai A, Kobayashi S, Oiyama I (1991) Cryopreservation of nucellar cells of navel orange (*Citrus sinensis* Osb.) by a simple freezing method. Plant Sci 74:243–248

Skriver K, Mundy J (1990) Gene expression in response to abscisic acid and osmotic stress. Plant Cell 2:503–512

Steponkus PL, Langis R, Fujikawa S (1992) Cryopreservation of plant tissues by vitrification. In: Steponkus PL (ed) Advances in low temperature biology, vol 1. JAI Press, London, pp 1–61

Suzuki M, Niino T, Akiyama T (1994) Cryopreservation of shoot tips of kiwifruit seedlings by the alginate encapsulation-dehydration technique. Plant Tissue Cult Lett 11:122–128

Taniguchi M, Nomura R, Kamihara M, Kijima I, Kobayashi T (1988) Effective utilization of horseradish (*Armoracia Iapathifolia*) and wasabi (*Wasabia japonica*) by treatment with supercritical carbon dioxide. J Ferment Bioeng 66:347–353

Thinh NT (1997) Cryopreservation of germplasm of vegetatively propagated tropical monocots by vitrification. PhD Thesis, Department of Agronomy, Kobe University, Kobe

Widholm JM (1972) The use of fluorescein diacetate and phenosafranine for determining viability of cultured plant cells. Stain Tech 47(4):189–194

Yamada K, Haruki K (1992) Mass propagation of wasabi (*Wasabia japonica* Matsumura) through shoot apex culture. Bull Shimane Agric Exp Stn 26:85–95

Yoshimatsu K, Yamaguchi H, Shimomura K (1996) Traits of *Panax ginseng* hairy roots after cold storage and cryopreservation. Plant Cell Rep 15:555–560

Section III
Woody Species

III.1 Cryopreservation of Somatic Embryos from *Aesculus hippocastanum* L. (Horse Chestnut)

Z. Jekkel[1,2], J. Kiss[1], G. Gyulai[1], E. Kiss[1], and L.E. Heszky[1]

1 Introduction

1.1 Origin and Importance of Horse Chestnut

Horse chestnut is the common name for *Aesculus hippocastanum*. The family *Hippocastanaceae* contains about 15 species in two genera, *Aesculus* and *Billia*, and occurs in the North Temperate Zone. Of the 15 species, one is European, and the others are native to Eurasia and North America. The trees are dicots, characterized by large winter buds covered with resinous, sticky scales; opposite, palmately compound, 10–25 cm long leaves with 5–7 obovate leaflets; large clusters of attractive yellow, red or whitish irregular flowers of four or five petals; and bark that exfoliates in gray plates to show orange-brown inner bark. The fruits are leathery, three-valved capsules containing large, brown seeds (Fig. 1). Many members of the genus *Aesculus*, commonly called buckeyes, are popular ornamental and shade trees.

The economic and ornamental importance of *Billia*, with only two species (*B. colombiana* and *B. hippocastanum*), is modest compared with *Aesculus*. The most economically important species are *Aesculus hippocastanum*, *A.* × *carnea*, *A. octandra*, *A. pavia* and *A. parviflora*.

Aesculus hippocastanum was native to the whole of Europe before glacial times. Its present "actual area" has been significantly narrowed to Northern Greece, Albania and Bulgaria. It is typically a highland species, which for instance in Northern Greece grows at 1000–1300 m above sea level in mixed populations together with *Abies*, *Tilia* and *Quercus* species. Horse chestnut was first brought to Hungary and Austria from Turkey by Clusius in 1576, and spread very fast throughout Europe due to its ornamental value. By the end of the 19th century, the growing area of *Aesculus* extended as far as 60° latitude. Nowadays, horse chestnut is the most frequently planted ornamental tree in central Europe due to its high tolerance to air pollution and wide phenotypic variation.

Horse chestnut is a fast-growing species with significant wood production; however, due to the unfavorable trunk morphology and wood quality, its importance in forestry and wood industry is low. Nevertheless, it is present in

---

[1] Department of Genetics and Plant Breeding, St. István University, 2103 Gödöllö, Hungary
[2] Aventis CropScience Hungary, 1036 Budapest, Hungary

Biotechnology in Agriculture and Forestry, Vol. 50
L.E. Towill and Y.P.S. Bajaj (Eds.) Cryopreservation of Plant Germplasm II
© Springer-Verlag Berlin Heidelberg 2002

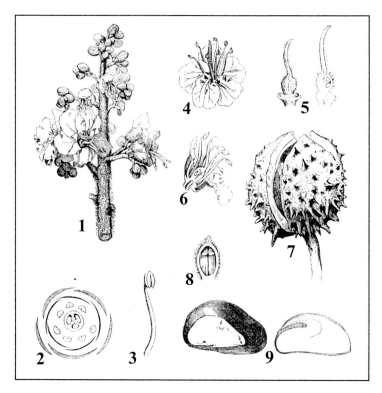

Fig. 1. Flower and seed morphology of *Aesculus hippocastanum* L. *1* Inflorescence; *2* cross section of the flower bud; *3* stamen; *4* opened flower; *5* pistil, entire and cut vertically; *6* flower cut vertically; *7* dehiscing three-valved capsule with seed; *8* sterile cell, open; *9* seed, entire and cut vertically

Hungarian forests and planted as an ornamental and shade tree along roads and in small populations for wild forage.

Under optimal conditions, the maximal height and cutting diameter can reach 30 and 1 m, respectively. Its maximum lifetime is around 150 years in central Europe, but when it reaches the age of about 100 years, crown-tip drying occurs frequently. It is mostly propagated by seeds, as its ability to reproduce from suckers is low.

Various parts of the tree can be used in pharmacology; these include bark and fruit (tonic, anti-odemic, anti-inflammatory, astringent; venous vasoconstrictor; febrifuge), seeds (tonic and decongestant) and pericarp (peripheral vasoconstrictor).

The fruits contain 60–70% starch, 5–10% sucrose and approx. 10% saponins. After extracting the saponins, seeds can be used for forage or alcohol production. The bark of the tree contains gallic and tannic acid, which are used for dying (black) silk and cotton. The soft wood of horse chestnut and buckeyes is used for paper pulp and woodenware and in carpentry.

1.2 Genetics and Breeding of *Aesculus* Species

The species are grouped into five series. Some are shrubby, while others form large trees. Most have large, showy flowers and are useful as ornamentals, but they are rarely managed for timber. Genetic work has been confined to the study of natural hybrids.

One natural hybrid (*A.* × *carnea* = *A. hippocastanum* × *A. pavia*) is between species belonging to different series. This particular hybrid is a true breeding allotetraploid. It presumably arose as a result of pollination of an unreduced egg cell by an unreduced pollen grain, a very rare occurrence in trees.

The other reported hybrids involve four American species having neighboring and slightly overlapping ranges and belonging to the same series. The four species could be combined in six different ways, and five of these combinations have been reported. In only one combination (*A. pavia* × *A. sylvatica*) has there been extensive introgression. In the other cases, hybrids were rare and reported only from arboreta, or hybrid swarms were small and confined to the immediate zone of overlap or to disturbed habitats. Apparently, more extensive introgression has been prevented by sterility in hybrids and selection against hybrids on anything but an intermediate habitat (Wright 1976).

While rare genetic variation can be observed in the natural habitat, a number of cultivars have been produced by breeding. One of the most popular varieties is *A. hippocastanum* 'Baumanni', with reduced height and pure white, mostly sterile double flowers blooming in late May. Having almost no fruit, it is not exposed to the damage caused by seed harvesting, and does not endanger road traffic. There are significant differences in flushing and blooming time among individual trees.

1.3 Pests and Diseases

One of the most threatening recent pests of *Aesculus hippocastanum* in Hungary is the horse chestnut leaf miner (*Cameraria ohridella*), which attacks in the early summer and eventually may cause lethal damage to populations. There are only a few other diseases and pests that are specific to *Aesculus*. *Anthracnose* (fungus) can cause brown blotches along the veins of the leaves. These spots expand and eventually cover most of the leaf area, and, finally, the leaves drop. Due to leaf scorch and leaf blotch, leaves turn yellow or reddish-brown at the edges and between the veins. These leaves wither, die and drop. On trees in woods, and sometimes in parks and avenues, leaves are often skeletonized by hairy caterpillars such as *Noctua aceris* and *Geometra aescularia*. In cities with hot and dry climates, aphids in large populations can attack trees, causing early browning and dropping of leaves. In natural horse-chestnut stands, *Septoria aesculi* may cause brown leaf spots. Older or damaged trees are susceptible to putrefactive mushrooms such as *Polyporus sqamosus*.

2 Cryopreservation

2.1 Cryopreservation of Horse Chestnut

The large seeds of the horse chestnut have a high moisture content and therefore do not survive well under conventional long-term seed storage (Bajaj 1995). Cryopreservation in liquid nitrogen (LN) can minimize damaging events, but limited work has been done on cryopreservation of horse-chestnut seeds. Jörgensen (1990) found that freezing horse-chestnut embryos in liquid nitrogen after cryoprotective treatment in dimethyl sulfoxide (DMSO) and sucrose followed by slow prefreezing (0.5–1.0 °C/min) to –40 or –50 °C before immersion in LN was suitable for storage. A similar pretreatment and cooling regime resulted in the survival of somatic embryos of *Citrus sinensis* (Marin and Duran-Vila 1988). In these experiments using cryoprotectants, cell dehydration was accomplished through freeze-desiccation in which the cooling rate was controlled precisely by a programmable biological freezer. Recently, cell dehydration has been achieved by desiccation in air (from 20 min to several hours), in which embryos are held in the air current of a laminar flow cabinet. Assy-Bah and Engelmann (1992) cryopreserved embryos of *Cocos nucifera* by desiccation and cryoprotective treatments before direct immersion in LN. However, there are many papers reporting success without cryoprotective treatment, in which embryos were plunged into LN immediately after desiccation. This has been reported for *Elaeis guineensis* (Dumet et al. 1993), *Fraxinus excelsior* (Brearley et al. 1995), *Musa* sp. (Abdelnour-Esquivel et al. 1992a), *Coffea* sp. (Abdelnour-Esquivel et al. 1992b; Normah and Vengadasalam 1992; Hatanaka et al. 1994), *Corylus avellana* (Gonzalez-Benito and Perez 1994) and *Vigna sesquipedalis* (Normah and Vengadasalam 1992).

Although there is evidence for heat shock induced freeze- and desiccation-tolerance in different plant tissues (Collins et al. 1993; Jennings and Saltveit 1994), its effect on the cryotolerance of cellular systems has not been studied extensively. The viability of rapidly frozen *Saccharomices cerevisiae* increased 20–30-fold when heat shock was incorporated in the cryopreservation protocol (Kaul et al. 1992). Similarly, heat shock treatments before a direct immersion in LN increased the survival rate of tobacco cells (Reinhoud et al. 1995).

In this review, cryopreservation methods for horse-chestnut somatic embryos are outlined. In order to obtain as high a recovery rate by rapid cooling as by slow cooling, several experiments were conducted to compare the viability of horse-chestnut somatic embryos after different cryopreservation methods, such as: cryopreservation with cryoprotectants at slow and fast cooling rates, cryopreservation by desiccation at a fast cooling rate and cryopreservation by heat shock treatments at slow and fast cooling rates.

2.2 Tissue Culture

Somatic embryos used for cryopreservation were induced from different explants. These were initiated, maintained, propagated and converted to plants using the methods described below.

2.2.1 Zygotic Embryo Culture

Developing fruits (200 explants/tree) were collected from the central part of the crown of four horse-chestnut trees in the Botanical Garden of the St. Istvàn University (Gödöllö) 1 month after flowering. The explants were surface-sterilized and the proembryos in globular developmental stage (1–2 mm in size) were isolated and placed on nutritive medium. The semisolid MS basal medium was supplemented with 2,4-D (4.4 µM), NAA (5.4 µM), sucrose (3%) and coconut milk (8%). The cultures were incubated at 26–28 °C with a light intensity of 44 µmol m^{-2}s^{-1} and with a 16-/8-h photoperiod.

2.2.2 Anther and Filament Culture

To induce embryogenic callus and somatic embryos, green flower buds (2–3 mm in size) were isolated, and after sterilization, anthers and filaments of 1–2 mm were excised. These were cultured on a semisolid MS basal medium, supplemented with 2,4-D (4.4 µM), NAA (5.4 µM), sucrose (3%) and coconut milk (8%). Cultures were kept in the dark at 28 °C. Somatic embryos developing directly from filament tissue were transferred to semisolid B5 medium (Gamborg et al. 1968) containing coconut milk (8%), BA (4.4 µM), NAA (2.7 µM) and sucrose (3%).

Calli derived from filaments were transferred to semisolid and liquid B5 medium supplemented with BA (8.8 µM), coconut milk (7%) and sucrose (3%). They were incubated in light (44 µ mol m^{-2}s^{-1}) with a 16-/8-h photoperiod at 26 °C.

2.2.3 Secondary Embryogenesis

To induce secondary embryogenesis, somatic embryos (2–10 mm in size) were transferred to MS medium containing NAA (2.7 µM), BA (2.2 µM) and sucrose (3%). One week after transfer, several secondary somatic embryos started to develop at the radicula of the cultured somatic embryos. These reached the globular stage on the 10th day (0.5–1 mm diameter) and the torpedo stage on the 15th day. On average, 20–30 secondary somatic embryos developed at one primary embryo pole. The developing secondary somatic embryos were transferred to B5 medium, supplemented with BA (2.2 µM), NAA (5 µM), coconut milk (5%) and sucrose (3%). The medium for plantlet regeneration from secondary somatic embryos was E1 (Gamborg et al. 1983), supplemented by BA (44 μM) and sucrose (3%). The other culture conditions were identical to those used in the embryo and callus cultures.

The organogenesis of secondary somatic embryos was analyzed using the scanning electron microscopy (SEM) technique.

2.2.4 Plant Regeneration

To induce plant regeneration, both somatic and secondary somatic embryos were transferred to culture tubes containing growth-regulator free MS and WPM medium with 1/2 amount of macroelements. The culture conditions were identical to those used in the embryo and callus cultures. After germination, plantlets (4–5 cm) were planted to pots and transferred to the greenhouse.

2.3 Cryopreservation Protocol

2.3.1 Preconditioning of Secondary Embryo Cultures for Cryopreservation

Adventitiously developing embryos were maintained in liquid E1B5 (Kiss et al. 1992) medium containing E1 (Gamborg et al. 1983), macro-, B5 (Gamborg et al. 1968) microelements, 5 μM NAA, 4.4 μM BA and 3% sucrose. Cultures were incubated in Erlenmeyer flasks (100 ml) on a rotary shaker (120 rpm) under a photoperiod of 16 h light/8 h dark with a light intensity of 44 μ mol m^{-2}s^{-1} at 26 °C and transferred weekly.

For cryopreservation, secondary embryos at the globular developmental stage (1.5–3.0 mm in diameter) were harvested on the 4th day of the subculture cycle. Pretreatment was carried out on a solid E1B5 medium containing filter-sterilized abscisic acid (ABA; analytical grade, mixed isomers, Sigma) (0.75, 7.5, 75.0 μ M). The embryos were cultured for 4 days on this medium (40 embryos/9-mm Petri dish, with 30 ml medium). ABA pretreatment was identical for each experiment; however, in the experiment where cryopreservation was preceded by heat shock treatments, only 0.75 μM ABA was used.

2.3.2 Cryoprotectant Treatments

Somatic embryos pretreated on nutritive medium containing ABA (0.75, 7.5 and 75 μM) were placed into 2-ml cryovials (20 embryos/vial) and 600 μ l cryoprotectant mixture (1.0 M sucrose, 0.5 M glycerol, 0.5 M DMSO) were added to the embryos. Embryos were incubated for 1 h in an ice bath. Slow cooling was carried out by holding samples for 10 min at 0 °C, cooling to –35 °C at 1 °C/min, holding for 30 min at this temperature and immersion in liquid nitrogen. In addition to the slowly cooled samples, additional vials with embryos in cryoprotectant at 0 °C were rapidly cooled by plunging directly into LN.

2.3.3 Air Desiccation Treatment

Somatic embryos pretreated on a medium containing ABA (0.75 µM, 7.5 and 75 µM) were transferred onto the surface of one layer of dry filter paper in open, empty Petri dishes (40 embryos/9-mm Petri dish) and desiccated in the air current of a laminar flow cabinet for 2, 3 and 4 h. These desiccated embryos were then packed in 2-ml cryovials and immersed directly in LN without cryoprotective treatment. The water content of the desiccated embryos was calculated according to weight before and after drying in an oven at 104 °C for 48 h.

2.3.4 Heat Shock Treatment

Somatic embryos preconditioned in Erlenmeyer flasks in 0.75 µM ABA containing nutritive medium in the culture were subjected to a heat shock treatment for 30–90 min immediately before freezing, with the immersion of the flasks in a water bath at 40 °C. After heat incubation, the somatic embryos were placed into 2-ml cryovials (20 embryos/vial) and filled with 600 µl cryoprotectant mixture (1.0 M sucrose, 0.5 M glycerol, 0.5 M DMSO), then cooled slowly or rapidly as described in Section 2.3.2.

2.3.5 Post-Thaw Treatments and Embryo Recovery

After cooling, cultures were stored in LN for 3–5 days, then thawed in a 40 °C water-bath until complete melting (40–50 s) had taken place. For recovery, embryos were transferred onto filter paper discs placed on solid E_1B_5 medium. These cultures were incubated in the dark at 25 °C for 2 weeks, and the filter paper discs with embryos were transferred to new medium on the 3rd day after thawing. Cultures were returned to the dark. After the dark period, embryos were transferred to new E1B5 medium again and placed under light conditions (44 µ $Em^{-2}s^{-1}$, with a 16-/8-h photoperiod) at 25 °C.

Viability was estimated by recovery on solid E1B5 medium. Recovery was the percentage of treated embryos showing cotyledon development (direct regrowth) and those producing secondary somatic embryos (Kiss et al. 1992) 6 weeks after thawing. Five Petri dishes per treatment were evaluated for embryo recovery. Post-thaw treatments were identical for all of the cryopreservation methods described.

3 Results and Discussion

Cryopreserved somatic embryos showed a high level of morphogenetic competence for secondary embryogenesis, with certain aberrations especially in the numbers of cotyledons (Fig. 2). Embryos first turned brown in culture, then

Fig. 2. Secondary embryogenesis and plant regeneration from cryopreserved zygotic embryos of horse chestnut (*Aesculus hippocastanum* L.). **A** Horse chestnut trees, 35–38 years old; **B** light micrograph (LM) of zygotic embryo development on WPM medium on *1* the 10th and *2* the 20th day of incubation; **C** scanning electron micrograph (SEM; 45×) of the secondary embryogenesis at the early cotyledonary stage; **D** SEM (75×) of the secondary somatic embryo at *1* the scutellar stage and *2* an aberrant embryo with three cotyledons; **E** SEM (75×) of the secondary somatic embryo at the late cotyledonary stage; **F** SEM (75×) of the aberrant secondary embryo with five cotyledons; **G** SEM (100×) of a properly developed shoot apical meristem of the secondary somatic embryo; **H** LM of plant regeneration

began to recover in 2–3 weeks after thawing (Fig. 3A). Some embryos developed normally and directly into plants (Fig. 2H). Other embryos first produced adventitious somatic embryos (Fig. 3), which, after a subsequent subculture, developed into plants at a low frequency due to the high level of embryo 'overdevelopment' (Fig. 3D). This undesired change in normal embryo development occurred in the form of enormous (up to 8–10 mm) elongation of cotyledons (Fig. 3C) and the abortion of embryonic shoot apical meristems. The percentage of direct embryo regrowth was higher than those of secondary embryos (Tables 1 and 2).

3.1 Viability of Somatic Embryos after Cryopreservation with Cryoprotectants at Slow and Rapid Cooling Rates

Embryos precultured on ABA-free medium did not show any recovery after cryopreservation when cooled rapidly or slowly (Table 1). The recovery rate was highest (43%) at 0.75 µM ABA concentration after slow freezing. Increasing ABA concentrations to 7.5 or 75.0 µM reduced survival. When fast cooling was applied, a few embryos pretreated with ABA survived cooling, but there were no significant differences among ABA treatments.

Similar to this observation, ABA treatments improved survival after cryogenic treatments for meristems in raspberry and blackberry (Reed 1993), zygotic embryos in wheat (Kendall et al. 1993), cacao bean (Pence 1991) and calli in asparagus (Uragami 1991).

3.2 Viability of Somatic Embryos after Cryopreservation with Desiccation at a Rapid Cooling Rate

The initial relative embryo moisture content (91%, fresh weight basis, when cultured in liquid medium) fell to about 80% when the embryos were precultured for 4 days on the solid medium (data not shown) independently of the ABA concentration in the medium. Desiccation in the laminar flow hood further reduced water content. After desiccation for 2, 3 and 4 h, the water content of embryos decreased to 53, 22 and 13%, respectively (Table 2).

While ABA had no effect on the reduction in water content, it provided a significant increase in recovery when the vials with embryos were immersed directly in LN (Table 2). The highest recovery (46%) was obtained when the water content of the embryos was reduced to 13% (after 4 h of desiccation in this system) following pretreatment on a medium containing 0.75 µM ABA. The 7.5 and 75.0 µM ABA pretreatments also significantly increased survival compared with embryos not treated with ABA. Maximum viability was observed at similar water contents in cryopreserved somatic embryos of *Coffea canephora* (13%, Hatanaka et al. 1994), zygotic embryos of *Coffea liberica* (15.5%, Normah and Vengadasalam 1992), *Coffea arabica* (16.4%, Abdelnour-Esquivel et al. 1992b) and *Fraxinus excelsior* (12%, Brearley et al. 1995). In the present study, desiccation for 2 h reduced moisture content to

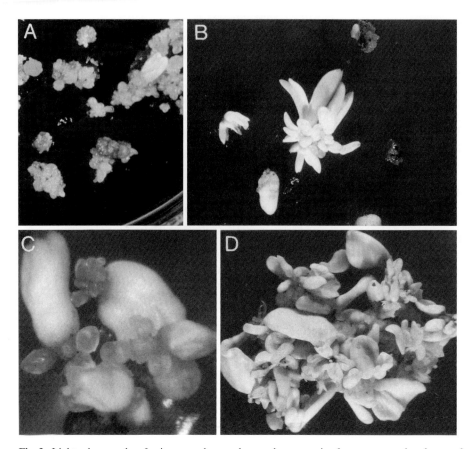

Fig. 3. Light micrographs of primary and secondary embryogenesis of cryopreserved embryos of horse-chestnut (*Aesculus hippocastanum* L.) trees. From the cyopreserved embryos **A** secondary somatic embryos are developing **B** mainly from the hypocotyl region **C** of the primary embryos. **D** Embryo overdevelopment with elongated cotyledons

Table 1. Effect of ABA in the pretreatment medium on the recovery of horse-chestnut somatic embryos after cryoprotectant treatment and slow or rapid cooling

	PERCENT RECOVERY[1]							
	ABA (µM)							
	0		0.75		7.5		75.0	
	slow	rapid	slow	rapid	slow	rapid	slow	rapid
Direct regrowth	0	0	40 ± 2	9 ± 2	20 ± 2	2 ± 0	5 ± 1	6 ± 2
Secondary embryos	0	0	3 ± 1	6 ± 2	9 ± 1	8 ± 0	0	1 ± 0
Total	0	0	43 ± 3a	15 ± 4b	29 ± 4c	10 ± 1d	5 ± 1e	7 ± 2e

The recovery data[1] show the percentage of the number of recovering embryos as the mean of 100 embryos from each treatment ± standard deviation. Total recovery rate of uncooled ABA-pretreated embryos varied in the range of 94–98%, and no significant differences were found in different ABA treatments.

Values followed by the same letter do not differ significantly ($P = 0.05$).

Table 2. Effect of desiccation period and ABA in the pretreatment medium on the recovery of horse-chestnut somatic embryos rapidly cooled

Desiccation duration (h)	Moisture content (% of fresh weight)		PERCENT RECOVERY[1]				
			Uncooled control	ABA μM			
				0	0.75	7.5	75
2	53	direct regrowth	68 ± 4	6 ± 2	6 ± 2	8 ± 1	13 ± 1
		secondary embryos	28 ± 2	2 ± 1	0	3 ± 1	8 ± 3
		total	96 ± 6a	8 ± 2j	6 ± 2j	11 ± 2g	21 ± 4f
3	22	direct regrowth	49 ± 3	13 ± 3	22 ± 1	18 ± 2	20 ± 3
		secondary embryos	27 ± 2	1 ± 0	12 ± 1	4 ± 1	6 ± 2
		total	76 ± 5b	14 ± 3	34 ± 2e	32 ± 3e	26 ± 5f
4	13	direct regrowth	33 ± 5	11 ± 2	37 ± 4	30 ± 2	36 ± 2
		secondary embryos	12 ± 1	0	9 ± 1	10 ± 3	8 ± 1
		total	45 ± 5c	11 ± 2g	46 ± 5.2c	40 ± 5d	44 ± 3c

The recovery data[1] show the percentage of the number of recovering embryos as the mean of 100 embryos from each treatment ± standard deviation.

Values followed by the same letter do not differ significantly ($P = 0.05$).

Desiccation was carried out in the air current of a laminar flow cabinet immediately before cooling on embryos precultured on ABA media without cryoprotectant mixture (Paragraph 2.3.4.). The recovery was determined in the 6th week after thawing.

53%, but these embryos displayed low survival. An adjustment of the desiccation period to between 3 and 4h (i.e., to a moisture content less than 22% but more than 13%) might enhance viability, but in somatic embryos of coffee, a rapid decrease in survival rate occurred at water contents higher than 13% (Hatanaka et al. 1994).

3.3 Viability of Somatic Embryos After Cryopreservation with Heat Shock at Slow and Rapid Cooling Rates

Heat shock for 30 and 60min at 40 °C had no effect, while a 90-min period somewhat decreased the viability of slowly cooled somatic embryos (Table 3). In contrast, various heat shock durations significantly increased the percentage of recovering embryos after cryopreservation by rapid cooling. The longer heat shock periods (i.e., 60 and 90min) resulted in 37 and 34% embryo recovery, respectively.

There are indications that heat shock treatments increase tolerance to the chilling in plant cellular systems (Harrington and Alm 1988; Andarajah et al. 1991; Collins et al. 1993). However, heat shock treatments have not generally

Table 3. Effect of heat shock duration at 40 °C on the recovery of horse-chestnut somatic embryos preconditioned in 0.75 µM ABA in the pretreatment medium and either slowly and rapidly cooled

Heat shock duration (min)		PERCENT RECOVERY[1]		
		Uncooled control	Slow cooling	Rapid cooling
0	direct regrowth	72 ± 4	31 ± 2	9 ± 3
	secondary embryos	28 ± 3	12 ± 2	6 ± 1
	total	100a	43 ± 3c	15 ± 4f
30	direct regrowth	69 ± 6	25 ± 3	16 ± 2
	secondary embryos	31 ± 5	17 ± 1	5 ± 1
	total	100a	42 ± 4c	21 ± 4e
60	direct regrowth	68 ± 4	35 ± 3	25 ± 3
	secondary embryos	30 ± 1	9 ± 1	12 ± 1
	total	98 ± 5a	44 ± 4c	37 ± 3d
90	direct regrowth	67 ± 3	15 ± 5	28 ± 3
	secondary embryos	29 ± 3	11 ± 2	6 ± 2
	total	96 ± 6b	26 ± 7e	34 ± 4d

The recovery data[1] show the percentage of the number of recovering embryos as the mean of 100 embryos from each treatment ± standard deviation.

Values followed by the same letter do not differ significantly ($P = 0.05$).

Different heat shock periods were applied immediately before cooling on embryos precultured on ABA media and treated with cryoprotectant mixture (Paragraph 2.3.4.). The recovery was determined in the 6th week after thawing.

improved cryotolerance of plant tissue cultures. Cell suspension cultures of tobacco reached their highest viability after a heat shock treatment at 37 °C for 2 h, when the cells were preconditioned on medium containing mannitol (Reinhoud et al. 1995).

4 Summary and Conclusions

This study compared three freezing protocols for the cryopreservation of somatic embryos from a species, *Aesculus hippocastanum,* possessing recalcitrant seed. Recovering embryos (43%) were obtained after 4-day preculture on medium containing 0.75 µM ABA, followed by cryoprotective treatments and slow cooling. The slow cooling step could be substituted by a 4-h desiccation period in the air flow of a laminar cabinet, or by a 60-min heat shock treatment at 40 °C following a pretreatment on ABA medium.

In these experiments, both desiccation and heat shock treatments increased the cryotolerance of horse-chestnut somatic embryos, but, more likely, their protective effect probably follows different pathways. While desiccation-induced stress genes synthesize proteins with direct cryoprotective effect, HS proteins by heat shock treatments could renaturate polypeptides that have been denaturated during the freeze-thaw cycle (Skowyra et al. 1990).

Different ABA pretreatments did not change the moisture content of the embryos, but increased their recovery. From this observation, it may be concluded that stress response regulation in horse chestnut follows the ABA-dependent pathway.

Acknowledgements. The authors wish to thank the National Research Fund (OTKA 1571, 1572) for financial support.

References

Abdelnour-Esquivel A, Mora A, Villalobos V (1992a) Cryopreservation of zygotic embryos of *Musa acuminata* (AA) and *M. balbisiana* (BB). Cryo Lett 13:159–164

Abdelnour-Esquivel A, Villalobos V, Engelmann F (1992b) Cryopreservation of zygotic embryos of *Coffea* spp. Cryo Lett 13:297–302

Andarajah K, Kott K, Beversdorf WD, McKersie BD (1991) Induction of desiccation tolerance in microspore-derived embryos of *Brassica napus* L. by thermal stress. Plant Sci 77:119–123

Assy-Bah B, Engelmann F (1992) Cryopreservation of mature embryos of coconut (*Cocus nucifera* L.) and subsequent regeneration of plantlets. Cryo Lett 13:117–126

Bajaj YPS (1995) Cryopreservation of plant cell, tissue and organ culture for the conservation of germplasm and biodiversity. In: Bajaj YPS (ed) Biotechnology in agriculture and forestry, vol 32. Cryopreservation of plant germplasm I. Springer, Berlin Heidelberg New York, pp 3–18

Brearley J, Henshaw GG, Davey C, Taylor NJ, Blakesley D (1995) Cryopreservation of *Fraxinus excelsior* L. zygotic embryos. Cryo Lett 16:215–218

Collins GG, Nie X, Saltveit ME (1993) Heat shock increases chilling tolerance of mung bean hypocotyl tissue. Physiol Plant 89:117–124

Dumet D, Engelmann F, Chabrillange N, Duval Y (1993) Cryopreservation of oil palm (E*laeis guineensis* Jack.) somatic embryos involving a desiccation step. Plant Cell Rep 12:352–355

Gamborg OL, Miller RA, Ojima K (1968) Nutrient requirements of suspension cultures of soybean root cells. Exp Cell Res 50:151–158

Gamborg OL, Davies BP, Stahlhut RW (1983) Cell division and differentiation in protoplasts from cell cultures of G*lycine* species and leaf tissues of soybean. Plant Cell Rep 2:213–215

Gonzalez-Benito ME, Perez C (1994) Cryopreservation of embryonic axes of two cultivars of hazelnut (*Corylus avellana* L). Cryo Lett 15:41–46

Harrington HM, Alm DM (1988) Interaction of heat and salt stress in cultured tobacco cells. Plant Physiol 88:618–625

Hatanaka T, Yasuda T, Tamaguchi T, Sakai A (1994) Direct regrowth of encapsulated somatic embryos of coffee (*Coffea canephora*) after cooling in liquid nitrogen. Cryo Lett 15:47–52

Jennings P, Saltveit ME (1994) Temperature and chemical shocks induce chilling tolerance in germinating *Cucumis sativus* (Cv. Poinsett 76) seeds. Physiol Plant 91:703–707

Jörgensen J (1990) Conservation of valuable gene resources by cryopreservation in some forest tree species. Plant Physiol 136:373–376

Kaul SC, Obuchi K, Iwahashi H, Komatsu Y (1992) Cryoprotection provided by heat shock treatment in S*accharomyces cerevisiae*. Cell Mol Biol 38:135–143

Kendall EJ, Kartha KK, Qureshi JA, Chermak P (1993) Cryopreservation of immature spring wheat zygotic embryos using an abscisic acid pretreatment. Plant Cell Rep 12:89–94

Kiss J, Heszky LE, Kiss E, Gyulai G (1992) High efficiency adventive embryogenesis on somatic embryos of anther, filament and immature proembryo origin in horse-chestnut (A*esculus hippocastanum* L.) tissue culture. Plant Cell Tissue Org Cult 30:59–64

Marin ML, Duran-Vila N (1988) Survival of somatic embryos and recovery of plants of sweet orange (*Citrus sinensis* L.) after immersion in liquid nitrogen. Plant Cell Tissue Org Cult 14: 51–57

Normah MN, Vengadasalam M (1992) Effects of moisture content on cryopreservation of *Coffea* and *Vigna* seeds and embryos. Cryo Lett 13:199–208

Pence VC (1991) Cryopreservation of immature embryos of *Theobroma cacao*. Plant Cell Rep 10:144–147

Reed BM (1993) Responses to ABA and cold acclimation are genotype dependent for cryopreserved blackberry and raspberry meristems. Cryobiology 30:179–184

Reinhoud PJ, Schrijnemakers WM, van Iren F, Kijne W (1995) Vitrification and heat shock treatment improve cryopreservation of tobacco cell suspension compared to two step freezing. Plant Cell Tissue Org Cult 42:261–267

Skowyra D, Georgopoulos C, Zylicz M (1990) The E. *coli* dnaK product, the hps homolog, can reactivate heat inactivated RNA polymerase in ATP hydrolysis manner. Cell 62:939–944

Uragami A (1991) Cryopreservation of asparagus (A*sparagus officinalis* L.) cultured in vitro. Res Bull Hokkaido Natl Agric Exp Stn 156:1–37

Wright JW (1976) Introduction to forest genetics. Academic Press, New York, 335 pp

III.2 Cryopreservation of Neem (*Azadirachta indica* A. Juss.) Seeds

Dominique Dumet and Patricia Berjak

1 Introduction

1.1 Plant distribution and Importance

Neem, Azadirachta indica (A. Juss.), is a member of the Meliaceae which is believed to have emerged from the Indian sub-continent and Myanmar (Kundu and Tigersted 1997). These authors describe the introduction of this evergreen tree to different parts of the world, mainly in the drier tropical and sub-tropical zones of Asia, Africa, the Americas, Australia and the South Pacific Islands. Nowadays, neem introductions are so old and widely spread that knowledge of its original distribution has become uncertain (Farooqui et al. 1998).

Neem as been described as "the tree for solving problems" (National Research Council 1992). Indeed, besides the use of the whole tree in the prevention of soil erosion, improvement of soil microclimate, provision of shade, fuel and timber, over 250 different compounds have been extracted from neem, some of them being highly valuable and used (inter alia) as biopesticides (Farooqui et al. 1998). Azadirachtin, the major bioactive chemical, is found in the neem kernel (Ley et al. 1993) and its concentration may vary considerably depending on plant ecotype and environment (Ermel et al. 1987; Singh 1987).

Conventionally, neem is propagated by seeds derived from cross-pollination with the disadvantage of potential loss of highly productive genotypes. In order to propagate selected elite trees, vegetative in vitro propagation of neem has been developed via shoot initiation as well as somatic embryogenesis from, respectively, leaf explants (Eeswara et al. 1998) and seed or hypocotyl tissues (Shrikhande et al. 1993; Su et al. 1997; Murthy and Saxena 1998).

1.2 Neem Storage

Field and seed gene banks still provide the main means for conserving agricultural plant germplasm. However, when stored in the field, neem, like any

School of Life and Environmental Sciences, University of Natal, Durban 4041, South Africa
Present address: D. Dumet, 3 rue du Marché, 36 700 Chatillon sur Indre, France

Biotechnology in Agriculture and Forestry, Vol. 50
L.E. Towill and Y.P.S. Bajaj (Eds.) Cryopreservation of Plant Germplasm II
© Springer-Verlag Berlin Heidelberg 2002

other species, is submitted to biotic and abiotic stress. History has shown that many disasters have occurred as a result of the narrow genetic base of crops that offer little resistance to certain diseases (Chin 1994). It was recently suggested by Farooqui and coworkers (1998), who studied variations in RAPD profiles amongst provenances, that neem does have a narrow genetic base. Consequently, it seems essential to combine both field and seed gene banking to ensure safe genetic resource management.

Seed banking feasibility depends on seed post-harvest behaviour. Indeed, when seeds are desiccation- and cold-tolerant, i.e. orthodox, long-term storage can be achieved following the recommendations of IBPGR (1976), hermetic storage at $-18\,°C$ after drying down to 5–1% water on a fresh weight basis, although the arbitrary use of such low moisture contents is now debatable (Walters and Engels 1998). In contrast, long-term storage of intermediate (less desiccation-tolerant than orthodox seeds, and chilling-sensitive in the case of certain species; Hong and Ellis 1996) and recalcitrant (desiccation-sensitive and sometimes cold-sensitive; Roberts 1973) seeds require the use of alternative strategies. Neem seeds have been classified variously as orthodox, intermediate or recalcitrant (Chaudhury and Chandel 1991; Bellefontaine and Audinet 1993; Berjak et al. 1995; Gamene et al. 1996). These categorisations were based on seed germination potential after slow desiccation, or when water content was maintained at or near the level at harvest and with storage at various temperatures for different periods of time. The diversity of the reported responses of neem seeds to desiccation and temperature are not yet understood. In addition to the provenance and the genetic variation that may exist between mother trees, other factors, such as harvesting methods, dehydration procedures and seed maturity at the time of the experiment, may result in different seed sensitivity towards desiccation and chilling temperatures (Berjak et al. 1995; Gamene et al. 1996).

Recent studies of Sacandé and collaborators (1998, 2000) on neem seed longevity did not reveal any difference in storage behaviour between seed lots harvested in Asia or in Sahelian Africa. Instead, the developmental stage at harvest appeared critical to seed longevity; seeds originating from mature, yellow fruits survived longer than those originating from younger, green or older, brown fruits. Moreover, the ability of those neem seeds to accumulate putatively protective compounds (oligosaccharides, glutathione and phospholipids) during slow drying, led those authors to conclude that neem seeds show features of orthodox, rather than recalcitrant, storage behaviour. Nevertheless, significant viability loss of neem seeds upon drying and cold storage distinguishes them from the orthodox category originating from temperate climates (Sacandé et al. 1998).

For neem seeds reported as orthodox, long-term storage should be unconstrained. In contrast, for intermediate and recalcitrant types (e.g. Berjak et al. 1995), long-term preservation requires the use of plant conservation biotechnology such as tissue culture (embryo/micro-plants) and/or cryopreservation. To our knowledge, the former approach has not yet been exploited.

2 Cryopreservation

Chaudhury and Chandel (1991) first reported successful cryopreservation of artificially desiccated Indian neem seeds. In our experiment (described below), seeds were harvested in the Gede region of Mombasa, Kenya, and cryopreservation experiments were performed on both seeds and isolated embryonic axes. It should be noted that seeds of this provenance are apparently unamenable to cold storage even at undiminished water contents, and lose viability within 4 months when equilibrated to ambient relative humidity (Berjak et al. 1995).

2.1 Material and Methods

Seeds were first hand-sorted from freshly harvested and depulped neem fruit in Kenya. Extracted seeds, each enclosed by the endocarp, were transported by air to our laboratory in Durban, South Africa. Upon arrival, the endocarps were cracked by hand and the seeds removed. Each seed (approximately 10 × 7 mm) was surrounded by a soft brown testa, enclosing the embryo consisting of a small axis between two relatively large yellow-green cotyledons.

Seeds. Dehydration was performed by placing the seeds in a monolayer upon a layer of activated silica gel in glass Petri dishes. These were maintained at ambient temperature (20–25 °C) for 1 and 2 days. Seeds were then introduced into a cryovial (5 seeds/cryovial) and cooled by plunging the cryotube into liquid nitrogen (LN). The cryotubes were then maintained in liquid nitrogen for 1 h, 1 day, 1 month and 4 months. Samples were warmed by placing them at ambient temperature and waiting 24 h prior to cryotube opening and seed transfer onto moistened filter paper.

Embryonic Axes. Using a stereomicroscope, embryonic axes were excised from their cotyledons before being surface sterilised in 1% sodium hypochlorite for 20 min. They were then rinsed in sterile water and briefly blotted before being introduced into 1-ml cryotubes containing 0.6 g activated silica gel (5 axes/cryotube). Desiccation, which took place within the cryotube, was allowed to proceed for a period of up to 2 h at ambient temperature. Cryovials were then directly plunged into LN where they were kept for 48 h. Warming was performed rapidly by transferring the tubes to a 40 °C water bath for 2 min. Embryonic axes were then transferred to a germination medium comprising half strength MS salts and vitamins (Murashige and Skoog 1962) containing 3% sucrose and 0.8% agar.

Water content of seeds and isolated axes [expressed as g water per g dry weight (g g^{-1}DW)] was determined gravimetrically after drying at 80 °C for 4 and 2 d, respectively.

Seeds were recorded as viable when showing radicle elongation (30 mm) and greening of the cotyledon 3 weeks after placing on moistened filter paper within Petri dishes. For isolated embryonic axes, survival was scored when they showed elongation from the original 2 mm to approximately 10 mm within 2–3 weeks.

2.2 Results

The totality of seed germination at the original water content ($0.61\,g\,g^{-1}$ DW) in this experiment was 92% (Table 1), when none of these seeds could tolerate freezing under the conditions used. It had also been previously shown that neem seeds from this provenance would lose viability within 4 months when stored at ambient relative humidity (Berjak et al. 1995). Although seed water contents dropped to $0.09\,g\,g^{-1}$ DW after 1 day of dehydration, 75% were still able to germinate. Moreover, this degree of dehydration allowed viable seeds to survive the temperature of liquid nitrogen for short, as well as more extended periods. Indeed, whether the seeds were stored in liquid nitrogen for 1 h or 1–4 months, their survival rate remained high (70–75%). Surprisingly, the 1-day cryostored condition showed a low germination rate (40%). Considering that for the other conditions tested the exposure time did not affect seed germination, it was concluded that the fungi associated with the 1-day cryopreserved sample may have significantly affected viability. Indeed, associated fungi are prevalent on and in the tissues of recalcitrant (and presumably) of intermediate seeds and are known to play a crucial role in viability loss of such seeds (Mycock and Berjak 1990). The presence of only one infected seed in the confines of the closed container in which the germination test was conducted, is likely to have affected the germination rate of them all, considering the rate of fungal proliferation under these conditions.

Germination of 2-day desiccated seeds remained high (90%) despite a very low water content ($0.06\,g\,g^{-1}$ DW; Table 1). When such seeds were cryopreserved, a decline in viability was observed when they were maintained in liquid nitrogen for longer than 1 h, although the reasons for this are not clear. It is possible that seeds were able to recover the stress of a short excursion at −196 °C better than an extended one. Similar observations were previously demonstrated for dehydrated embryonic axes of tea (Wesley-Smith et al. 1992).

The initial water content of the embryonic axes was relatively high in comparison to the whole seed, 2.33 vs. $0.61\,g\,g^{-1}$DW (Tables 1 and 2). After

Table 1. Survival of neem seeds after desiccation and/or cryopreservation (8 to 24 seeds were assessed each time)

Desiccation duration	Water content ($g\cdot g^{-1}$DW)	No Cryostorage	Survival rate after different cryostorage durations			
			1 hour	1 day	1 month	4 months
None	0.61 ± 0.003	92	0	Not tested	Not tested	Not tested
1 day	0.09 ± 0.02	75	75	40	75	70
2 days	0.06 ± 0.01	90	85	55	45	65

Table 2. Survival of embryonic axes after desiccation and/or cryopreservation (10 to 20 embryonic axes were assessed each time)

Desiccation duration	Water content (g·g⁻¹DW)	Survival rate after	
		Desiccation	Desiccation and cryopreservation
0	2.33 ± 0.49	100	0
1 hour	0.23 ± 0.03	100	100
2 hours	0.19 ± 0.06	79	85

1 h of desiccation over silica gel, axes had lost almost 90% of their initial water ($0.23\,\mathrm{g\,g}^{-1}$ DW). However, all the axes survived this rapid desiccation and subsequent freezing in liquid nitrogen did not affect their viability; indeed, all survived after cryopreservation. Further desiccation of embryos allowed only the removal of a small fraction of water while substantially lowering their survival after both this water loss and subsequent cryopreservation (Table 2).

Very high survival rates after cryopreservation were obtained with both seeds and isolated embryonic axes. Optimal pre- and post-freezing treatment conditions were seemingly attained for the embryonic axes, as 100% survival after cryopreservation was recorded when axes were desiccated to what appeared to be an optimal water content. Complete survival may also occur with the whole seeds if no fungi are present. This contention is borne out by the fact that when axes were grown under aseptic conditions in vitro, 100% survival was recorded, as long as dehydration was not extreme (Table 2). In contrast, fungal contamination was commonly observed when the whole seeds were ready to germinate.

3 Conclusions

Long-term conservation of neem germplasm, at least that deriving from a humid coastal provenance in Kenya, can be achieved via cryopreservation of either the whole seeds or embryonic axes. Whether exposed for a short period to liquid nitrogen or cryostored for up to 4 months, their recovery remained satisfactory. Despite the fact that better survival rates after cryopreservation (up to 100%) were obtained with embryonic axes, the simplicity of the cryopreservation process used for seeds suggests this choice as a first approach for the establishment of neem gene banks. This is because while embryo cryopreservation implies several time-consuming steps such as excision, sterilisation and later, in vitro culture, these procedures are not necessary when dealing with whole seeds, which only require endocarp removal prior to desiccation. Considering their water content at harvest and their desiccation tolerance during, and following, rapid dehydration of the seeds used in this experiment, they might be classified as intermediate. However, as previously mentioned, intermediate behaviour is not invariable for neem. Furthermore, in terms of the fact that if recalcitrant – or any non-orthodox – seeds are able to be dehydrated rapidly (as is the case with neem), then they can tolerate remarkably low water contents (Pammenter et al. 1998). However, it is doubtful whether such dehydrated seeds would survive for any significant period under ambient or refrigerated conditions, making their cryopreservation imperative. For those seeds which have been classified as recalcitrant, slow dehydration of the whole seed has been suggested to result inevitably in lethal damage at a relatively high water content (Pammenter et al. 1998). However, isolated axes of such seeds are able to sustain far lower water contents, and, in this condition, are amenable to cryopreservation, e.g. those of *Camellia sinensis* (Wesley-Smith et al. 1992) and *Quercus robur* (Berjak et al. 1999a,b). For recalcitrant seeds that cannot lose water rapidly, embryonic axes would provide the only appropriate explants for cryobanking, if the objective were to conserve heterozygosity.

The danger of not conserving the genetic diversity of a species has recently been vividly illustrated for neem, where most of the resources are represented by field gene banks. An alarming report by Shankara Bhat et al. (1998) describes a new disease which has already affected many neem trees, of all ages and sizes, in Karnataka State in Southern India. The die-back disease, which is thought to be caused by an unknown species of the fungal pathogen, *Phomopsis*, resulted in the entire loss of fruit production (Shankara Bhat et al. 1998) and is a major threat to neem. Such observations suggest the urgent need for a cryogene banking programme for this important species.

References

Bellefontaine R, Audinet M (1993) Les problèmes de semences forestières notamment en Afrique. In: Some L, de Kam M (eds) Tree seed problems, with special reference to Africa. Backhuis, Leiden, pp 268–274

Berjak P, Campbell GK, Farrant JM, Omondi-Oloo W, Pammenter NW (1995) Responses of seeds of *Azadirachta indica* (neem) to short-term storage under ambient or chilled conditions. Seeds Sci Technol 23:779–792

Berjak P, Kioko JI, Norris M, Mycock DJ, Wesley-Smith J, Pammenter NW (1999a) Cryopreservation – an elusive goal? In: Marzalina M, Khoo KC, Jayanthi N, Tsan FY, Krishnapillay B (eds) Recalcitrant seeds. Proceeding of the IUFRO Seed Symposium, Forest Research Institute, Kuala Lumpur, Malaysia, pp 96–109

Berjak P, Walker M, Watt MP, Mycock DJ (1999b) Experimental parameters underlying failure or success in plant germplasm cryopreservation: a case study on zygotic axes of *Quercus robur* L. Cryo Lett 20:251–262

Chaudhury R, Chandel KPS (1991) Cryopreservation of desiccated seeds of neem (*Azadirachta indica* A. Juss) for germplasm conservation. Ind J Plant Genet Resour 4:67–72

Chin HF (1994) Seedbanks: conserving the past for the future. Seed Sci Technol 22:385–400

Eeswara JP, Stuchbury T, Allan EJ, Mordue AJ (1998) A standard procedure for the micropropagation of the neem tree (*Azadirachta indica* A. Juss). Plant Cell Rep 17:215–219

Ermel K, Pahlich E, Schmutterer H (1987) Azadirachtin content of neem kernels from different geographical locations and its dependance on temperature, relative humidity and light. In: Schmutterer H, Ascher KRS (eds) Natural pesticides from the neem tree (*Azadirachta indica* A. Juss) and other tropical plants. Proceedings of the Third International Neem Conference, Nairobi. German Agency for Technical Cooperation, Eschborn, pp 171–184

Farooqui N, Ranade S, Sane PV (1998) RAPD profile variation amongst provenances of neem. Biochem Mol Biol Int 45:931–939

Gamene CS, Kraak HL, Van Pijlen JG, de Vos CHR (1996) Storage behaviour of neem (*Azadirachta indica*) seeds from Burkina Faso. Seed Sci Technol 24:441–448

Hong TD, Ellis RH (1996) A protocol to determine seed storage behaviour. IPGRI, Rome

International Board for Plant Genetic Resources (1976) In: Report of IBPGR Working Group on engineering, design and cost aspects of long term storage facilities. Rome, Italy

Kundu S, Tigersted PMA (1997) Geographical variation in seed and seedling traits of neem (*Azadirachta indica* A. Juss) amongst ten populations studied in growth chamber. Silvae Genet 46:129–137

Ley SV, Denholm AA, Wood A (1993) The chemistry of *azadirachtin*. Nat Product Rep 10:109–157

Murashige T, Skoog F (1962) A revised medium for rapid growth and bioassay with tobacco cultures. Physiol Plant 15:473–497

Murthy BNS, Saxena PK (1998) Somatic embryogenesis and plant regeneration of neem (*Azadirachta indica* A. Juss). Plant Cell Rep 17:469–475

Mycock DJ, Berjak P (1990) Fungal contaminants associated with several homoiohydrous (recalcitrant) seed species. Phytophylactica 22:413–418

National Research Council (1992) Neem: a tree for solving global problems. National Academy Press, Washington, DC

Pammenter NW, Greggains V, Kioko JI, Wesley-Smith J, Berjak P, Finch-Savage WE (1998) Effects of differential drying rates on viability retention of recalcitrant seeds of E*kebergia capensis*. Seed Sci Res 8:463–472

Roberts EH (1973) Predicting the storage life of seeds. Seed Sci Technol 1:499–541

Sacandé M, Hoekstra FA, van Pijlen JG (1998) A multifactorial study of conditions influencing longevity of neem (*Azadirachta indica*) seeds. Seed Sci Res 8:473–482

Sacandé M, Hoekstra FA, van Aelst AC, De Vos CHR (2000) Is oxidative stress involved in the loss of neem seed viability? Seed Sci Res 10:381–392

Shankara Bhat S, Sateesh MK, Devaki NS (1998) A new destructive disease of Neem (*Azadirachta indica*) incited by *Phomopsis azadirachtae*. Curr Sci 74:17–19

Shrikhande M, Thengane SR, Mascarenhas AF (1993) Somatic embryogenesis and plant regeneration in *Azadirachta indica* A. Juss. In Vitro Cell Dev Biol Plant 29:38–42

Singh RP (1987) Comparison of antifeedant efficacy and extract yields from different parts and ecotypes of neem (*Azadirachta indica* A. Juss) tree. In: Schmutterer H, Ascher KRS (eds) Natural pesticides from the neem tree (*Azadirachta indica* A. Juss) and other tropical plants. Proceeding of the 3rd International Neem Conference, Nairobi. German Agency for Technical Cooperation, Eschborn, pp 171–184

Su WW, Hwang WI, Kirn SY, Sagawa Y (1997) Induction of somatic embryogenesis in *Azadirachta indica*. Plant Cell Tissue Organ Culture 50:91–95

Walters C, Engels J (1998) The effect of storing seeds under extremely dry conditions. Seed Sci Res 8:3–8

Wesley-Smith J, Vertucci CW, Berjak P, Pammenter NW, Crane J (1992) Cryopreservation of desiccation-sensitive axes of *Camellia sinensis* in relation to dehydration, freezing rate and the thermal properties of tissue water. J Plant Physiol 140:596–604

III.3 Cryopreservation of *Coffea* (Coffee)

S. Dussert[1], N. Chabrillange[2], F. Engelmann[1,2], F. Anthony[3], N. Vasquez[3], and S. Hamon[1]

1 Introduction

Coffee trees belong to the tribe *Coffeae* in the *Rubiaceae* (Bridson and Verdcourt 1988). The genus *Coffea* L. is subdivided in two subgenera: *Coffea* and *Baracoffea*. Approximately 100 taxa have been identified so far in the subgenus *Coffea* (Charrier and Berthaud 1985). All species are woody, ranging from small-sized shrubs to robust trees and originate from the inter-tropical forests of Africa and Madagascar. Commercial coffee production relies on two species only, *C. arabica* L. and *C. canephora* Pierre, but many *Coffea* species form a valuable gene reservoir for different breeding purposes (Berthaud and Charrier 1988). *C. arabica* is a natural allotetraploid and autogamous, while other species are diploid and generally self-incompatible (Charrier and Berthaud 1985).

Although *C. arabica* seeds can withstand desiccation down to 0.08–0.10 g $H_2O g^{-1} DW$ water content (Becwar et al. 1983; Ellis et al. 1990; Dussert et al. 1999; Eira et al. 1999), they cannot be considered orthodox because they remain cold-sensitive (Van der Vossen 1977; Couturon 1980; Ellis et al. 1990) and desiccation does not increase their longevity (Van der Vossen 1977; Ellis et al. 1990). Fully hydrated seeds stored at 19 °C with 100% relative humidity remained viable for 36 months for *C. arabica* and 15 months for *C. canephora* and *C. stenophylla* (Couturon 1980). Because of their intermediate storage behaviour, coffee seeds cannot be used for long-term conservation and coffee genetic resources are conventionally conserved as trees in field gene banks.

However, the maintenance of field gene banks has significant drawbacks: (1) genetic erosion in some species or genetic groups due to their poor adaptation to the local environment and to attacks by pests and pathogens; and (2) high labour costs and large space requirements. Thus, research for alternative methods to field conservation for coffee genetic resources became a priority and attention focused on the use of in vitro culture techniques (Berthaud and Charrier 1988).

Numerous in vitro techniques have been developed for medium-term storage of coffee germplasm (Dussert et al. 1997b). The establishment of an

[1] IRD (previously ORSTOM), 911 Av. Agropolis, BP 5045, F-34032, Montpellier, France
[2] IPGRI, Via dei Tre Denaci, 4721a, 00057 Maccarese, Italy
[3] CATIE, Ap. 59, 7170 Turrialba, Costa Rica

Biotechnology in Agriculture and Forestry, Vol. 50
L.E. Towill and Y.P.S. Bajaj (Eds.) Cryopreservation of Plant Germplasm II
© Springer-Verlag Berlin Heidelberg 2002

in vitro coffee core collection was thus initiated in 1991 at IRD; however, a few years later, the limits of this technique became apparent with the occurrence of some genotypic selection and intraspecific genetic drift (Dussert et al. 1997c). This stressed the importance of developing cryopreservation protocols for long-term conservation of coffee germplasm. Research for the development of cryopreservation techniques was performed with seeds, zygotic embryos, apices and somatic embryos.

2 Cryopreservation of Seeds

2.1 Cryopreservation of *C. arabica* Seeds

2.1.1 Importance of Water Content and Cooling Process (Rapid vs. Slow)

In *C. arabica*, when cryopreserved seeds are rehydrated above water immediately after warming, seedlings can be recovered only if they are dehydrated to a very narrow range of water contents, $0.20–0.22\,g\,H_2O\,g^{-1}$ DW (Dussert et al. 1997a, 1998; Eira et al. 1999), and slowly cooled before immersion into liquid nitrogen (LN) (Dussert et al. 1997a, 1998). Optimal water content $(0.2\,g\,H_2O\,g^{-1}$ DW) corresponds to the unfreezable water content of *C. arabica* seeds, as determined by differential scanning calorimetry (DSC) analysis (Dussert et al. 2001). It can be achieved either using silica gel (Dussert et al. 1998) or by equilibrating seeds for 3 weeks under 78% relative humidity (RH), obtained using a saturated solution of NH_4Cl (Dussert et al. 1997a). This second method is strongly recommended for routine use in coffee gene banks since, in the case of coffee seeds, there is no detrimental effect of a slow drying process in this range of water contents and optimal water content for cryopreservation is achieved in a very reproducible manner (in comparison with the procedure involving silica gel where the drying duration to achieve $0.2\,g\,H_2O\,g^{-1}$ DW is highly dependent on initial seed water content).

When cooled rapidly (200 °C/min) and rehydrated rapidly immediately after warming, germination *sensu stricto* was observed in 53% of seeds at $0.2\,g\,H_2O\,g^{-1}$ DW, while at other water contents, germination was very low or nil (Table 1). However, none of these germinated seeds produced normal seedlings and, after 6 months in culture, no further development was noted. By contrast, after desiccation to $0.2\,g\,H_2O\,g^{-1}$ DW, if seeds were cooled to −50 °C at 4 °C/min prior to immersion in LN (slow cooling), 12% of seeds developed into normal seedlings.

After warming, when rehumidifying seeds in a 90% RH chamber at 25 °C for 24 h, Eira et al. (1999) observed that 10–30% of seeds cooled rapidly after desiccation to $0.2\,g\,H_2O\,g^{-1}$ DW produced normal seedlings. Vasquez (unpubl. results) obtained similar percentages of seedling recovery with seeds cooled rapidly and osmo-conditioned (see Sect. 2.1.4) after warming. As shown hereafter with slowly cooled seeds, controlling seed rehydration after warming in

Table 1. Germination (%) of *C. arabica* seeds desiccated to various water contents and cooled, or not (desiccation control), slowly or rapidly, and development of normal seedlings (%). Slow cooling consisted of precooling seeds to $-50\,^{\circ}$C at $4\,^{\circ}$C min^{-1} prior to immersion in liquid nitrogen. Rapid cooling consisted of directly immersing seeds in LN (average cooling rate $200\,^{\circ}$C min^{-1}). From Dussert *et al.* (1998)

Water content ($g\,H_2O\,g^{-1}$ dw)	Treatment	Germination (%)	Normal seedlings (%)
0.37	Desiccation control	98	94
	Slow cooling	0	0
	Rapid cooling	0	0
0.28	Desiccation control	97	90
	Slow cooling	4	0
	Rapid cooling	0	0
0.20	Desiccation control	90	85
	Slow cooling	63	12
	Rapid cooling	53	0
0.14	Desiccation control	90	79
	Slow cooling	35	0
	Rapid cooling	4	0

order to avoid imbibitional injury appears to be a key factor to achieve seedling recovery after rapid cooling.

2.1.2 Effect of the Cooling Rate

A significant effect of the cooling rate on the production of normal seedlings was observed in *C. arabica* seeds cooled to $-50\,^{\circ}$C before immersion in LN (Dussert et al. 1998). Seeds cooled at $20\,^{\circ}$C/min did not survive, but 13.0 and 24.2% of seeds developed into normal seedlings when cooled at 4 and $2\,^{\circ}$C/min, respectively. Lowering the cooling rate to $1\,^{\circ}$C/min did not increase significantly the percentage of normal seedlings recovered from treated seeds.

2.1.3 Effect of Cooling Temperature

The effect of the cooling temperature was investigated after desiccation to $0.2\,g\,H_2O\,g^{-1}$ DW: seeds were cooled to 0, -20, -50 and $-100\,^{\circ}$C at $1\,^{\circ}$C/min, then warmed directly (cooling controls) or immersed in LN before warming.

There was no detrimental effect of cooling alone to temperatures of 0, -20 and $-50\,^{\circ}$C; in all cases, around 95% of the seeds developed into normal seedlings (Fig. 1). However, a drastic decline in viability was observed when seeds were cooled to $-100\,^{\circ}$C, since only 24% of them produced normal seedlings.

For cooling to temperatures above $-50\,^{\circ}$C, viability of seeds then immersed in LN after cooling was always significantly lower than that of

Fig. 1. Effect of cooling temperature on the rate of normal seedlings produced (%) from *C. arabica* cooling control (○) and cryopreserved seeds (□). Points followed by the same letter were not significantly different at the 0.05 probability level as determined by Ryan's test. (Dussert et al. 1997a)

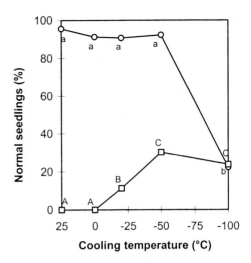

cooling controls. By contrast, when seeds were cooled to −100 °C, there was no difference in survival between cooling controls and cryopreserved seeds. However, the percentage of normal seedlings produced from cryopreserved seeds increased with lowering cooling temperatures down to −50 °C, from 0% without cooling up to 30% for a cooling temperature of −50 °C, and then decreased down to 25% for seeds cooled to −100 °C.

Thus, −50 °C appeared to be the optimal temperature for transfer of *C. arabica* seeds to LN as the optimal combination of two distinct effects: that of cooling and that of immersion in LN after cooling.

2.1.4 Effect of Post-warming Osmo-conditioning

Even if they are cryopreserved under optimal conditions, the percentage of cryopreserved seeds which develop into normal seedlings remains low, about 17% on average (Dussert et al. 2000a), and the growth rate of these seedlings is dramatically reduced in comparison with unfrozen controls. Moreover, varying the warming rate had no beneficial effect on both viability and vigour of cooled seeds (Dussert et al. 1998). Thus, in order to improve recovery of normal seedlings after cryopreservation, various post-warming treatments were investigated. It was shown for the first time that seed osmo-conditioning – controlled rehydration of seeds in a solution with low osmotic potential – could have a dramatic beneficial effect on seedling recovery after cryopreservation (Dussert et al. 2000b).

The effect of an osmo-conditioning treatment after warming was first investigated with seeds of two *C. arabica* varieties, Bourbon and Typica. Bourbon seeds were treated for 1 or 2 weeks using solutions with osmotic potentials of −1, −2 and −4 MPa, while osmotic potentials of −1, −1.25 and −1.5 MPa in combination with osmo-conditioning durations of 2, 4 or 6 weeks were tested with Typica seeds. Osmo-conditioning was carried out at 27 °C in the dark by placing batches of 10 seeds in Petri dishes sealed with Parafilm ribbon on a thin layer of cotton wool imbibed with 20 ml of aqueous PEG 6000 solution.

With Bourbon seeds, the beneficial effect of osmo-conditioning after warming was observed both on the final percentage of seedlings recovered and their mean germination time (Fig. 2). The time to produce half of the final percentage of normal seedlings was about three-fold lower with osmo-conditioned seeds than with non-osmo-conditioned seeds (12–13 vs. 36 days). Moreover, after a 2-week osmo-conditioning treatment with solutions with osmotic potential of –1 and –2 MPa, the percentage of seedlings recovered from cryopreserved seeds was 34–39%, against 13% only for cryopreserved seeds that were not osmo-conditioned.

The beneficial effect of osmo-conditioning on the percentage of seedlings produced by cryopreserved Typica seeds was significant in all treatments using PEG solutions of –1 and –1.25 MPa: the percentage of seedlings produced from treated seeds ranged from 50–74%, against 16% for untreated seeds (Table 2). Moreover, the percentage of seedlings obtained increased in line with increasing durations of osmo-conditioning treatment with seeds treated with a –1.25 MPa solution.

In the framework of an IPGRI-funded project, the possibility of using the protocol described in Fig. 3 as a standard protocol for cryopreservation of *C. arabica* seeds at CATIE (Costa Rica) was tested with 25 accessions of the CATIE's *C. arabica* germplasm collection (Vasquez, unpubl. results). Four categories of accessions could be defined with regard to seedling recovery after the three treatments studied: desiccation alone, desiccation and cryopreservation, followed or not by an osmo-conditioning treatment (Table 3). For 23 accessions (categories A and B), the individual statistical analysis showed that seedling production from osmo-conditioned cryopreserved seeds was signifi-

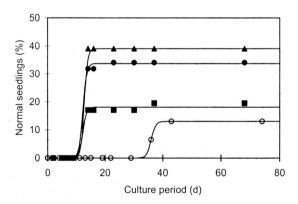

Fig. 2. Germination of cryopreserved seeds of *C. arabica* var. Bourbon after *a* 1-week or *b* 2-week osmo-conditioning treatment with PEG solutions with osmotic potentials of –1 (▲), –2 (●) and –4 (■) MPa or without osmo-conditioning treatment (○). (Dussert et al. 2000b)

Table 2. Percentage of normal seedlings recovered from cryopreserved seeds of *C. arabica* variety Typica after a 2-, 4- and 6-week osmo-conditioning treatment with PEG solutions with osmotic potentials of –1, –1.25 and –4 MPa or without osmo-conditioning treatment. Percentages of normal seedlings followed by the same letter were not significantly different at the P = 0.05 level. From Dussert *et al.* (2000b)

Post-warming treatment	Osmotic potential (MPa)	Normal Seedlings (%)
Control	0	16[a]
2-week osmo-conditioning	–1	52[b]
	–1.25	50[b]
	–1.5	2[a]
4-week osmo-conditioning	–1	54[b]
	–1.25	62[b]
	–1.5	22[a]
6-week osmo-conditioning	–1	64[b]
	–1.25	74[b]
	–1.5	10[a]

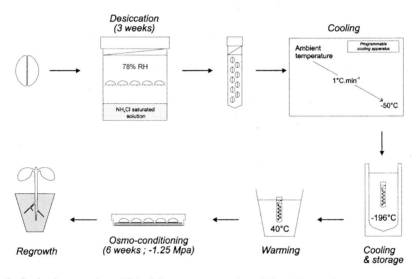

Fig. 3. Optimal protocol established for cryopreservation of *C. arabica* seeds

cantly higher than from cryopreserved seeds that were not osmo-conditioned after warming. Moreover, for 18 (category A) out of these 23 accessions, the percentage of seedlings recovered from cryopreserved and osmo-conditioned seeds was not significantly different from that of the desiccation controls. No beneficial effect of the osmo-conditioning treatment could be detected for one (category C) of the 25 accessions tested because of the very low detrimental effect of cryopreservation observed with this accession. For the remaining

Table 3. Categorization of 25 accessions of CATIE's *C. arabica* collection according to survival after cryopreservation, followed or not by osmo-conditioning, in comparison with desiccation controls: number of accessions in each category, *n*, and mean *(minimum-maximum)* percentage of normal seedlings recovered from seeds after: i) desiccation to $0.2\,g\,H_2O\,g^{-1}$ dw; ii) desiccation and cryopreservation (cooling to $-50\,°C$ at $1\,°C\,min^{-1}$ prior to immersion in liquid nitrogen); and, iii) desiccation, cryopreservation and osmo-conditioning (6-week treatment with a $-1.25\,MPa$ PEG solution)

Category	*n*	Seedling recovery (%)		
		Desiccation	Desiccation + cryopreservation	Desiccation + cryopreservation + osmo-conditioning
A	18	57 *(20–85)*	21 *(0–64)*	58 *(24–82)*
B	5	76 *(52–93)*	12 *(4–30)*	43 *(28–64)*
C	1	100	74	78
D	1	44	0	0

accessions (category D), no seedlings could be recovered after cryopreservation, with or without osmo-conditioning treatment after warming.

This study confirmed the efficiency of the protocol employed and its applicability for long-term conservation of *C. arabica* germplasm. A duplicate of CATIE's *C. arabica* core collection, which comprises 74 accessions, is currently being cryopreserved and it is planned that a duplicate of the whole core collection will be conserved for the long-term under cryopreservation by the end of the year 2000.

2.2 Cryopreservation of *Coffea* spp. Seeds in Relation to Calorimetric Properties of Tissue Water and Lipid Composition

The effect of exposure to LN temperature on viability of seeds desiccated to various water contents was investigated in nine coffee species (Dussert et al. 2001). Three groups of species could be distinguished based on seed survival after LN exposure. In group 1 species (*C. brevipes, C. canephora, C. liberica* and *C. stenophylla*), no seedling production could be obtained after LN exposure due to endosperm injury. In group 2 species (*C. arabica* and *C. eugenioides*), recovery was very low or nil after rapid cooling and only moderate after slow cooling. In group 3 species (*C. pseudozanguebariae, C. racemosa* and *C. sessiliflora*), very high percentages of seedling development were observed after both rapid and slow cooling. A high interspecific variability for the high moisture freezing limit was observed within the species of groups 2 and 3 since it ranged from $0.14–0.26\,g\,H_2O\,g^{-1}$ DW. A very highly significant correlation was found for those species between the unfreezable water content, as determined from DSC analysis, and the high moisture freezing limit of their seeds. No significant correlation was found between seed lipid content, which varied from 9.8–34.6% DW, and survival after LN exposure. However, a negative relationship was found between seed unfreezable water content and lipid content. Interspecific differences in fatty acid composition of seed lipids resulted in a high variability in the percentage of unsaturated fatty acids which ranged from 28.7–54.4% among the nine species studied. For all species studied, a highly significant correlation was found between the percentage of unsaturated fatty acids and the percentage of seedling recovery after rapid or slow cooling.

3 Cryopreservation of Excised Zygotic Embryos

3.1 Zygotic Embryos Excised from Seeds Before Cryopreservation

Successful cryopreservation of zygotic embryos, extracted from mature seeds, was achieved for *C. liberica* (Normah and Vangadasalam 1992), *C. arabica* (Abdelnour-Esquivel et al. 1992; Florin et al. 1993), *C. canephora* and the interspecific hybrid arabusta (Abdelnour-Esquivel et al. 1992). The simple protocol established by Abdelnour-Esquivel et al. is presented in Fig. 4.

With all species studied, partial dehydration of excised embryos was sufficient to obtain high survival rates after cryopreservation. Cooling was carried out by direct immersion in liquid nitrogen and embryos were warmed rapidly by plunging the cryotubes in a 40 °C water-bath.

Desiccation could be either rapid (Normah and Vangadasalam 1992; Abdelnour-Esquivel et al. 1992), by placing embryos under the air current of

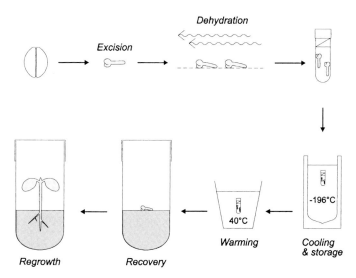

Fig. 4. Cryopreservation of coffee zygotic embryos involving their excision before desiccation as described by Abdelnour-Esquivel et al. (1992). (Adapted from Dussert et al. 1997b)

a laminar flow cabinet, or slow, i.e. 5 days in a chamber with a relative humidity of 43% (Florin et al. 1993). The optimal desiccation duration depended on both the initial water content of embryos and on the desiccation method employed. For *C. arabica*, the maximal survival (96%) was achieved after 0.5 h of rapid desiccation, from 43% moisture content (MC) down to 16% MC. However, no embryos survived if they were rapidly desiccated to 7.7% MC, whereas a survival of 63% was obtained when embryos were slowly dehydrated to 6.5% MC (Florin et al. 1993). Maximal survival of *C. liberica* embryos also achieved at 16% MC was 60% (70% for the dehydrated control). Survival rates of 84% (27% MC) and 42% (29% MC) were obtained for the arabusta hybrid and *C. canephora*, respectively, after 0.5 h rapid drying, suggesting that survival could be improved if MC were further reduced.

If monitoring the moisture content appeared to be the key factor for successful cryopreservation of zygotic embryos, other factors could still influence survival. When placing embryos on a culture medium containing 0.6 M sucrose during the 5 days of slow dehydration, survival increased from 63% to 93% of the dehydrated controls (Florin et al. 1993). Moreover, the formulation of the recovery medium may be sub-optimal and, thus, germination would not reflect the potential survival rate of cryopreserved embryos (Engelmann et al. 1995). Immature embryos of *C. arabica* displayed lower survival than mature ones after cooling when recovery took place on the standard medium (Abdelnour-Esquivel et al. 1992). Addition of $100 \, \text{mg} \, \text{l}^{-1}$ GA_3 to the recovery medium increased the survival of cryopreserved immature embryos of *C. arabica* from 52 to 83%, a value similar to that obtained with mature embryos (Abdelnour-Esquivel et al. 1992).

3.2 Zygotic Embryos Excised from Seeds After Cryopreservation

Although their efficiency has been demonstrated for coffee species, cryopreservation protocols developed for zygotic embryos present several constraints and disadvantages for routine use in coffee gene banks: (1) the need to work under aseptic conditions before cryopreservation; (2) the difficulty in achieving reproducible desiccation conditions when using the sterile airstream of a laminar flow cabinet or silica gel: since the desiccation periods are generally very short, they need to be very precise and are highly dependent on the initial moisture content of the samples and of the characteristics of the air-flow or the silica gel used; and (3) the impossibility to treat simultaneously large amounts of material since the time needed to extract one embryo (1–2 min for a single coffee seed) is very long compared with the optimal desiccation period (e.g., 30 min under the laminar flow for coffee embryos with an initial MC of 60% (fresh weight basis), as reported by Abdelnour-Esquivel et al. (1992).

We have thus developed a new and simple approach which consists of drying *C. arabica* seeds to $0.2 \, \text{g} \, H_2O \, \text{g}^{-1}$ DW under controlled RH, cooling them by direct immersion in LN and extracting embryos from the seed after warming (Fig. 5). This new method avoids many of the problems encountered

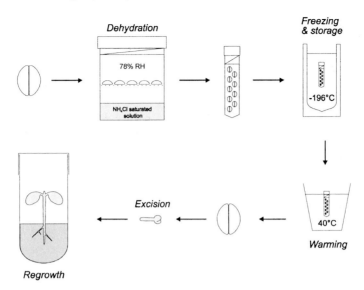

Fig. 5. Cryopreservation of *C. arabica* zygotic embryos involving their excision from seeds after warming as described in Dussert et al. (1997a)

with traditional zygotic embryo cryopreservation protocols for all coffee species and, possibly, for other species which produce intermediate seeds of relatively small size. Equilibrating coffee seeds under 78% RH allowed the seeds to reach an optimal water content for cryopreservation without any viability loss, in a very easy and reproducible manner. This also allowed the processing of large amounts of seeds at the same time. Moreover, aseptic conditions were required only after warming. This method was very efficient since, when embryos were extracted from seeds after cooling, no decrease in viability was observed after cryopreservation (97% survival) in comparison with uncooled controls. The only drawback of this approach, compared with classical protocols, is that it requires a larger volume for storing seeds in LN containers, in comparison with excised embryos.

4 Cryopreservation of Apices

Cryopreservation of coffee apices has only been reported by Mari et al. (1995). Experiments were carried out on *C. racemosa* and *C. sessiliflora*. The successive steps of the method developed by these authors are illustrated in Fig. 6.

Nodal cuttings, isolated from in vitro plantlets of the coffee germplasm collection, were placed on standard medium supplemented with $0.5\,mg\,l^{-1}$ 6-benzyladenine (BA) and $0.2\,mg\,l^{-1}$ naphthaleneacetic acid (NAA) to induce growth of axillary shoots. After 3 weeks, apices, consisting of the meristematic

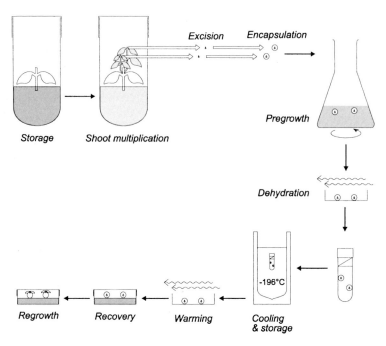

Storage   Shoot multiplication

Fig. 6. Cryopreservation of coffee apices as described by Mari et al. (1995). (Reproduced from Dussert et al. 1997b)

dome, the subapical zone and one or two leaf primordial, were excised from axillary shoots. Apices were left overnight on the standard medium to recover from dissection stress, encapsulated in 3% calcium alginate beads, pretreated in liquid medium with high sucrose concentration for various durations, then partially dehydrated and cryopreserved. Apices of *C. sessiliflora* required a 3–10 day growth in liquid medium containing 0.75 M sucrose to achieve survival after cryopreservation, whereas those of *C. racemosa* required a progressive increase of the sucrose concentration from 0.5 to 1 M. Survival was 27 and 38% for *C. racemosa* and *C. sessiliflora*, respectively. Growth recovery of frozen apices took place either directly with the production of foliar primordia, or through callusing, with a frequency varying according to the pregrowth treatment.

5 Cryopreservation of Somatic Embryos

Research on cryopreservation of somatic embryos was carried out with the two cultivated species *C. arabica* and *C. canephora*. Various techniques were employed successfully including conventional slow freezing (Bertrand-Desbrunais et al. 1988), simplified freezing (Tessereau 1993), encapsulation-dehydration (Hatanaka et al. 1994) and desiccation (Tessereau et al. 1994; Mycock et al. 1995).

The conventional protocol developed by Bertrand-Desbrunais et al. (1988) for somatic embryos of *C. arabica* consisted of cultivating clumps of globular embryos for 24h on a medium enriched with sucrose, pretreating them with liquid medium containing sucrose and dimethyl sulfoxide (DMSO), and then cooling them slowly with a programmable freezing apparatus. Recovery was observed in 50% of cryopreserved samples through secondary embryogenesis only and no direct regrowth of embryos could be achieved.

Recovery through secondary embryogenesis was also observed by Tessereau (1993) with heart-stage *C. canephora* embryos cryopreserved using a simplified freezing process which consisted of placing the samples in the −20 °C compartment of a commercial refrigerator for freezing. This protocol ensured very high survival since adventitious embryos could be regenerated from up to 100% of the cryopreserved material.

The encapsulation-dehydration technique gave direct regrowth of heart- and torpedo-stage somatic embryos of *C. canephora* (Hatanaka et al. 1994). After pregrowth on media containing progressively increased sucrose concentrations, somatic embryos were encapsulated in alginate beads, pregrown for 1 day in liquid medium containing 0.5 M sucrose, dehydrated until bead water content reached 13% (fresh weight basis) and cooled rapidly. Under these conditions, 63% of embryos remained alive and half of them developed into whole plantlets.

Somatic embryos of *C. canephora* and *C. arabica* could also withstand cryopreservation and directly regenerate whole plants using the desiccation technique. *C. canephora* somatic embryos were submitted to a 12-week freeze-hardening treatment consisting of a culture on media containing high sucrose concentrations and abscisic acid followed by a 7-day desiccation period under 75% RH (Tessereau et al. 1994). After rapid cooling, 64% of cryopreserved embryos developed directly into plantlets. Mycock et al. (1995) obtained comparable results with *C. arabica* embryos using desiccation but under very different conditions. After a short pretreatment with a mixture of glycerol and sucrose followed by rapid desiccation and cooling, 70% of *C. arabica* embryos regenerated plantlets.

6 Conclusions

Different cryopreservation protocols are now available for coffee seeds, zygotic embryos, apices and somatic embryos. They have been successfully applied to a relatively large range of coffee species. The choice of the most appropriate material to cryopreserve will depend on the species and on the objective of conservation. Seeds represent the material of choice for the long-term conservation of *C. arabica* genetic diversity, since this species is autogamous. For other species that are allogamous, cryopreserving seeds will ensure gene conservation. Even though the technique still must be optimised, it is already possible to apply it for the establishment of cryopreserved *C. arabica* germplasm collections.

Apices and somatic embryos will be employed for the conservation of particular genotypes. The techniques developed for cryopreserving somatic embryos are sufficiently advanced to envisage their routine utilisation in the near future. However, their applicability will be restricted in terms of germplasm conservation, since the capacity to produce somatic embryos is highly genotype-specific, and embryogenic lines have only been obtained from a limited number of genotypes. As regards apices, the preliminary results obtained have shown that they can be successfully cryopreserved using the encapsulation-dehydration technique. However, before routine cryopreservation of apices can be envisaged, additional research is needed to improve the encapsulation-dehydration technique, test other cryopreservation techniques and assess their efficiency with a large number of species and genotypes.

In conclusion, considerable progress has been made over the last few years in the development of cryopreservation for coffee. It can be envisaged that, in the near future, cryopreservation will be routinely used among other techniques to ensure the long-term and cost-effective conservation of coffee genetic resources.

References

Abdelnour-Esquivel A, Villalobos V, Engelmann F (1992) Cryopreservation of zygotic embryos of *Coffea* spp. Cryo Lett 13:297–302

Berthaud J, Charrier A (1988) Genetic resources of coffee. In: Clarke RJ, Macrae R (eds) Coffee, vol 4. Agronomy. Elsevier, London, pp 1–42

Becwar MR, Stanwood PC, Lehonhardt KW (1983) Dehydration effects on freezing characteristics and survival in liquid nitrogen of desiccation-tolerant and desiccation-sensitive seeds. J Am Soc Hortic Sci 108:613–618

Bertrand-Desbrunais A, Fabre J, Engelmann F, Dereuddre J, Charrier A (1988) Reprise de l'embryogenèse adventive à partir d'embryons somatiques de caféier (*Coffea arabica* L.) après leur congélation dans l'azote liquide. CR Acad Sci Paris 307:795–801

Bridson DM, Verdcourt B (1988) Flora of tropical East Africa – *Rubiaceae*, part 2. Polhill RM (ed). Balkema, Rotterdam, 227 pp

Charrier A, Berthaud J (1985) Botanical classification of coffee. In: Clifford MN, Willson KC (eds) Coffee. Botany, biochemistry and production of beans and beverage, Am edn. Westport, Connecticut, pp 13–47

Couturon E (1980) Le maintien de la viabilité des graines de caféiers par le contrôle de leur teneur en eau et de la température de stockage. Café Cacao Thé 1:27–32

Dussert S, Chabrillange N, Engelmann F, Anthony F, Hamon S (1997a) Cryopreservation of coffee (*Coffea arabica* L.) seeds: importance of the precooling temperature. Cryo Lett 18:269–276

Dussert S, Chabrillange N, Engelmann F, Anthony F, Noirot M, Hamon S (1997b) In vitro conservation of coffee (*Coffea* spp.) germplasm. In: Razdan MK, Cocking EC (eds) Conservation of plant genetic resources in vitro. M/s Science Publishers, Enfield, USA, pp 287–305

Dussert S, Chabrillange N, Anthony F, Engelmann F, Recalt C, Hamon S (1997c) Variability in storage response within a coffee (*Coffea* spp.) core collection under slow growth conditions. Plant Cell Rep 16:344–348

Dussert S, Chabrillange N, Engelmann F, Anthony F, Louarn J, Hamon S (1998) Cryopreservation of seeds of four coffee species (*Coffea arabica*, *C. costatifructa*, *C. racemosa* and *C. sessiliflora*): importance of water content and cooling rate. Seed Sci Res 8:9–15

Dussert S, Chabrillange N, Engelmann F, Hamon S (1999) Quantitative estimation of seed desiccation sensitivity using a quantal response model: application to nine species of the genus *Coffea* L. Seed Sci Res 9:135–144

Dussert S, Chabrillange N, Engelmann F, Anthony F, Hamon S (2000a) Cryopreservation of coffee (*Coffea arabica* L.) seeds: toward a simplified protocol for routine use in coffee genebanks. In: Engelmann F, Takagi H (eds) Cryopreservation of tropical plant germplasm – current research progress and application. JIRCAS, Tsukuba, Japan and IPGRI, Rome, Italy, pp 161–166

Dussert S, Chabrillange N, Vasquez N, Engelmann F, Anthony F, Guyot A, Hamon S (2000b) Beneficial effect of post-thawing osmoconditioning on the recovery of cryopreserved coffee (*Coffea arabica* L.) seeds. Cryo Lett 21:47–52

Dussert S, Chabrillange N, Rocquelin G, Engelmann F, Lopez M, Hamon S (2001) Tolerance of coffee (*Coffea* spp.) seeds to ultra-low temperature exposure in relation to calorimetric properties of tissue water, lipid composition and cooling procedure. Physiol Plant 112:495–504

Eira MTS, Walters C, Caldas LS, Fazuoli LC, Sampaio JB, Dias MC (1999) Tolerance of *Coffea* spp. seeds to desiccation and low temperature. Rev Brasil Fisiol Veg 11:97–105

Ellis RH, Hong TD, Roberts EH (1990) An intermediate category of seed storage behaviour? I. Coffee. J Exp Bot 41:1167–1174

Engelmann F, Dumet D, Chabrillange N, Abdelnour-Esquivel A, Assy-Bah B, Dereuddre J, Duval Y (1995) Factors affecting the cryopreservation of coffee, coconut and oil palm embryos. Plant Gen Res Newslett 103:27–31

Florin B, Tessereau H, Pétiard V (1993) Conservation à long terme des ressources génétiques de caféier par cryoconservation d'embryons zygotiques et somatiques et de cultures embryogènes. In: Proceedings of the 15th ASIC, Montpellier, France, pp 106–113

Hatanaka T, Yasuda T, Yamagushi T, Sakai A (1994) Direct regrowth of encapsulated somatic embryos of coffee (*Coffea canephora*) after cooling in liquid nitrogen. Cryo Lett 15:47–52

Mari S, Engelmann F, Chabrillange N, Huet C, Michaux-Ferrière N (1995) Histo-cytological study of apices of coffee (*Coffea racemosa* and *C. sessiliflora*) in vitro plantlets during their cryopreservation using the encapsulation-dehydration technique. Cryo Lett 16:289–298

Mycock DJ, Wesley-Smith J, Berjak P (1995) Cryopreservation of somatic embryos of four species with and without cryoprotectant pre-treatment. Ann Bot 75:331–336

Normah MN, Vengadasalam M (1992) Effects of moisture content on cryopreservation of *Coffea* and *Vigna* seeds and embryos. Cryo Lett 13:199–208

Tessereau (1993) Development of a simplified method for the cryopreservation of plant tissues and somatic embryos and study of the acquisition of freezing tolerance. PhD Thesis, University of Paris VI, France

Tessereau H, Florin B, Meschine C, Thierry C, Pétiard V (1994) Cryopreservation of somatic embryos: a tool for germplasm storage and commercial delivery of selected plants. Ann Bot 74:547–55

Van der Vossen HAM (1977) Methods of preserving the viability of coffee seed in storage. Kenya Coffee 45:31–35

III.4 Cryopreservation of *Eucalyptus* sp. Shoot Tips by the Encapsulation-Dehydration Procedure

M. Pâques, V. Monod, M. Poissonnier[1], and J. Dereuddre[2]

1 Introduction

Cryopreservation allows long-term storage of plant cells, tissues or organs in a stable state without physiological alterations (Kartha 1985). It is successfully applied not only to genetic resources conservation (Pâques et al. 1997; Poissonnier et al. 1997; Zhao et al. 1999), but also to facilitate breeding, especially within the framework of woody species improvement programs.

Cryopreservation is particularly useful to maintain woody species in a juvenile status; otherwise, they quickly lose rooting ability with age. Indeed selected genotypes in the juvenile stage can be stored in liquid nitrogen and as long as necessary for the estimation of the tree value under field test conditions (Smith 1997). It also is a technique of interest for retarding aging of shoot tips from stock plants grown either in vitro or in situ.

For these different reasons, AFOCEL has developed a research program on the cryopreservation of trees of industrial interest as well as endangered broad-leave trees. Among those, Eucalyptus is attractive for rapid production of considerable woody biomass. However, its culture in France is limited by its cold sensitivity. A large program of selection and genetic improvement (Marien 1982) was undertaken to develop fast growing clones adapted to French climatic conditions. Eucalyptus was chosen as a model in this study to set up bud cryopreservation for two main reasons: in vitro culture of Eucalyptus is now well handled and produces a large number of buds for experimentation; moreover, the fast growth of Eucalyptus allows in 2–3 years the evaluation of the eventual impact of cryopreservation on the recovered "cryo-plants" grown in a field test.

A method of encapsulation-dehydration has recently been proposed for cryopreservation of shoot tips (Dereuddre et al. 1990; Plessis et al. 1992). This procedure has two major advantages: it is easy to perform and does not require a sophisticated cooling apparatus. It involves encapsulation of organs in calcium alginate beads, preculture with sucrose as a single cryoprotective agent, dehydration over dry silica gel, direct cooling in liquid nitrogen

[1] AFOCEL, Station de Biotechnologies, Domaine de l'Etançon, 77370 Nangis, France e-mail: biotech@afocel.fr – Personal address: Dr. M. Pâques 49 qua Franklin Roosevelt 77920 Sahois/Seiwe

[2] Université P. et M. Curie, Laboratoire de Cryobiologie Végétale, 12 rue Cuvier, 75230 Paris cedex 05, France

(LN), and slow rewarming in air at room temperature. This procedure was successfully applied to a cold tolerant clone of *Eucalyptus gunnii* (Poissonnier et al. 1992). In order to define optimal conditions for the cryopreservation of Eucalyptus shoot tips, several preculture conditions were studied: sucrose concentration, preculture duration, and preculture with other cryoprotective agents.

The optimal conditions developed for *Eucalyptus gunnii* were applied to the cryopreservation of two other cold-sensitive Eucalyptus species: *Eucalyptus globulus* and *Eucalyptus gunnii* × *Eucalyptus dalrympleana*.

2 Materials and Methods

2.1 Plant Material and Culture Conditions

Shoot tips excised from *Eucalyptus gunnii* (clone 034) in vitro plantlets were used to determine the best cryopreservation protocol. Improved methods were then applied to two other species: *Eucalyptus globulus* (clone 010) and a hybrid *Eucalyptus gunnii x Eucalyptus dalrympleana* (clone 208).

In vitro stock cultures were maintained on a modified Murashige and Skoog (MS) medium (Murashige and Skoog 1962) containing growth regulators (6-benzylaminopurine and naphthaleneacetic acid), $30\,g\,l^{-1}$ sucrose, and gelled with $7\,g\,l^{-1}$ Bacto-agar (Difco, Detroit, USA). The pH of the medium was adjusted to 5.6 before autoclaving at $110\,°C$ for 30 min. Cultures were placed under a light intensity of $60\,\mu mol\,s^{-1}\,m^{-2}\,PAR$, with a 16-h light/8-h dark photoperiod at a temperature of $+25\,°C$ (day) and $+20\,°C$ (night).

2.2 Cryopreservation Procedure

Steps described below were performed aseptically. The different solutions were autoclaved at $110\,°C$ for 30 min before use.

For cryopreservation, shoot tips (0.5–1 mm in length) were excised from in vitro plantlets at the end of a 3-week subculture. They consisted of the apical dome, two or four primordial leaves, and a small tissue base. They were trapped in calcium alginate beads and transferred to stepwise sucrose-enriched liquid medium. Preculture was performed on a rotary shaker (100 rpm) under the same temperature and light conditions as described above for stock cultures.

Several concentrations of sucrose (0.3–1.5 M) were tested. Sucrose was progressively added every 48 h (slow increase in sucrose concentration) to obtain successive concentrations of 0.3, 0.5, 0.75, 1, 1.25 and 1.5 M. The preculture duration depended on the final sucrose concentration and ranged from 2 days for 0.3 M, to 12 days for 1.5 M.

In order to decrease the total preculture duration, shorter preculture durations (2 or 3 days) were tested with some of the concentrations (0.5–1.25 M) allowing tolerance to liquid nitrogen (LN). After overnight preculture with 0.3 M sucrose, sucrose concentration was increased to 0.5 M and 0.75 M during the following day, and then to 1 and 1.25 M during the second day. Once the final concentration had been reached, encapsulated shoot tips were maintained in the preculture medium for 16 h (for 0.75 and 1.25 M) or 24 h (for 0.5 and 1 M). Under these conditions, preculture duration was reduced to 1 day for 0.1 and 0.3 M, to 2 days for 0.5 and 0.75 M, and to 3 days for 1 and 1.25 M. Different periods of incubation (1–6 days) were also tested, with 1 M final sucrose concentration. Control shoot tips were incubated for 2 days in medium containing sucrose at standard concentration (0.1 M).

Once the final concentration had been reached, encapsulated shoot tips were maintained in the preculture medium for 16 h (for 0.75 and 1.25 M) or 24 h (for 0.5 and 1 M). Different periods of incubation (1–6 days) were also tested, with 1 M final sucrose concentration. Control shoot tips were incubated for 2 days in medium containing sucrose at standard concentration (0.1 M).

Other sugars were tested in preculture in comparison with sucrose: a trisaccharide (raffinose), a disaccharide (trehalose), and a polyol (sorbitol). They were added progressively over 2 days to the preculture medium to a final concentration of 1 M (i.e., 0.1 M sucrose + 0.9 M cryoprotectant). Shoot tips were then maintained in this medium for 4 days before dehydration and cooling treatments.

After preculture, encapsulated shoot tips were dehydrated over silica gel at room temperature (+23 °C) for 6 h, in order to obtain 0.22 ± 0.02 g residual water in beads/g dry weight. Dehydrated beads were then placed in cryovials and immediately immersed in LN for 1 h. They were rewarmed in air at room temperature for 15 min before transfer to standard culture medium in Petri dishes. After 7 days of culture, shoot tips were extracted from beads and placed on fresh culture medium. Bead water content, expressed in g water/g dry weight, was estimated for 20 beads after encapsulation, preculture and dehydration. Bead dry weight (DW) was assessed after oven drying at +70 °C.

Survival of shoot tips was assessed after the different steps of the cryopreservation procedure (encapsulation, preculture, dehydration and freezing). Results were expressed as percentage of shoot recovery: i.e., apices developing new axillary shoots 6 weeks after plating. Three replicates were performed for each experiment, using 20–30 shoot tips per set of conditions.

3 Results

3.1 Effect of Final Sucrose Concentration in Preculture Medium

3.1.1 Slow Increase in Sucrose Concentration

The water content in beads was initially 12.13 g H_2O/g DW. It was not modified by preculture with 0.1 M sucrose, the standard sucrose concentration in culture medium (Table 1). After stepwise preculture in medium enriched with

Table 1. Effects of final sucrose concentration (M) in the preculture medium on the water content in beads (g H₂O/g DW) after preculture and after 6 hours of dehydration. Values in brackets represent standard deviation

Treatment	Final sucrose concentration (M) in preculture medium						
	0.1	0.3	0.5	0.75	1.0	1.25	1.5
preculture	**12.13**	**5.93**	**4.14**	**2.76**	**1.98**	**1.51**	**1.19**
	(0.95)	*(0.45)*	*(0.03)*	*(0.21)*	*(0.05)*	*(0.01)*	*(0.00)*
dehydration	**0.19**	**0.20**	**0.21**	**0.21**	**0.22**	**0.24**	**0.23**
	(0.04)	*(0.09)*	*(0.01)*	*(0.01)*	*(0.01)*	*(0.01)*	*(0.00)*

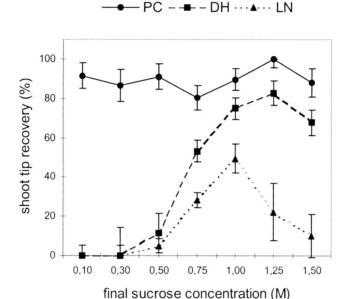

Fig. 1. Shoot tip recovery versus the final sugar concentration slowly increased in the preculture medium. Recovery (in % of treated shoot tips) was determined at different steps of the cryopreservation procedure: preculture (*PC*), dehydration (*DH*) and cooling in liquid nitrogen (*LN*). *Vertical bars* represent standard deviation

sucrose, the water content in beads decreased as a function of the final sugar concentration, from 5.93 (0.3 M) to 1.19 g H₂O/g DW (1.5 M).

During dehydration with silica gel, the water content in beads decreased sharply to between 0.19 and 0.23 g H₂O/g DW. During dehydration, water loss was greater for lower preculture medium sucrose concentrations; beads precultured with 0.1 M sucrose lost 11.94 g H₂O/g DW compared with 0.96 g for beads treated with 1.5 M sucrose.

After encapsulation, 80–100% of shoot tips resumed growth. These percentages were not affected by the preculture, whatever the final sucrose concentration in the preculture medium (Fig. 1).

Tolerance to dehydration (12% recovery) was first noted after preculture with 0.5 M sucrose. Shoot recovery increased progressively with sucrose concentration, reached a maximum (83%) at 1.25 M, and decreased to about 65% at the highest sucrose concentration tested (1.5 M).

Like tolerance to dehydration, tolerance to cooling in LN required high sucrose levels in the preculture medium. Shoot recovery after freezing increased with sucrose concentration in preculture medium, from 5% (for 0.5 M) to 1 M (49%), and then decreased at higher concentrations. After preculture with 1.5 M sucrose, 10% of shoot tips survived and grew after freezing.

Control shoot tips precultured for 2 days on 0.1 M sucrose did not withstand dehydration and subsequent cooling in LN.

3.1.2 Rapid Increase in Sucrose Concentration

The water content in precultured beads more rapidly exposed to the final sucrose concentration did not differ significantly from values obtained after slow increase in sucrose concentration. Similar values after 6 h of dehydration were also obtained (data not shown). The recovery of precultured control shoot tips remained above 80% (Fig. 2). A rapid increase in sucrose concentration in the preculture medium did not affect shoot tip survival.

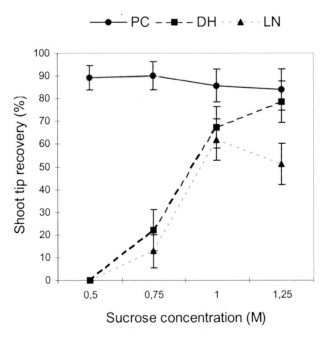

Fig. 2. Shoot tip recovery versus the final sucrose concentration (M) quickly increased in the preculture medium. Recovery (in % of treated shoot tips) was determined at different steps of the cryopreservation procedure: preculture (*PC*), dehydration (*DH*) and cooling in liquid nitrogen (*LN*). *Vertical bars* represent standard deviation

Shoot survival after dehydration and after cooling in LN depended on the sucrose concentration in the preculture medium. As with the slow increase in sucrose concentration in the preculture medium, the maximum recovery rate was obtained for 1 M sucrose. The survival after LN exposure was 62 and 49%, respectively, for the slow and rapid increase in sucrose concentration in the preculture medium. For 1.25 M sucrose in the preculture medium, the survival rate after cooling reached 52% for the quick procedure, while it decreased drastically to about 26% for the slower sugar increase in the preculture medium. A concentration of 1 M sucrose was therefore selected for further experiments.

3.2 Effects of Preculture Duration with 1 M Sucrose Concentration

After a 2-day increase in sucrose concentration in the preculture medium, encapsulated shoot tips were maintained for 1, 2, 4 or 6 days in 1 M sucrose preculture medium. Recovery of precultured shoot tips ranged from 83–93% (Fig. 3), and was not altered by an extended period of incubation with 1 M sucrose. After dehydration to 18% residual water, shoot tip recovery increased

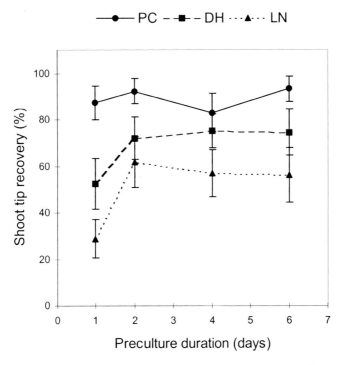

Fig. 3. Effects of the culture duration (days) in 1 M sucrose preculture medium on shoot tips recovery (in % of treated shoot tips) determined after preculture (*PC*), dehydration (*DH*) and cooling in liquid nitrogen (*LN*). *Vertical bars* represent standard deviation

from 52 to 72% when the incubation period was extended from 1 to 2 days. Recovery of LN-cooled shoot tips increased likewise from 29 to 62%. Further extension of preculture with 1 M sucrose (4 or 6 days) did not significantly change shoot recovery, which remained above 55%.

A 2-day preculture with increasing sucrose concentrations up to 1 M and a 4-day incubation with 1 M sucrose were chosen for subsequent experiments. These conditions will be referred to as the standard preculture.

3.3 Effects of Different Cryoprotectants

The osmotic pressure of the final preculture medium at 1 M concentration was similar for sucrose (1105 mOsmol/kg H_2O), trehalose (1110 mOsmol/kg H_2O) and sorbitol (1100 mOsmol/kg H_2O), and higher for raffinose (1360 mOsmol/kg H_2O). Nevertheless, the osmotic dehydration due to the preculture greatly depended on the cryoprotectant used. The highest water content in beads (3.77 g H_2O/g DW) was obtained with sorbitol, the lowest (1.23 g H_2O/g DW) with raffinose, a triholoside. Intermediate water contents were obtained with the disaccharides, 2.22 and 1.75 g H_2O/g DW for trehalose and sucrose, respectively. After 6 h of dehydration with silica gel, a similar low water content was obtained: 0.23 ± 0.02 g H_2O/g DW for all carbohydrate-treated beads.

Preculture with these different cryoprotective agents had no deleterious effect on the viability of precultured control shoot tips (Table 2). However, tolerance to dehydration and to freezing in liquid nitrogen depended on the carbohydrate used in the preculture medium. Sorbitol was ineffective in permitting survival after dehydration and freezing. Shoot recovery was only obtained with sugars as a cryoprotectant.

Highest recovery was observed with sucrose. Shoot tips precultured with trehalose showed low recovery after dehydration (40%), and after cooling (13%). With raffinose, 77% of shoot tips resumed growth after dehydration, but only 34% survived after cooling in LN. These results could not be related to different levels of dehydration since with all cryoprotectants tested the residual water content in beads was similar after 6 h of dehydration.

Table 2. Effects of cryoprotectant in the preculture medium (cryoprotectant 0.9 M + sucrose 0.1 M) on shoot tip recovery (in %) after preculture, dehydration and cooling in LN. Values in brackets represent standard deviation

Treatment	Cryoprotectant			
	Sucrose	Trehalose	Raffinose	Sorbitol
Preculture	**95.7**	**93.7**	**88.4**	**92.4**
	(11.35)	*(2.15)*	*(2.08)*	*(4.05)*
Dehydration	**82.5**	**39.8**	**77.1**	**0.0**
	(8.27)	*(24.77)*	*(10.39)*	*(0.0)*
Liquid nitrogen	**73.2**	**13.0**	**34.1**	**0.0**
	(7.15)	*(7.12)*	*(17.21)*	*(0.0)*

Table 3. Shoot tip recovery (in %) after cooling in LN. Three Eucalyptus species were tested: *E. gunnii* (clone 034), *E. gunnii x E. dalrympleana* (clone 208) and *E. globulus* (clone 010). The sucrose concentration in the preculture medium was increased slowly over 7 days (A) or rapidly over 2 days (B). Values in brackets represent standard deviation

Eucalyptus species	Slow increase in sucrose concentration	Rapid increase in sucrose concentration
E. gunnii (clone 034)	**40.1** *(17.9)*	**63.1** *(12.0)*
E. gunnii x dalrympleana (clone 208)	**10.9** *(5.4)*	**43.2** *(6.3)*
E. globulus (clone 010)	**8.7** *(4.8)*	**12.7** *(11.9)*

3.4 Application to Other Clones of *Eucalyptus*

The cryopreservation procedure was assessed for two other *Eucalyptus* genotypes using fast and slow sucrose exposure in the preculture medium (final preculture time of 4 d in 1 M sucrose). Shoot tips from *E. gunnii dalrympleana* (clone 208) and *E. globulus* (clone 010), respectively, known for their tolerance and sensitivity to frost, were compared to those from *E. gunnii* (Table 3).

Between 80 and 100% of untreated shoot tips (as control) of the three clones resumed growth. Tolerance to freezing varied with clone and preculture procedure. Survival was observed for each of the three Eucalyptus genotypes whatever their cold tolerance to natural frost. A rapid increase in sucrose concentration during the preculture gave the best results: 63–13% according to the genotypes incubation with 1 M sucrose. Whatever the procedure, shoot tips of *E. globulus* showed the lowest tolerance of cooling and *E. gunnii* the highest.

4 Discussion

In vitro cryopreservation of Eucalyptus shoot tips by the encapsulation-dehydration procedure was presented in a previous paper (Poissonnier et al. 1992). In that work we showed that low water content in beads and preculture with high sucrose concentrations were required for post-cryopreservation survival. Best results were obtained with 1 M sucrose and 0.2 g residual water/g DW in beads. However, survival remained relatively low and variable. In the present work, different preculture conditions were studied in order to improve *Eucalyptus* shoot tip recovery after cooling in LN.

High sucrose concentrations are usually required for cryopreservation of shoot tips: 0.7 M for *Asparagus* (Uragami et al. 1990), 0.75 M for *Pyrus* (Dereuddre et al. 1990), *Solanum* (Fabre and Dereuddre 1990), *Cichorium* (Vandenbussche et al. 1993) and *Saccharum* (Gonzales et al. 1993), 1 M for *Vitis*

(Plessis et al. 1992) and *Morus* (Niino et al. 1992a). As with *Vitis,* Eucalyptus shoot tips were sensitive to direct preculture with 1M sucrose, and the final concentration was therefore obtained progressively by gradual addition of sucrose to the preculture medium. This final concentration could be reached in 2 or 7 days without a significant change in *Vitis* shoot tip tolerance to cooling in LN. By contrast, Eucalyptus shoot recovery was higher after a 2-day increase in sucrose concentration.

During stepwise preculture, the progressive increase in sugar concentration favored dehydration of beads containing the shoot tips. The fact that sucrose enters cells cannot be ignored (Fabre 1991). During preculture with 0.75 M sucrose, plasmolysis of potato meristematic cells occurred during the first hours, and decreased during the days that followed: the increase in tolerance to dehydration and LN cooling was linked to a decline in plasmolysis due to sucrose permeation into cells and/or sugar synthesis (Fabre 1991). During the dehydration process, sucrose concentration in beads with shoot tips increases significantly and may exceed solubility limits (Paul 1994). The liquid phase in the bead and cell sap may enter a supersaturated state that allows vitrification during rapid freezing. Thermal analysis has shown that sucrose pretreatment associated with marked dehydration is necessary to obtain glass transition in cells of carrot (Dereuddre et al. 1991) and oil palm somatic embryos (Dumet et al. 1993). Dehydration of material is a key step to induce freezing tolerance. However, this treatment proved lethal for shoot tips precultured with low sucrose concentrations or with sorbitol. For such shoot tips within beads, the quantity of water released during the 6-h dehydration was substantial and induced serious damage. High sucrose concentration favors dehydration tolerance. For numerous species, sucrose, or a combination of sucrose, raffinose and stachyose, apparently plays an important role in the acquisition of desiccation tolerance (Hoekstra and van Roekel 1988; Koster and Leopold 1988) and freezing tolerance (Skre 1988; Sauter and Van Cleve 1991; Hinesley et al. 1992). Sucrose may ensure membrane integrity during dehydration by forming hydrogen bonds between sugar hydroxyl groups and polar head groups of phospholipids (Anchordoguy et al. 1987; Hoekstra et al. 1992; Crowe and Crowe 1993). It also ensures the stability of isolated proteins in the same way (Carpenter 1993). This effect can be improved in the presence of oligosaccharides such as raffinose or stachyose, which prevent sucrose from crystallizing (Caffrey et al. 1988).

In this study, tolerance of Eucalyptus shoot tips to dehydration increased with the final sucrose concentration in the preculture medium. While good recovery after dehydration was observed for a large range of concentrations (1–1.5 M), the best tolerance to cooling was only obtained for 1 M sucrose preculture. Higher concentrations (1.25 and 1.5 M) did not ensure good tolerance to cooling in LN, although shoot recovery remained high after dehydration. Similar results were observed with shoot tips in beads precultured with raffinose. After subsequent dehydration, sugar may crystallize or precipitate with detrimental effects on shoot tips during the cooling-warming cycle, since, in some experiments performed with raffinose, crystallization occurred after dehydration. This may be due to the low solubility of raffinose, as described by Koster and Leopold (1988). The sugar composition of a cryoprotective medium may play an important role in preventing sugar crystallization during dehydration as well as during cooling. However, after rehydration on culture medium, most shoot tips trapped in these beads remained green and resumed growth. Nevertheless, all shoot tips encapsulated in such beads were killed after cooling in LN. Different mechanisms may be involved in tolerance of dehydration and of cooling.

The role of sucrose as a cryoprotectant may not be purely thermodynamic. Sucrose might act directly on shoot metabolism, as suggested for potato shoot tips. Tolerance to cooling in LN was improved by increasing preculture duration and intracellular accumulation of starch (Fabre 1991). For Eucalyptus, preculture gave low survival after cryopreservation. Indeed, for preculture with final sucrose concentrations of 0.5 and 0.75 M, cryopreserved shoot tip recovery decreased when preculture duration was reduced to 2 day (faster increase in sucrose concentration). The extent of dehydration of the beads was unchanged. Shoot tolerance to cooling increased, however, when incubation with 1 M sucrose was extended from 1 to 2, 4 or 6 day. During preculture, sucrose might be incorporated by cells, enhance carbohydrate metabolism, and thus modify cellular sugar composition in favor of better cryoprotection. From this point of view, results obtained with trehalose are interesting. Trehalose is known to play a similar role to sucrose in membrane stabilization in yeast, fungi, and microscopic animals (Crowe and Crowe 1993). However, trehalose does not seem to be synthesized by plant cells, and its cryoprotective effect on Eucalyptus shoot tips therefore appears to be a simple colligative effect. This points out the importance of preculture with sucrose to stimulate sugar metabolism. However, these poor results could also be attributed to low plasma membrane permeability to trehalose. Further research on shoot tip metabolic changes during preculture may cast light on this question.

The choice of an adequate sucrose concentration and the extension of the incubation period improved cryopreservation tolerance of shoot tips from the three clones tested. Nevertheless, shoot tip recovery depended on the clone and seemed to be a function of their winter cold tolerance. In field tests, clone 034 of *Eucalyptus gunnii* is considered highly cold-tolerant (Cauvin 1988). Good tolerance to cooling in LN was observed for this clone whatever the procedure used. On the other hand, clone 208 of *E. gunnii dalrympleana* is considered cold-sensitive (Cauvin 1988) and the recovery rate for shoot tips of this clone was always lower than for clone 034. *E. globulus* (clone 010) is known to be a very cold-sensitive species of value for its vigor and tolerance to chalky soil (Marien 1982). Shoot tips of clone 010 showed the lowest tolerance to cryopreservation. However, the number of clones tested was low and this apparent relationship between cold tolerance and shoot tip recovery after freezing may be fortuitous. Differences in cryopreservation tolerance between genotypes have often been reported, whatever the procedure used (Tyler et al. 1988; Niino et al. 1992b; Gonzales-Arnao et al. 1993). This observation could not always be linked to species cold hardiness, as shown for *Pyrus* shoot tips (Reed 1990). For some species, such as *Solanum* (Towill 1981), carnation (Fukai et al. 1991b), and *Chrysanthemum* (Fukai et al. 1991a), unsuitable regeneration medium explained this variation. However, in the present study, control shoot tip recovery of the three clones tested was high and similar. Variation in freezing tolerance could be due to differences in sensitivity to sucrose treatment, in the degree of injury caused by the freeze/thaw cycle, and in the physiological state of these three clones cultured in vitro.

In conclusion, the encapsulation-dehydration procedure can be considered efficient for cryopreservation of Eucalyptus shoot tips excised from in vitro plantlets. The choice of an adequate sucrose concentration and the extension of the incubation period in the final preculture medium improved tolerance of shoot tips of *Eucalyptus gunnii*. This method will be extended to shoot tips from plants grown in the nursery and to clones obtained after selection and controlled crossing. We have underlined the complexity of the mechanisms involved in the acquisition of tolerance to freezing in liquid nitrogen. Further research is needed to elucidate the phenomena involved and, in particular, the role of sucrose.

References

Anchordoguy TJ, Rudolph AS, Carpenter JF, Crowe JH (1987) Modes of interaction of cryoprotectants with membrane phospholipids during freezing. Cryobiology 24:324–331

Caffrey M, Fonseca V, Leopold AC (1988) Lipid sugar interactions. Plant Physiol 86:754–758

Carpenter JF (1993) Stabilization of proteins during freezing and dehydration: application of lessons from Nature. Cryobiology 30:220–221

Cauvin B (1988) Eucalyptus – les tests de résistance au froid. Ann AFOCEL 1987:161–195

Crowe JH, Crowe LM (1993) Evidence for direct interaction between disaccharides and dry phospholipids. Cryobiology 30:226–227

Dereuddre J, Scottez C, Arnaud Y, Duron M (1990) Résistance d'apex caulinaires de vitroplants de Poirier (*Pyrus Communis*) enrobés dans l'alginate à une déshydratation puis une congélation dans l'azote liquide: effet d'un endurcissement préalable au froid. CR Acad Sci Paris Ser III 310:317–323

Dereuddre J, Hassen N, Blandin S, Kaminski M (1991) Resistance of alginate-coated somatic embryos of carrot (*Daucus carota* L.) to desiccation and freezing in liquid nitrogen. 2. Thermal analysis. Cryo Lett 12:135–148

Dumet D, Engelmann F, Chabrillange N, Duval Y, Dereuddre J (1993) Importance of sucrose for the acquisition of tolerance to desiccation and cryopreservation of oil palm somatic embryos. Cryo Lett 14:243–250

Fabre J (1991) Cryoconservation d'apex de Solanacées tubérifères après encapsulation et déshydration. Etude des modifications cellulaires au cours du prétraitement. Thèse Univ Picardie et M Curie, 123 pp

Fabre J, Dereuddre J (1990) Encapsulation-dehydration: a new approach to cryopreservation of *Solanum* shoot tips. Cryo Lett 11:413–426

Fukai S, Goi M, Tanaka M (1991a) Cryopreservation of shoot tips of *Chrysanthemum morifolium* and related species native to Japan. Euphytica 54:201–204

Fukai S, Goi M, Tanaka M (1991b) Cryopreservation of shoot tips of Caryophyllacae ornamentals. Euphytica 56:149–153

Gonzales Arnao MT, Engelmann F, Huet C, Urra C (1993) Cryopreservation of encapsulated apices of sugarcane: effect of freezing procedure and histology. Cryo Lett 14:303–308

Hinesley LE, Pharr DM, Snelling LK, Funderburk SR (1992) Foliar raffinose and sucrose in four conifer species: relationship to seasonal temperature. J Am Soc Hort Sci 117:852–855

Hoekstra FA, Van Roekel T (1988) Desiccation tolerance of *Papaver dubium* L. pollen during its development in the anther. Plant Physiol 88:626–632

Hoekstra FA, Crowe JH, Crowe LM, Van Roekel T, Vermeer E (1992) Do phospholipids and sucrose determine membrane phase transitions in dehydrating pollen species? Plant Cell Environ 15:601–606

Kartha KK (1985) Meristem culture and germplasm preservation. In: Kartha KK (ed) Cryopreservation of plant cells and organs. CRC Press, Boca Raton, pp 115–134

Koster KL, Leopold AC (1988) Sugars and desiccation tolerance in seeds. Plant Physiol 88:829–832

Marien JN (1982) Les taillis d'eucalyptus. Culture de biomasse ligneuse, taillis à courte rotation. AFOCEL, Paris, pp 73–117

Murashige T, Skoog F (1962) A revised medium for rapid growth and bioassays with tobacco tissue cultures. Physiol Plant 15:473–497

Niino T, Sakaï A, Yakuwa H (1992a) Cryopreservation of dried shoot tips of mulberry winter buds and subsequent plant regeneration. Cryo Lett 13:51–58

Niino T, Sakaï A, Yakuwa H, Nojiri K (1992b) Cryopreservation of in vitro-grown shoot tips of apple and pear by vitrification. Plant Cell Tissue Organ Culture 28:261–266

Pâques M, Poissonnier M, Dumas E, Monod V (1997) Cryopreservation of dormant and non-dormant broad-leaved trees. Acta Hortic 447:491–497

Paul H (1994) Contribution à la régénération par embryogenèse somatique, et à la cryoconservation d'apex axillaires de vitroplants et d'embryons somatiques, chez le Pommier (*Malus domestica Borkh*). Thèse Univ Picardie, 167 pp

Plessis P, Leddet C, Dereuddre J (1992) Résistance à la déshydratation et à la congélation dans l'azote liquide d'apex enrobés de vigne (*Vitis vinifera* L. cv. chardonnay). CR Acad Sci Paris Ser III, 313:373–378

Poissonnier M, Monod V, Pâques M, Dereuddre J (1992) Cryoconservation dans l'azote liquide d'apex d'*Eucalyptus gunnii* cultivé' in vitro après enrobage et déshydratation. Ann AFOCEL 1991:5–23

Poissonnier M, Monod V, Accart F, Dumas E, Pâques M (1997) Un nouvel outil pour la conservation des espèces en danger: la cryoconservation. Les Colloques de l'INRA 84:321–330

Reed B (1990) Survival of in vitro-grown apical meristems of *Pyrus* following cryopreservation. HortScience 25:111–113

Sauter JJ, Van Cleve B (1991) Biochemical and ultrastructural results during starch-sugar-conversion in ray parenchyma cells of *Populus* during cold adaptation. J Plant Physiol 139:19–26

Skre O (1988) Frost resistance in forest trees: a literature survey. Medd Nor Inst Skogforsk 40:1–35

Smith DR (1997) The role of in vitro methods in pine plantation establishment: the lesson from New Zealand. Plant Tissue Culture Biotechnol 3:63–73

Towill LE (1981) Survival at low temperature of shoot tips from cultivars of *Solanum tuberosum* group tuberosum. Cryo Lett 2:373–382

Tyler NJ, Stushnoff C, Gusta LV (1988) Freezing in water in dormant vegetative apple buds in relation to cryopreservation. Plant Physiol 87:201–205

Uragami A, Sakaï A, Nagaï M (1990) Cryopreservation of dried axillary buds from plantlets of *Asparagus officinalis* grown in vitro. Plant Cell Rep 9:328–331

Vandenbussche B, Demeulemeester MAC, de Proft MP (1993) Cryopreservation of alginate-coated in vitro grown shoot-tips of chicory (*Cichorium intybus* L.) using rapid freezing. Cryo Lett 14:259–266

Zhao Y, Wu Y, Engelmann F, Zhou M, Zhang D, Chen S (1999) Cryopreservation of apple shoot tips by encapsulation-dehydration: effect of preculture, dehydration and freezing procedure on shoot regeneration. Cryo Lett 20:103–108

III.5 Cryopreservation of *Guazuma crinita* Mart. (Guazuma)

E. Maruyama

1 Introduction

1.1 Botany, Geographical Distribution and Importance

Guazuma crinita Mart., which belongs to the Sterculiaceae family (chocolate tree family), is a medium-sized tree native to South America in the Amazon forest region of Peru, Brazil and Ecuador (Freytag 1951; Encarnación 1983). This species has a soft-light wood with good working properties that has been used for light construction, panelling, interior joinery, mouldings, cases, matches and packing.

Guazuma (locally know as "bolaina blanca") is a versatile fast-growing tree with excellent adaptability to a wide range of sites including areas with poorly drained heavy clay soils which are widespread throughout the tropics (Maruyama et al. 1996). Fast initial growth rates of up to 3 m in height per year with a final rotation age of 10 to 15 years have been reported in the tropical forest of Peru-Amazon (Fig. 1). It has been planted alone or in association with mahogany (*Swietenia macrophylla*), cedro (*Cedrela odorata*), tornillo (*Cedrelinga catenaeformis*), ishpingo (*Amburana cearensis*), and other valuable tree species, or in association with agricultural crops in agroforestry systems, especially as living poles for pepper plant cultivation (Fig. 2).

1.2 Methods for Storage and Need for Cryopreservation

Guazuma germplasm is traditionally conserved in field gene banks in natural forests or in its cultivation areas. Propagation of this species is by seeds and seed storage is the most popular and economical method for germplasm conservation. However, propagation and conservation by vegetative means are desirable for better preservation of true-to-type genetic characteristics. The conservation of desirable genetic materials using conventional in situ or ex situ methods, such as a field gene bank, is costly and risky due to diseases, pest attacks, natural disasters, and human intervention.

Bio-Resources Technology Division, Forestry and Forest Products Research Institute, Box 16, Tsukuba Norinkenkyudanchi-nai, Ibaraki 305-8687, Japan

Biotechnology in Agriculture and Forestry, Vol. 50
L.E. Towill and Y.P.S. Bajaj (Eds.) Cryopreservation of Plant Germplasm II
© Springer-Verlag Berlin Heidelberg 2002

Fig. 1. Five-year-old open field plantation of *G. crinita* (Pucallpa, Peru)

Although the use of in vitro gene banks alleviates these problems, the maintenance of large collections under conventional in vitro storage systems requires much handling and is expensive because most cultures need subculturing at regular intervals to prevent browning and loss of viability. In addition, the risks of contamination and somaclonal variation increase with time.

Recently, medium-term storage of in vitro cultured material based on the principles of slow growth or minimum growth conditions has been achieved. Guazuma shoot tips have been stored successfully for more than 12 months at 25 °C without considerable loss of viability using alginate-encapsulation techniques (Maruyama et al. 1997a). In vitro shoots were stored in a culture room at about 25 °C for 18 months on WPM supplemented with 10 µM ABA (Maruyama, unpubl.).

Fig. 2. Plantation of *G. crinita* in association with pepper crop (Pucallpa, Peru)

Cryopreservation of in vitro cultured materials may be a reliable method for long-term conservation of plant genetic resources without apparent risk of genetic alterations and using minimum space and requiring less labor and maintenance costs.

2 Cryopreservation of in Vitro Cultured Materials

2.1 Materials and Methods

2.1.1 Plant Material

Shoot tips, nodal segments and root tips excised from 2–3-month-old plantlets, regenerated in vitro by the method described by Maruyama et al. (1996), were cut into segments about 2 mm long. Adventitious bud clusters obtained from petiole culture by the method described by Maruyama et al. (1997b) were cut into small (1.0–1.5 mm^3) or large (3.0–4.0 mm^3) cubic segments.

2.1.2 Cryoprotectant Mix

The following cryoprotectant solutions (w/v), modified from Sakai et al. (1991) and Towill (1990), were tested: Solution A: 20% glycerol and 15% sucrose; Solution B: 30% glycerol, 15% sucrose, 15% ethylene glycol, and 15% dimethyl sulfoxide (DMSO); Solution C: 25% glycerol, 15% sucrose, 15% ethylene glycol, 13% DMSO, and 2% polyethylene glycol (PEG); Solution D: 35% ethylene glycol, 10% DMSO, and 5% PEG. The aqueous volume added consisted of woody plant medium (WPM) (Lloyd and McCown 1980) without sucrose and growth regulators. All the cryoprotectant solutions were filter-sterilized.

2.1.3 Cryopreservation Methods

2.1.3.1 Simple Cooling

Shoot tip, nodal segment, and root tip explants in 1.5-ml cryotubes were treated with solution A at 25 °C for 5, 10, 15, 20, 30, 45, and 60 min and then cooled in a freezer at −30 °C for 1 h prior to immersion in liquid nitrogen (LN) and kept there for at least 1 h.

2.1.3.2 Rapid Cooling

Shoot tip, nodal segment, and root explants in 1.5-ml cryotubes were treated with solutions A and B at 25 °C for 0, 10, 20, 30, and 45 min. Adventitious bud cluster explants were treated with solutions B, C, or D for 5, 15, 30, 45, 60, and 90 min. Then the cryotubes were directly immersed in LN and kept there for at least 1 h.

2.1.3.3 Slow Cooling

Shoot tip, nodal segment, and root tip explants in 1.5-ml cryotubes were treated with solutions A and B at 25 °C for 0, 10, and 20 min and then cooled to −40 °C at a rate of 0.5 °C/min prior to immersion in LN and held there for at least 1 h.

2.1.3.4 Encapsulation/Dehydration

Explants were progressively cultured at 5 °C by successive daily transfer onto solidified WPM containing 5, 10, and 20% (w/v) sucrose. Then they were encapsulated by the method described by Maruyama et al. (1997c) in 3% (w/v) alginate-coated beads containing 20% (w/v) sucrose. Encapsulated (con-structed beads about 5 mm in diameter containing one shoot tip or nodal segment) and non-encapsulated explants were treated with the same medium supplemented with 30% (w/v) sucrose for 16 h at 5 °C. After treatments with sucrose they were dehydrated at 25 °C for 0–24 h in a laminar flow cabinet or inside Petri dishes (9 cm in diameter) containing about 50 g silica gel sterilized by heating at 110 °C for 16 h. Explants were placed in cryotubes and then cooled in LN by the slow cooling or by the rapid cooling method.

2.1.4 Survival and Plant Regeneration

Explants were warmed from LN by rapid transfer of cryotubes to a water bath at 37 °C. After warming, cryoprotectant solutions were drained from the cryotubes and replaced with a medium containing 40% (w/v) sucrose and kept for 20 min. Then, shoot tips and nodal segments were transferred onto solidified WPM containing 10 μM zeatin and cultured at 25 °C under a photon flux density of 65 μmol m^{-2} s^{-1}. Root tips and adventitious bud clusters were transferred into liquid WPM containing 10 μM zeatin and cultured on a bio-shaker at 75 rpm under a photon flux density of 25 μmol m^{-2} s^{-1}. All the explants were cultured at 25 °C and under a 16-h photoperiod provided by cool white fluorescent lamps (100 V, 40 W; Toshiba Co.). Explants were examined for survival at weekly intervals. An explant was defined as "surviving" if it turned green and produced leaves or adventitious buds. The percentage of explants that developed into plantlets was recorded as plant regeneration.

2.1.5 Acclimatization of Plantlets

Plantlets regenerated from cryopreserved explants were transplanted into plastic pots filled with vermiculite after washing the roots with tap water to free any adhering agar. They were acclimatized in a growth cabinet at 25–30 °C at a photon flux density of 35 μmol m^{-2} s^{-1} under a 16 h photoperiod. Plantlets were acclimatized under high relative humidity (about 90–95%) during the first 2 weeks in plastic boxes with transparent covers. Covers were opened gradually during the next 2 weeks (covering about 99% of plastic boxes during the first 5 days, and then about 97, 90, and 80% over 3 days for each condition). Thereafter covers were completely removed. Plantlets were irrigated with water for the first 2 weeks and then with 0.1% (v/v) Hyponex 5-10-5 plant-food solution (Hyponex Co., Inc.; w/v: 5.00% N, 4.36% P, and 4.15% K).

2.2 Results and Discussion

The effects of cryopreservation methods on survival and plant regeneration of different explants are shown in Table 1. Survival of explants after immersion in LN was not achieved in any of the treatments tested by the simple cooling and encapsulation/dehydration methods. Only the use of rapid cooling and slow cooling methods resulted in plant recovery after storage in LN. These results suggest that the explants tested were not suitable materials or were not in a suitable physiological state for cryopreservation by simple cooling since

Table 1. Effects of different cryopreservation methods on survival (S) and plant regeneration (PR) of *G. crinita* explants after cooling in LN

Explants	Simple cooling method[a]		Rapid cooling method[b]		Slow step cooling method[c]		Encapsulation/ dehydration method[d]	
	S (%)	PR (%)	S (%)	PR (%)	S (%)	PR (%)	S (%)	PR (%)
Shoot tip	0	0	0	0	50	15	0	0
Nodal segment	0	0	0	0	5	0	0	0
Root tip	0	0	30	5	30	5	0	0
Adv. bud cluster			85	25				

[a] Explants were treated with cryoprotectant solution and then cooled in a freezer at –30 °C for 1 h prior to immersion in LN.

[b] Explants were treated with cryoprotectant solution and then cooled by direct immersion in LN.

[c] Explants were treated with cryoprotectant solution and then cooled to –40 °C at a rate of 0.5 °C/min prior to immersion in LN.

[d] Explants, with or without alginate encapsulation, were treated with a medium enriched with sucrose before dehydration in a laminar flow cabinet or inside petri dishes containing silica gel, and then cooled in LN by slow step cooling or by the rapid cooling method.

Shoot tip, nodal segment, and root tip explants were cut into segments about 2 mm long. Adventitious bud cluster explants were cut into about 1.0–1.5 mm³ cubic segments.

this method had been successfully applied to cultured cells (Sakai 1995). Treatments in media with a high concentration of sucrose by the encapsulation/dehydration method, successful for some species (Niino and Sakai 1992; Suzuki et al. 1994; Sakai 1995), caused damage to explants apparently because this species does not tolerate drastic dehydration processes. Osmotic dehydration damage was observed in explants that were not cooled; almost all surviving explants did not develop into plantlets, and a dehydration period of more than 4 h was inhibitory to the survival of shoot tips and nodal segments (data not shown).

Slow cooling was the best method for cryopreservation of shoot tips (Fig. 3A). A survival rate of 50% and 15% plant regeneration were achieved when shoot tips were treated with cryoprotectant solutions A and B for 20 and 10 min, respectively, and then cooled to –40 °C at a rate of 0.5 °C/min prior to storage in LN. For root tip explants no difference between the rapid cooling method and the slow step cooling method was found (Table 1). As in the case of shoot tip explants, plant regeneration from cryopreserved root tips was obtained through adventitious bud formation (Fig. 3B) and subsequently shoot differentiation on medium supplemented with 10 and 1 μM ZEA, respectively. Although this method is time-consuming and laborious, and requires controlled cooling equipment and complicated procedures, it is the most commonly used method for the cryopreservation of meristems/shoot tips, cell cultures and somatic embryos. Several species have been successfully cryopreserved following slow cooling methods (Kartha 1985; Withers 1985; Chen and Kartha 1987; Kartha and Engelmann 1994; Grout 1995; Sakai 1995).

In the present study, although rapid cooling failed to cryopreserve the shoot tip and nodal segment explants, this method was effective for the

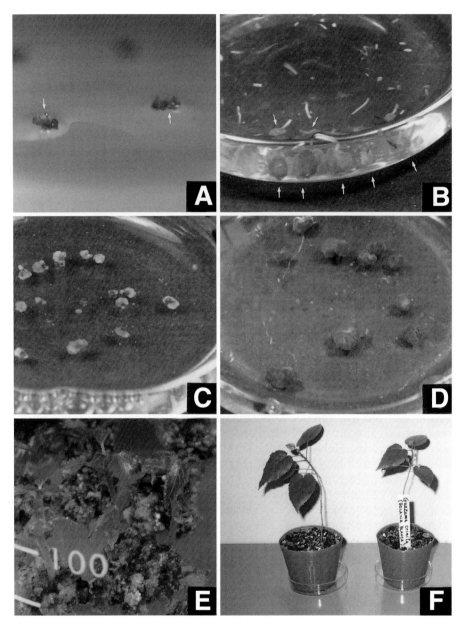

Fig. 3A–F. Cryopreservation of *G. crinita*. **A** Shoot tips showing areas of growth (*arrows*) after storage in liquid nitrogen (LN) by slow step cooling method. **B** Adventitious bud cluster formation (*arrows*) on root tips after storage in LN by rapid cooling. **C** Cryopreserved adventitious bud cluster segments after 15 days of culture in recovery growth medium. **D** Cryopreserved adventitious bud cluster segments after 45 days of culture in recovery growth medium. **E** Shoot development from cryopreserved adventitious bud cluster segments after 60 days of culture onto WPM supplemented with 1 µM zeatin. **F** Acclimatized plants regenerated from cryopreserved explants

cryopreservation of adventitious bud clusters of *G. crinita*. Survival of adventitious bud clusters after storage in LN varied depending on the size of explant, cryoprotectant solution and on the duration of the cryoprotectant treatment. High survival percentages (73–85%) were achieved for small cubic segments (1.0–1.5 mm^3) pretreated with cryoprotectant solution B or C. In contrast, large cluster explants (3.0–4.0 mm^3) and cryoprotectant solution D treated bud clusters did not survive after storage in LN (Fig. 4A). These results suggest that (1) the large size of the explants is not favorable for the dehydration-cryoprotective action of the vitrification solution, or that (2) the duration of the cryoprotectant treatments for large cluster explants was insufficient and that (3) glycerol and sucrose are necessary components of the cryoprotectant solution for cryopreservation of bud clusters of *G. crinita* by the rapid cooling method. Based on these results, large-sized cluster explants and cryoprotectant solution D were not used for further experiments.

Survival of small cluster explants did not differ when they were treated with either solution B or C at 25 °C for various periods of time prior to direct immersion in liquid nitrogen. The highest rate of survival was obtained with the explants treated with solution B for 15–45 min or with solution C for 15–60 min, respectively. In both cryoprotectant treatments, a pretreatment of 5 min in a cryoprotectant solution was insufficient, and pretreatment for 90 min was inhibitory to the survival of explants (Fig. 4B). The original green color of the bud cluster segments was evident immediately following cooling but was lost within 24–48 h after warming. However, the bud cluster segments that ultimately survived regained their green color within 1–2 weeks after transfer to recovery growth medium (Fig. 3C). After 45 days of culture, clumps of numerous bulbous structures about 5 mm in diameter were formed (Fig. 3D). When transferred onto agar-solidified WPM containing 1 μM zeatin, after 60 days, about 30% of the surviving cryopreserved explants formed shoots (Fig. 3E). No differences were observed among the rates of shoot development from untreated control and surviving cryopreserved explants. Regenerated plantlets were successfully acclimatized and all survived and grew well (Fig. 3F). No morphological abnormalities were observed in the plants regenerated from cryopreserved explants.

Rapid cooling is the simplest method of cryopreservation and does not require sophisticated and expensive controlled cooling equipment and complicated cryoprotective procedures. Cells and meristems are directly immersed in LN (Withers 1985; Towill 1990; Kohmura et al. 1992; Matsumoto et al. 1994; Sakai 1995).

Although several authors have indicated that cold-hardening and/or pre-culturing with a high concentration of sugar medium are essential to successful cryopreservation of in vitro cultured plant materials (Reed 1988; Niino et al. 1992; Kartha and Engelmann 1994), adventitious bud clusters of *G. crinita* cryopreserved by the rapid cooling method showed high percentages of survival without any cold-hardening and/or pre-culturing treatments. Similar results were reported in the vitrified bud clusters of asparagus (Kohmura et al. 1992).

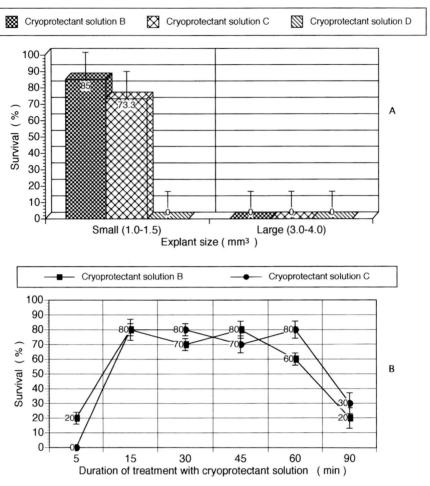

Fig. 4A,B. Cryopreservation of adventitious bud cluster explants of *G. crinita* cultured in vitro. **A** Effects of explant size and cryoprotectant solution on survival after storage in liquid nitrogen (LN). Explants were treated with a cryoprotectant solution for 45 min at 25 °C and then immersed in LN. **B** Effects of duration of treatment with cryoprotectant solution prior to cooling on survival after storage in LN. Explants placed in 1.5-ml cryotubes were treated with a cryoprotectant solution for different periods of time at 25 °C and then directly immersed into LN. Ten segments were treated for each of four replicates. *Bars* indicate standard error

Cold-hardening and/or pre-culturing with sugar-enriched medium were not effective in inducing dehydration tolerance in *G. crinita*. Attempts at using several cold-hardening and/or pre-culturing treatments modified from Niino and Sakai (1992), Suzuki et al. (1994), Matsumoto et al. (1994), and Brison et al. (1995), to enhance the survival and plant regeneration rate (data not presented), were not successful.

3 Summary and Conclusions

Vegetatively propagated plants are preferable sources for germplasm conservation. In our institute, in vitro cultured plants of Guazuma have been maintained by conventional tissue culture methods and by slow growth or minimum growth conditions. However, both methods can ensure only medium-term storage and practical problems exist. Thus, cryogenic storage techniques for germplasm of *G. crinita*, using different in vitro cultured materials, were examined.

The results of the cryopreservation experiments performed may be summarized as follows: (1) slow cooling was the best method for cryopreservation of shoot tips, (2) rapid cooling was not effective for the cryopreservation of shoot tips and nodal segments; however, this method was effective for adventitious bud clusters, (3) for root tip explants, there was no difference between the rapid cooling and slow step cooling methods, (4) the simple cooling and encapsulation/dehydration method failed to cryopreserve all explants tested, and (5) cold-hardening and/or pre-culturing treatments were not effective in enhancing survival or plant recovery after storage in LN.

Although plant recovery percentages need to be improved for a more effective cryopreservation system, these results suggest a usable system since more than 800,000 shoots can be regenerated per year from only one shoot tip explant (Maruyama et al. 1996). Thus, further propagation from a few cryopreserved surviving explants is really possible.

Cryopreservation of in vitro cultured materials by using both the slow cooling and rapid cooling methods are suitable for the long-term storage of *G. crinita* germplasm.

References

Brison M, Bocaut MT, Dosba F (1995) Cryopreservation of in vitro grown shoot tips of two interspecific *Prunus* rootstocks. Plant Sci 105:235–242

Chen THH, Kartha KK (1987) Cryopreservation of woody species. In: Bonga JM, Durzan DJ (eds) Cell and tissue culture in forestry, vol 2. Specific principles and methods: growth and developments. Martinus Nijhoff, Dordrecht, pp 305–319

Encarnación F (1983) Nomenclatura de las Especies Forestales Comunes en el Perú. Proyecto PNUD/FAO/PER/81/002 Fortalecimiento de los Programas de Desarrollo Forestal en la Selva Central, Documento de Trabajo no 7, Lima, 149 pp

Freytag GF (1951) A revision of the genus *Guazuma*. Ceiba (Hond) 1:193–225

Grout B (ed) (1995) Genetic preservation of plant cells in vitro. Springer, Berlin Heidelberg New York

Kartha KK (ed) (1985) Cryopreservation of plant cell and organs. CRC Press, Boca Raton

Kartha KK, Engelmann F (1994) Cryopreservation and germplasm storage. In: Vasil IK, Thorpe TA (eds) Plant cell and tissue culture. Kluwer, Dordrecht, pp 195–230

Kohmura H, Sakai A, Chokyu S, Yakuwa T (1992) Cryopreservation of in vitro cultured multiple bud clusters of asparagus (*Asparagus officinalis* L. cv. Hiroshimagreen (2n=30)) by the techniques of vitrification. Plant Cell Rep 11:433–437

Lloyd G, McCown B (1980) Commercially-feasible micropropagation of mountain laurel, *Kalmia latifolia,* by use of shoot-tip culture. Comb Proc Int Plant Prop Soc 30:421–427

Maruyama E, Ishii K, Kinoshita I, Ohba K, Saito A (1996) Micropropagation of bolaina blanca (*Guazuma crinita* Mart.), a fast-growing tree in the Amazon region. J For Res 1:211–217

Maruyama E, Kinoshita I, Ishii K, Ohba K, Saito A (1997a) Germplasm conservation of the tropical forest trees, *Cedrela odorata* L., *Guazuma crinita* Mart., and *Jacaranda mimosaefolia* D. Don., by shoot tip encapsulation in calcium-alginate and storage at 12–25 °C. Plant Cell Rep 16:393–397

Maruyama E, Ishii K, Kinoshita I, Ohba K, Saito A (1997b) Micropropagation of *Guazuma crinita* Mart. by root and petiole culture. In Vitro Cell Dev Biol Plant 33:131–135

Maruyama E, Kinoshita I, Ishii K, Shigenaga H, Ohba K, Saito A (1997c) Alginate-encapsulated technology for the propagation of the tropical forest trees: *Cedrela odorata* L., *Guazuma crinita* Mart., and *Jacaranda mimosaefolia* D. Don. Silvae Genet 46:17–23

Matsumoto T, Sakai A, Yamada K (1994) Cryopreservation of in vitro grown apical meristems of wasabi (*Wasabia japonica*) by the vitrification and subsequent high plant regeneration. Plant Cell Rep 13:442–446

Niino T, Sakai A (1992) Cryopreservation of alginate-coated in vitro grown shoot tips of apple, pear and mulberry. Plant Sci 87:199–206

Niino T, Sakai A, Nojiri K (1992) Cryopreservation of in vitro grown shoot tips of apple and pear by vitrification. Plant Cell Tissue Org Cult 28:261–266

Reed BM (1988) Cold acclimation as a method to improve survival of cryopreserved *Rubus* meristems. Cryo Lett 9:166–171

Sakai A (1995) Cryopreservation for germplasm collection in woody plants. In: Jain SM, Gupta PK, Newton RJ (eds) Somatic embryogenesis in woody plants, vol 1. History, molecular and biochemical aspects, and applications. Kluwer, Dordrecht, pp 293–315

Sakai A, Kobayashi S, Oiyama I (1991) Survival by vitrification of nucellar cells of naval orange (*Citrus sinensis* var. *brasiliensis* Tanaka) cooled to –196 °C. J Plant Physiol 137:465–470

Suzuki M, Niino T, Akihama T (1994) Cryopreservation of shoot tips of kiwifruit seedlings by the alginate encapsulation-dehydration technique. Plant Tissue Cult Lett 11:122–128

Towill LE (1990) Cryopreservation of isolated mint shoot tips by vitrification. Plant Cell Rep 9:178–180

Withers LA (1985) Cryopreservation and storage of germplasm. In: Dixon RA (ed) Plant cell culture, a practical approach. IRL Press, Oxford, pp 169–192

III.6 Cryopreservation of *Olea europaea* L. (Olive)

M.A. Revilla[1], D. Martínez[1], J.M. Martínez-Zapater[2],
and R. Arroyo García[2]

1 Introduction

1.1 Botany, Distribution and Importance of the Olive Tree

The olive tree, *Olea europaea* L., belongs to the family Oleaceae, which comprises plant species distributed through the tropical and temperate regions of the world. From about 29 genera of this family, only *Fraxinus*, *Jasminum*, *Ligustrum*, *Phillyrea*, Syringa and *Olea* have economic or horticultural interest (Heywood 1978).

Within the genus *Olea* there are about 35 species. *Olea europaea* L. includes cultivated and wild olive trees. It is generally considered that the cultivated olives belong to the subspecies *sativa* and wild olives to the subspecies *sylvestris*.

The olive is one of the most ancient cultivated plants, whose origin as a crop dates back to 4000–3000 B.C. A large number of varieties are currently recognized. Morphological characteristics or isoenzyme analysis have identified about 300 olive varieties in Spain alone. These are vegetatively propagated and have been collected for germplasm banks (Barranco 1997).

The olive is a long-lived evergreen tree cultivated mainly for oil (93%) and also for table olives (both black and green). It has economic relevance within the Mediterranean Basin, which supplies 97% of the world production. Olive oil ranks sixth in world production of fluid vegetable fats (Loussert and Brousse 1980) but is first in taste and nutritional qualities, due to the high content of oleic acid, a monounsaturated fat with beneficial effects on human cholesterol levels (Grande Covián 1989).

1.2 Storage of Olive Germplasm

The conservation of plant genetic resources for food and agriculture is a major element of any strategy to achieve sustainable agricultural development

[1] Dep. Biología de Organismos y Sistemas, Facultad de Biología, C/C. Rodrigo Uría s/n, Universidad de Oviedo, 33071 Oviedo, Spain
[2] Dep. de Genética Molecular de Plantas, Centro Nacional de Biotecnología, CSIC, Campus de la Universidad Autónoma de Madrid, Cantoblanco, 28049 Madrid, Spain

Biotechnology in Agriculture and Forestry, Vol. 50
L.E. Towill and Y.P.S. Bajaj (Eds.) Cryopreservation of Plant Germplasm II
© Springer-Verlag Berlin Heidelberg 2002

(Withers and Engelmann 1998). Maintenance and preservation of cultivated olives and their relatives is essential for breeding. Once new cultivars with good cultivation and yield qualities are released, the current major cultivars replaced by them may be lost.

Vegetatively propagated crops, including olive, are conventionally conserved in the field offering ready accession and observation, permeating detailed evaluation. However, there are certain drawbacks that limit its efficiency and threaten its security. The genetic resources are exposed to pests, diseases, weather damage and other unforeseen problems and, in addition, they are not in a condition that is readily usable for international germplasm exchange. Field gene banks are costly to maintain and, as a consequence, are prone to economic decisions that may limit the level of replication, the quality of maintenance, and even their survival in times of economic stringency (Withers and Engelmann 1998). Even under the best circumstances, field gene banks require considerable input of land, labor, management, and materials.

Ex situ conservation involves removing the plant genetic resources from their natural habitat and placing them under artificial storage conditions. The most familiar approach to ex situ conservation is seed storage. However, olive varieties are highly heterozygous and, therefore, seeds are not useful for the conservation of the specific varietal traits.

In vitro propagation through cuttings facilitates the perpetuation of a given genotype, particularly for traditionally vegetatively propagated material, has a low risk of introducing pathogens and also allows use for various processes in biotechnology. However, in vitro storage methods can ensure only short- or medium-term needs. Problems arise from loss of vigor in shoots, eventual microbial contamination and the task of maintaining all valuable collections. Thus, it is highly desirable to develop techniques for long-term preservation of germplasm.

1.3 Genetic Stability

Plant shoot tips are genetically stable for the preservation of germplasm, providing clonal propagation and production of disease-free plants. However, cryopreservation involves a series of stresses that may destabilize the plant material and lead to genetic modifications in recovered cultures and regenerated plants. Therefore, it is necessary to evaluate the genetic stability of material recovered from cryopreserved samples before this technique is routinely used for long-term conservation of plant germplasm.

Identification of olive plant material has usually been based on the morphological characteristics of the tree, flower or fruit (Barranco 1997) or on isoenzymatic variation (Trujillo et al. 1995). However, it would be highly desirable to use DNA molecular markers for olive plant materials to evaluate genome modifications as a result of the different steps followed to cryopreserve samples.

In the past few years, new markers like AFLPs (amplified fragment length polymorphism) have been added to the list of molecular markers to detect

1   2   3   4
ab ab ab ab

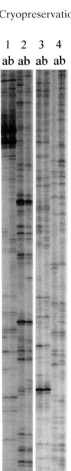

Fig. 1. Amplified fragment length polymorphism (AFLP) profiles, obtained with four different primer combinations (referred to as *1, 2, 3* and *4*), of two samples of different cultivars of olive. *a* cv. Arberquina; *b* cv. Manzanillo. Selective amplification of restriction fragments was performed using (^{33}P)-labelled *Eco*RI primers. We used 5 µl of the preamplification template for each PCR reaction. Samples amplified with different primer combinations were loaded into 4.5% denaturing polyacrylamide gels and electrophoresed for 2 h. Gels were dried on chromatography paper, and exposed to autoradiographic film. DNA was extracted from freeze-dried young leaf material using the kit Dneasy (Quiagen). AFLP analysis was performed according to Vos et al. (1995) with the modifications described by Cervera et al. (1998)

polymorphisms in the DNA sequence (Vos et al. 1995). AFLP is a PCR-based method that is able to generate complex banding patterns per reaction. AFLPs have been shown to be very useful in the analysis of biological diversity, in varietal identification and in genetic mapping. These markers are useful because they analyze a greater number of loci in each reaction and are much more reproducible than RAPDs (randomly amplified polymorphic DNA). Preliminary results presented in Fig. 1 show the utility of AFLPs in the identification of olive varieties. Two olive varieties from Spain (Cordoba), Arbequina and Manzanillo, were used in this study. This molecular method can provide potential markers to monitor genetic instability generated in olive in vitro cultures and to test the genetic integrity of cryopreserved specimens.

2 Cryopreservation of Olive Germplasm

Cryopreservation refers to the storage of biological materials at low temperatures such that viability is retained after thawing. Techniques of cryopreservation are still relatively new for plant tissues, and a great deal of empirical work must be done when attempting to cryopreserve a new line. There are certain prerequisites for the successful application of cryopreservation, such as efficient in vitro methods to regenerate complete plants from the explants used in cryopreservation, as well as an efficient cryopreservation protocol. Shoot tips or buds are recognized as the major explants for long-term germplasm storage for species that are maintained clonally. Olive micropropagation has been achieved either using nodal segments of juvenile (Cañas et al. 1992) or through micrografting shoot tips from mature origin plants (Revilla et al.

1996). However, as far as we know, the successful in vitro culture of olive shoot tips (from in vitro or ex vitro plants), comprising the meristematic dome and a few leaf primordia, has not yet been reported. This is a major handicap in developing a cryopreservation procedure for olive. Survival after cryopreservation has been recently reported for olive shoot tips from in vitro shoot cultures (Martínez et al. 1999), but the recovery of complete plants was not achieved and therefore more experiments need to be made in the regrowth phase after thawing. Olive embryo cryopreservation (González-Río et al. 1994) may be useful for experimental studies on genetic manipulation, but it is not desirable for maintenance of genotypes of olive in base collections due to their high heterozygosity.

2.1 Cryopreservation of Embryos

Plant material from seedlings is frequently used to carry out experiments on genetic manipulation aimed at the improvement of the olive tree (Rugini and Fedeli 1990). However, a loss of germination capacity together with an increase in the contamination levels of the embryos cultured in vitro have been observed during the storage of olive fruits for seed and embryo extraction (González-Río et al. 1994). This prevents the continuous availability of embryonic plant material through the year. Seed cryostorage provides a method for decreasing this loss of viability since at LN temperature (−196°C) biological processes are virtually stopped.

2.1.1 Material and Cryopreservation Procedure

Olive material was provided by the Department of Olivicultura and Arboricultura Frutal (C.I.D.A. Córdoba, Spain). The fruits of *Olea europaea* L. cv. Aberquina were collected in October and stored at 4°C until their use, about 4 months later. After removing the fleshy part of the olive (mesocarp), the endocarp was broken with the aid of a vice and the seeds were then soaked in water overnight. Subsequently, the embryo was excised by making a longitudinal excision in the endosperm with a scapel. Embryos were then surface sterilized with 70% ethanol for 2 min, followed by immersion in 1.2% active sodium hypochlorite solution for 10 min, rinsed with sterile water, cultured on solid MS medium ($30 g l^{-1}$ sucrose, without plant regulators, pH 5.8) and germinated at 26°C under a 16-h photoperiod.

Sterilized embryos were dried on filter paper in the air stream of a laminar flow cabinet, placed in 2-ml cryovials and directly immersed in LN. Warming was performed by placing the cryovials in a water bath at 45°C for 2 min.

Samples were counted after 15 days and scored as the percentage of embryos exhibiting green cotyledons (survival) and after 40 days as percentage of seeds germinated.

2.1.2 Results and Discussion

After sterilization, the embryos rapidly lost moisture in the air flow cabinet, the major loss occurring during the first 2.5 h (14% moisture content). Survival and germination percentages were not significantly lower than those of the control (non-desiccated embryos), except when embryo moisture content was reduced to 3% (18 h of desiccation). Embryos did not survive cooling in LN unless they were first desiccated. In the present study (González-Río et al. 1994), no significant differences after cryopreservation were found in the percentages of survival (70–80%) and germination (53%–70%) of olive embryos between 10 and 3% of moisture contents (13 and 18 h of desiccation, respectively). The germination of olive embryos began after 3–4 days in culture, when it was possible to observe the divergence of the cotyledons. After 1 week, the radicle elongated and the cotyledons became green. After 20–30 days of culture the first leaves of the seedling appeared.

Seedlings obtained from cryopreserved embryos were established in pots; these plants were morphologically uniform when cultivated in the greenhouse. The results of the present study indicated that excised embryos of *Olea europaea* are tolerant to desiccation. The desiccated embryos did not lose viability after rapid cooling and storage at the temperature of LN (−196 °C).

Further cryopreservation studies, performed on seeds with endocarps (olive stones) and seeds (embryo and endosperm) after breaking the endocarp, have shown germination percentages around 95% after immersion directly in LN and rapid rewarming (unpubl. data).

Many desiccation-tolerant (orthodox) seeds survive exposure to LN (Stanwood 1985). Embryo axes of several tree species, some from recalcitrant seeds, have been successfully stored in liquid nitrogen (Pence 1995). In many of those reports desiccation was employed as a cryoprotective treatment.

2.2 Cryopreservation of Shoot Tips

A simple cryopreservation method has been developed for apical shoot tips from in vitro shoots. This includes sucrose preculture, partial desiccation and immersion in liquid nitrogen. Differential scanning calorimetry (DSC) was used to assess the phase change behavior of water in the samples during cryopreservation.

2.2.1 Plant Material and Cryopreservation Procedure

Micropropagated shoots of *Olea europaea* var. Arbequina, originating from embryo material (González-Río et al. 1994), were multiplied on proliferation medium: DKW medium (Driver and Kuniyuki 1984), supplemented with 4.4 M benzylaminopurine (BAP), 0.05 M indole-3-butyric acid (IBA), 3%

sucrose and 0.75% agar (Roko, La Coruña) at pH 5.8. Growth conditions were 25 °C and 16 h light/8 h dark (40 μmol m^{-2} s^{-1}). Transfers of the micro-shoots to fresh medium were made every 40 days.

Experiments were undertaken with nodal segments (0.5 cm length), apical shoot tips (comprising the meristematic dome with one or two pairs of leaf primordia, 2–5 mm length) and axillary shoot tips (1 mm length) that were excised from 40-day-old micro-shoots under a stereomicroscope. All were precultured for 2 days on solid DKW medium supplemented with 0.75 M sucrose, and dehydrated at 20 °C in the air current of a flow cabinet by placing 20 shoot tips on top of filter paper in an open Petri dish containing 30 g of silica gel. Twenty shoot tips were placed in each 2-ml cryovial (Nalgene). Vials were cooled by direct immersion in LN and warmed in a flow cabinet at room temperature for 15 min. The shoot tips were then transferred to 90-mm Petri dishes containing proliferation medium. After 6 weeks of culture under the same growth conditions as described before, survival for control (sucrose preculture plus dehydration) and cooled samples was measured as percentage of shoot tips that regenerated new shoots.

For cold acclimation, in vitro shoot cultures (40 days after the last transfer) were maintained in dark conditions at 4 °C for 2 months. For differential scanning calorimetric analyses one shoot tip was placed in an aluminum pan of a Mettler TA 4000 system (DSC 30), and each analysis was performed on at least six samples to ensure reproducibility.

2.2.2 Results and Discussion

Three types of explants were assayed in the cryopreservation experiments: nodal segments, excised axillary and apical shoot tips. Axillary shoot tips were completely dry after exposure to the air flow of the cabinet and did not survive (data not shown). Nodal segments, the explants used for olive micropropagation, survived dehydration to a 11% of water content under the air current of the flow cabinet, but not the further cooling in LN (data not shown). Apical shoot tips gave the best survival in terms of percentages of shoot recovery after cooling (Fig. 2).

Very high survival percentages (ca. 90%) were obtained from non-dehydrated samples without cooling (controls). However, as the dehydration period increased, a gradual decrease in survival of the controls was observed (Fig. 2), and no survival was obtained after 4 h. When samples were cooled, survival was obtained from shoot tips dehydrated for periods between 1 and 3 h. Maximum recovery shoot percentages (ca. 35%) were obtained after 1.5–2 h of dehydration, which corresponded to moisture contents of 33–28%. If we consider survival of the LN-cooled samples relative to that of the dehydration controls, 60–77% of shoots were recovered following these treatments.

To determine the most effective dehydration period for cryopreservation of olive shoot tips, cooling and warming thermograms of apical shoot tips

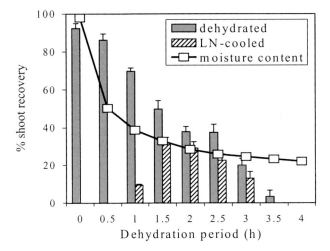

Fig. 2. Effects of the duration of dehydration on the percentage of shoot recovery from apical shoot tips either precultured for 2 days in 0.75 M sucrose and dehydrated for 0–4 h (controls) or subsequently cooled in LN, and on the percentage of moisture content. *Bars* represent the standard error of the mean. Five replicate batches (15 shoot tips each) were used per condition. (Martínez et al. 1999)

pretreated with 0.75 M sucrose were performed (Fig. 3) at the dehydration periods previously assayed (Fig. 2).

Peaks of crystallization and melting were displayed in samples dehydrated under the laminar air flow bench for 0, 0.5 and 1 h. These peaks were very small for samples dehydrated for 1.5 h. Shoot tips dried to 30% moisture content over a 2-h period exhibited a glass transition with mid points at −55.4 and −53.5 °C for cooling and rewarming, respectively (Fig. 3). These results confirm the absence of ice formation or recrystallization inside olive shoot tips dehydrated for 2 h during cooling and subsequent rewarming. These studies show the efficacy of dehydration treatments in producing protective vitrification events (Dereuddre et al. 1991; Dumet et al. 1993; Benson et al. 1996; Martínez et al. 1998).

Cold hardening of the in vitro cultures had no effect on the percentage of shoot recovery compared with non-acclimated shoot tips. Precultures for 2 days with different sucrose concentrations showed no significant differences between 0.3 and 0.75 M; 1.0 M sucrose was toxic.

The first indication of survival of apical shoot tips after cryopreservation was a greening of the explants (Fig. 4A), followed by elongation of the first leaf primordium within 7–15 days of culture on recovering medium (Fig. 4B). One-half to one-third of the samples showed further reactivation of the meristematic dome after another 30 days, which led to differentiation of new leaf primordia and subsequent formation of shoots (Fig. 4C). Simultaneously, a brown callus proliferated in the base of the explants. After 8 weeks of culture shoots had only three or four pairs of leaves and had stopped growing. Their

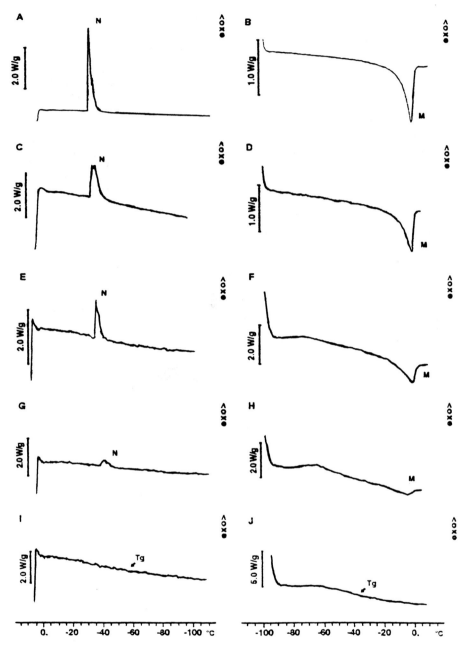

Fig. 3. Cooling (*left*) and warming (*right*) DSC thermograms of olive apical shoot tips, precultured for 2 days in 0.75 M sucrose and exposed to different dehydration periods in the air current of a laminar flow cabinet: **A, B** 0 h; **C, D** 0.5 h; **E, F** 1 h; **G, H** 1.5 h; **I, J** 2 h. *N* Nucleation. *M* melting. *Tg* transition temperature. (Martínez et al. 1999)

Fig. 4. Growth recovery from olive apical shoot tips with **a** two leaf primordial, **b** after 15 days, and **c** after 45 days of culture in proliferation medium. *Scale bar* 1 mm. (Martínez et al. 1999)

growth was not recoverable by transfer to fresh medium. Experiments to characterize a proper recovery medium and growth conditions to culture olive apical shoot tips are under way in our laboratory. Control shoot tips showed a similar growth pattern, but an actively growing green callus proliferated at the base of the explants. In many cases the callus covered the whole explant and inhibited the growth of the shoot.

3 Summary and Conclusions

Two simple cryopreservation methods are described in this chapter for embryos and shoot tips of *Olea europaea* L. var. Arbequina. After partial desiccation and cooling high germination percentages were achieved for cryopreserved embryos. However, lower survival percentages were observed for apical shoot tips after sucrose preculture, partial dehydration, and direct cooling in LN.

Shoot apices, a genetically stable plant material, are ideal candidates for long-term germplasm conservation. Indeed, the constituent cells of shoot meristems are small and undifferentiated, which usually is important for survival after cryopreservation in LN. However, olive shoot tips are prone to rapid oxidation and desiccation during the establishment of sterile cultures (Rugini et al. 1995; Revilla et al. 1996). As shown during the development of the present cryopreservation protocol, the shoot tips are sensitive to desiccation with the recovery rate progressively reduced as dehydration increased (Fig. 2).

Sucrose plays a role for cryoprotection at the cellular level due to the colligative action of relatively small molecules that depresses the freezing point. Nevertheless, other sugars or polyols such as trehalose, sorbitol or mannitol (Kartha and Engelmann 1994) may also provide protection against freezing, drying or both. Mannitol represents an important sugar fraction in *Olea europaea* (Flora and Madore 1993). A beneficial effect of mannitol on in vitro olive cultures (Leva et al. 1994) and stress protection by production of this osmolyte has been described for several plant species (Tarczynski et al. 1993; Stop and Pharr 1994). Several experiments on salt stress have recently been performed in our laboratory using olive callus derived from petiole. These experiments have shown that the use of mannitol in place of sucrose as a carbon source in the culture medium induced an increase in levels of proline (unpubl. results). The effects of proline in plant cell protection against stress situations are well documented in the literature (Delauney and Verma 1993). The use of mannitol in the preculture could improve resistance to dehydration of olive shoot tips and increase survival after cryopreservation.

Experiments using encapsulation of olive shoot tips in an alginate matrix resulted in very low survival rates to dehydration (data not shown). Successful cryopreservation by means of a sucrose preculture, followed by a direct plunge into liquid nitrogen without any additional cryoprotection, has been reported for shoot tips extremely sensitive to dehydration (Uragami et al. 1989; Panis et al. 1996). New strategies leading to increased olive shoot tip resistance to desiccation must be developed in future studies.

Absence of crystallization and melting peaks was observed in the thermograms obtained with naked olive shoot tips after sucrose preculture in samples with 30% moisture content (2-h dehydration). Hence, in the cryopreservation procedure described for olive tree apices, the problem of a relatively low survival rate relates more to their intolerance of dehydration than to sensitivity to LN treatment.

Acknowledgements. This work was supported by CICYT, project OLI96–2149

References

Barranco D (1997) Variedades y patrones. In: Barranco D, Fernández-Escobar D, Rallo L (eds) El cultivo del olivo. Junta de Andalucia, Sevilla, Spain, pp 57–79

Benson EE, Reed BM, Brennan RM, Clacher KA, Ross DA (1996) Use of thermal analysis in the evaluation of cryopreservation protocols for *Ribes nigrum* L. germplasm. Cryo Lett 17:347–362

Cañas LA, Avila J, Vicente M, Benbadis A (1992) Micropropagation of olive (*Olea europaea* L.). In: Bajaj YPS (ed) Biotechnology in agriculture and forestry, vol 18. High-tech and micropropagation II. Springer, Berlin Heidelberg New York, pp 493–505

Cervera MT, Cabezas JA, Sancha JC, Martínez de Toda F, Martínez-Zapater JM (1998) Application of AFLPs to the characterization of grapevine *Vitis vinifera* L genetic resources. A case study with accessions from Rioja (Spain). Theor Appl Genet 97:51–59

Delauney AJ, Verma DPS (1993) Proline biosynthesis and osmoregulation plants. Plant J 4:215–223

Dereuddre J, Hassen N, Blandin S, Kaminski M (1991) Resistance of alginate-coated somatic embryos of carrot (*Daucus carota* L.) to desiccation and freezing in liquid nitrogen: thermal analysis. Cryo Lett 12:135–148

Driver JA, Kuniyuki AH (1984) In vitro propagation of Paradox walnut rootstock. HortScience 19:507–509

Dumet D, Engelman F, Chanbrillange N, Dereuddre J (1993) Importance of sucrose for the acquisition of tolerance to desiccation and cryopreservation of palm somatic embryos. Cryo Lett 14:243–250

Flora LL, Madore MA (1993) Stachyose and mannitol transport in olive (*Olea europaea* L.). Planta 189:484–490

González-Río F, Gurriarán MJ, González-Benito E, Revilla MA (1994) Desiccation and cryopreservation of olive (*Olea europaea* L.) embryos. Cryo Lett 15:337–342

Grande Covián F (1989) El aceite de oliva en la prevención de las enfermedades cardiovasculares. II. Simposium ciéntifico del aceite de oliva. Expoliva 89, Jaén, Spain

Heywood HU (1978) Flowering plants of the world. Oxford Univ Press, London, 335 pp

Kartha KK, Engelmann F (1994) Cryopreservation and germplasm storage. In: Vasil IK, Thorpe TA (eds) Plant cell and tissue culture. Kluwer, Dordrecht, pp 195–230

Leva AR, Petruccelli R, Bartolini G (1994) Mannitol in vitro culture of *Olea europaea* L (cv. Maurino). Acta Hortic 356:43–46

Loussert R, Brousse G (1980) El olivo. Mundi-prensa, Madrid

Martínez D, Revilla MA, Espina A, Jaimez E, García, JR (1998) Survival cryopreservation of hop shoot tips monitored by differential scanning calorimetry. Thermochim Acta 317:91–94

Martínez D, Arroyo-García R, Revilla MA (1999) Cryopreservation of in vitro grown shoot-tips of *Olea europaea* L. var. Arbequina. Cryo Lett 20:29–36

Panis B, Totté N, Nimmen KV, Withers LA, Swennen R (1996) Cryopreservation of banana (*Musa* spp.) meristem cultures after preculture on sucrose. Plant Sci 121:95–106

Pence VC (1991) Cryopreservation of seeds of Ohio native plants and related species. Seed Sci Technol 19:235–251

Pence VC (1995) Cryopreservation of recalcitrant seeds. In: Bajaj YPS (ed) Biotechnology in agriculture and forestry, vol 32. Cryopreservation of plant germplasm I. Springer, Berlin Heidelberg New York, pp 29–50

Revilla MA, Pacheco J, Casares A, Rodríguez R (1996) In vitro reinvigoration of mature olive trees (*Olea europaea* L.) through micrografting. In Vitro Cell Dev Biol Plant 32:257–261

Rugini E, Fedeli E (1990) Olive (*Olea europaea* L.) as an oilseed crop. In: Bajaj YPS (ed) Biotechnology in agriculture and forestry, vol 10. Legumes and oilseed crops I. Springer, Berlin Heidelberg New York, pp 593–641

Rugini E, Pezza A, Muganu M, Caricato G (1995) Somatic embryogenesis in olive (*Olea europaea* L.). In: Bajaj YPS (ed) Biotechnology in agriculture and forestry, vol 30. Somatic embryogenesis and synthetic seed I. Springer, Berlin Heidelberg New York, pp 404–414

Stanwood PC (1985) Cryopreservation of seed germplasm for genetic conservation. In: Kartha KK (ed) Cryopreservation of plant cells and organs. CRC Press, Boca Raton, pp 199–226

Stop JMH, Pharr DM (1994) Mannitol metabolism in celery stresses by excess macronutrients. Plant Physiol 106:503–511

Tarczynski MC, Jensen RG, Bohnert HJ (1993) Stress protection of transgenic tobacco by production of the osmolyte mannitol. Science 259:508–510

Trujillo I, Rallo L, Arús P (1995) Identifying olive cultivars by isozyme analysis. J Am Soc Hortic Sci 120:318–324

Uragami A, Sakai A, Nagai M, Takahashi T (1989) Survival of cultured cells and somatic embryos of *Asparagus officinallis* cryopreserved by vitrification. Plant Cell Rep 8:418–421

Vos P, Hogers R, Bleeker M, Reijans M, vande Lee T, Hornes M, Frijtters A, Por J, Peleman J, Kuiper M, Zabeau M (1995) AFLP: a new technique for DNA fingerprinting. Nucleic Acids Res 23:4407–4414

Withers LA, Engelmann F (1998) In vitro conservation of plant genetic resources. In: Altman A (ed) Agricultural biotechnology. Marcel Dekker, New York, pp 57–88

III.7 Cryopreservation of Germplasm of *Populus* (Poplar) Species

Maurizio Lambardi

1 Introduction

1.1 Distribution and Important Species

Trees and shrubs of the family Salicaceae (poplars, aspens and willows) provide a wide range of industrial construction woods (sawtimber, peeler logs, pulpwood), fuel wood, poles, light packing material, plywood, matches, as well as being important in biomass and feedstock production. Furthermore, they are vital for protection of soil, crops, livestock and dwellings, and, as ornamental plants, they are used extensively in parks, gardens, streets and open spaces (Fig. 1). The species are dioecious, with male and female flowers on separate trees. Between the two genera forming the family, i.e., *Salix* and *Populus*, the latter is by far the more important from an economic point of view and includes both poplars and aspens. The genus *Populus* is characterized by a chromosome number of 2n=38, although triploid (2n=57) and tetraploid (2n=76) individuals have occasionally been recorded. The 34 species of the genus are botanically divided into five sections (Turanga, Leuce, Aigeiros, Tacamahaca and Leucoides) and these have an extremely wide geographical distribution throughout the Mediterranean, temperate, and colder regions of the northern hemisphere. While two sections (Turanga and Leucoides) do not include species of commercial value, the other three include several species of great economic importance.

Section Leuce contains *P. tremula*, *P. tremuloides* and *P. grandidentata* (aspens), *P. alba* and other minor species (white poplars). Aspens are found in North America, Europe, Western Asia and North Africa; white poplars are typical of the Mediterranean basin and Eastern Europe. An economically important hybrid between aspen and white poplar is the grey poplar (*P. alba* × *P. tremula* = *P. canescens*).

Section Aigeiros contains the true poplars, including *P. nigra* (black poplar) and *P. deltoides* (cottonwood), the former naturally occurring in the Mediterranean area and Western Asia, the latter native to North America. More than 90% of the cultivated poplars in the world belong to these two

National Research Council (CNR), Istituto sulla Propagazione delle Specie Legnose, via Ponte di Formicola 76, 50018 Scandicci (Firenze), Italy

Biotechnology in Agriculture and Forestry, Vol. 50
L.E. Towill and Y.P.S. Bajaj (Eds.) Cryopreservation of Plant Germplasm II
© Springer-Verlag Berlin Heidelberg 2002

Fig. 1. Use of white poplar (*Populus alba* L.) as ornamental plants along a river bank in Florence, Italy. (Lambardi, unpubl.)

species and to their natural and artificial hybrids (*P.* × *euramericana* syn. *P. canadensis*).

Section Tacamahaca (balsam poplars) is the largest section of the genus and includes species of major (*P. balsamifera* and *P. trichocarpa*, native to North America and Canada) and minor economic value (e.g., *P. ciliata*), the latter found in China, Korea, eastern Europe, and the Himalayas.

In central Asia, in eastern Europe and in the Mediterranean area, poplars (together with willows) have been closely associated with agriculture since antiquity, providing timber, fuel and leaves (used as forage for animals). Naturally occurring trees of black and white poplar were the first cultivated forms. However, during the 17th century, the introduction of *P. deltoides* from North

America into Europe led to a spontaneous hybridization with *P. nigra* (which originated the *P. × euramericana*) and, in time, caused a transformation in poplar cultivation. During the early years of the 20th century, poplar cultivation was already widespread in Europe with the support of industry, thanks also to easy propagation by cuttings. Through the exploitation of the natural characteristics of poplar, breeding work, over time, produced clones with desirable attributes (such as rapid growth, straightness, good fodder value, favourable branching habit) which can be utilized for cultivation in short rotation cycles, today even less than 10 years. Moreover, the availability of species and clones which can be grown in a wide range of climates and soils (including marginal and sub-marginal soils) has made the establishment of extensive plantations possible in several areas of the northern hemisphere.

Today, natural areas and uniform plantations of poplar are spread over a large part of the world. Canada reports large natural areas of mixed hardwood and hardwood/conifer stands, while natural and planted poplars in China cover 1.34 million ha. The most recent information on the main areas of poplar plantations are: France 245,000 ha, Romania 163,000 ha of which 80,000 ha are native stands, Hungary 162,000 ha, Turkey 157,000 ha, Iran 150,000 ha, Germany 103,000 ha, Spain 98,000 ha, Italy 71,000 ha, Argentina 55,000 ha, Belgium 45,000 ha, Egypt 40,000 ha, The Netherlands 31,000 ha, and India 26,400 ha (FAO 1996).

1.2 Various Methods for Storage of Poplar Germplasm

1.2.1 In Situ Conservation

Due to the large, naturally occurring genetic variability of *Populus*, the conservation of natural stands is the first option, allowing for natural evolution of the species. This approach is even more important when, as in the case of poplars and aspens, trees are unusually susceptible to diseases, pathogens and abiotic stress. However, at present, few coordinated programs of in situ preservation of poplar species are in progress in the world. Among them and worthy of note is a program of collaboration between European countries aimed at ensuring the conservation of natural stands of *P. nigra* (Turok et al. 1998).

1.2.2 Ex Situ Conservation

Seeds and Pollen. The establishment of collections of seeds and pollen should be regarded as a method for long-term ex situ preservation of poplar genetic resources. Both seed germination and pollen viability are affected in poplar by many factors, such as the time and method of collecting seeds and pollen, the time lapse between collection and storage, their moisture content and storage temperature. Poplar seeds are very small (1000 seeds weigh about 1 g) and their germinability starts to decrease 3 to 4 weeks after ripening time. Therefore, a higher germination rate is achieved when seed is collected in the proximity of ripening. Dehydration is fundamental for the storage of seeds and pollen: poplar seeds can be stored successfully for years at low temperatures if the moisture content is reduced to 6–8%, while a reduction to 10% is required for pollen. Following these rules, poplar seeds can be stored for years at temperatures ranging from −18 to −40 °C (Cagelli 1997). Pollen in glass ampoules under vacuum maintained its fertility for at least 5 years following storage at −18 °C (Hermann 1976).

In-Field Collections. In-field collections (clonal arboretums) are the traditional ex situ preservation method for woody species of high economic value. The establishment of in-field collections is relatively simple due to the ease in propagating the majority of poplar clones by cuttings (Zsuffa et al. 1993). The Italian in-field clonal collection, managed by the Poplar Research Institute (Istituto di Sperimentazione per la Pioppicoltura) of Casale Monferrato (Fig. 2), is the largest in Europe and contains almost 1800 genotypes from numerous *Populus* species and hybrids, among which are *P. nigra, P. alba, P. deltoides, P. × euramericana, P. simonii, P. trichocarpa, P. maximowiczii,* and *P × interamericana* (S. Bisoffi, pers. comm.).

Nowadays, an advanced and efficient system of ex situ conservation requires the production of appropriate databases that have the objective of

Fig. 2. Germplasm preservation of *Populus deltoides* in clonal arboretum. The trees form stock hedges and are used as a cutting source. (Courtesy of the Poplar Research Institute of Casale Monferrato, Italy)

providing an efficient tool for information, management, exchange of clones between countries, detection of duplications among national collections and identification of accessions. A European database of clones has been developed for black poplar by EUFORGEN (European Forest Genetic Resources Programme), and reports information on as many as 2000 clones maintained in the clonal arboretums of 18 European countries (Turok et al. 1998).

It should be noted that, because of the difficulty in controlling the numerous pests and diseases affecting poplar, conservation in the field of vegetatively propagated plants involves contamination hazards, and requires periodic and careful monitoring of trees to keep them disease-free.

1.2.3 In Vitro Conservation

Minimal Growth Storage. In vitro cloning has been widely studied in *Populus* over the past 20 years, and optimized micropropagation procedures have been established for several species and hybrids (Douglas 1989; Lubrano 1992; Chun 1993). These methods make poplar particularly suitable for medium-term germplasm preservation by means of the minimal growth storage technique. However, few reports have thoroughly explored poplar response to the storage at above freezing temperatures. Shoot cultures of *P. alba* × *P. grandidentata* were maintained for more than 5 years at 4 °C in darkness, following a 1-month prestorage culture on MS (Murashige and Skoog 1962) medium, containing 1.33 μM benzyladenine (BA). However, shoot regrowth declined from 70% after 2-year storage to 25% after 5 years (Son et al. 1991). Our experience with white poplar showed that better maintenance and prompt and superior regrowth can be obtained when shoots are cultured on Quoirin and Le Poivre (1977) medium, containing 2.5 μM BA, 0.3 μM gibberellic acid (GA_3) and 0.3 μM 1-naphthaleneacetic acid (NAA), and stored at 4 °C, under a 8-h photoperiod at a photon fluence rate of $30 \mu mol\, m^{-2} s^{-1}$ (unpubl. data).

Storage of Twigs and Buds. Ahuja (1988) investigated the potential for storing dormant buds of pure aspens (*P. tremula, P. tremuloides*) and hybrids (*P. tremula* × *P. tremuloides*) for medium-term germplasm preservation. The buds were stored either along with twigs or as isolated buds and, after different periods of storage, bud explants were cultured in vitro on a screening medium to monitor the growth and differentiation of microshoots. Of the five temperatures tested (ranging from 0 to −80 °C), best survival and microshoot differentiation after 2-year storage were observed from isolated buds, coated with wax and maintained under 190 Torr vacuum at −80 °C. Buds stored with twigs after 2 years showed extreme sensitivity to low temperatures. In general, the genotype greatly affected the storage potential of aspen buds.

1.3 Need for Conservation/Cryopreservation

Among woody species, poplar is one for which the necessity of a wide and coordinated program of germplasm preservation is absolutely necessary and urgent, in consideration of its large genetic variability (natural or from breeding activity). As regards natural stands, it is well known that riparian formations, in which the Salicaceae are often prominent components, are diminishing for reasons such as the artificialization of riverbanks, the conversion of riparian forest to agricultural land or to commercial poplar plantation. The International Poplar Commission (IPC) has recently highlighted the need for conservation of natural occurrences of riparian forest. Moreover, an active monitoring and conservation program is recommended for minor species (*P. euphratica, P. ilicifolia* = *P. euphratica* subsp. *Denhardtorum*) and other low-latitude species adapted to the warm and dry climates of sub-tropical countries, especially *P. yunnanensis* and *P. ciliata* in Asia, and species of the Tacamahaca and Aigeiros sections in Central America.

Because of the ease of crossing and cloning, the large genetic variability, the adaptability to a variety of soils and climates, and the superior growth rate of many hybrids, poplar has been intensely utilized in breeding programs carried out in the USA, Canada and several European countries, e.g., Germany, Italy, Hungary, Holland, Sweden, and Belgium (Zsuffa et al. 1993). At present, breeders are well aware of poplar disease problems and are actively selecting for resistance to numerous leaf, stem and root pathogens, such as *Marssonina brunnea* in Italy, which causes a 60% reduction of wood production in susceptible clones (FAO 1980). As a result of this intense breeding activity, hundreds of clones have been selected. In spite of this, only a small number are widely used and these account for over 50% of all poplar plantations in the world. Thus, there is a considerable risk of genetic diversity loss following introgression by a reduced number of cultivars.

Hence, the preservation of a broad genetic pool of the genus *Populus*, both natural and from breeding activity, should be assessed. The exploration of different approaches is urgently required since germplasm preservation carried out with traditional systems (e.g., clonal arboretums, from which to collect budwood, cuttings, etc.) in the near future will face serious problems due to the vast space required, the high costs of maintenance and the risks of diseases. Among them, cryopreservation of tissues and organs by vitrification has recently given encouraging results with several woody species (Sakai 1995, 2000; Lambardi et al. 2000b), and could represent an efficient and strategically important alternative for long-term poplar germplasm preservation.

2 Cryopreservation

2.1 Brief Review of Cryopreservation of Poplar

Various tissues and organs of aspen and poplar have been cryopreserved (Table 1). One-year-old twigs of several *Populus* species were not injured when cooled to −196 °C, if they were first held at low temperatures for 6–24 h. The effective prefreezing temperatures differed considerably among the tested

Table 1. Summary of reports on *Populus* cryopreservation (not including shoot-tip vitrification)

Species	Material frozen	Freezing and thawing method	Viability	Reference
P. siebodi	Twigs (15 cm length, 0.8 cm diameter)	PF: 24 h at −30°C; LN: 24 h; TH: unspecified	100%	Sakai (1960)
P. maximowiczii, P. nigra	Twigs (15 cm length, 0.5 cm diameter)	PF: 6 h at −15°C (*P.m.*) or −20°C (*P.n.*); LN: 30 min; TH: 4 h at −10°C + 16 h at 0°C + room temperature	Unspecified	Sakai (1965a)
P. maximowiczii, P. simonii, P. nigra, P. glaubka	Twigs (10 cm length, 0.6–0.7 cm diameter)	PF: 16 h at −15°C or −20°C (*P. glaubka*); LN: 15 min; TH: 1 h at −10°C + 16 h at 0°C	Unspecified	Sakai (1965b)
P. x euramericana	Callus from cambial tissue	PF: −30°C (5°C/day); LN: 2 h; TH: rapid	Unspecified	Sakai and Sugawara (1973)
P. tremula x, P. tremuloides	Seeds	LN: 6 days; TH: 2 min at 40°C + room temperature	99%	Ahuja (1986)
Hybrid (unspecified)	Callus, suspension cells	CP: 5% glycerol (callus), 5% glycerol + 5% DMSO + 4% sucrose (suspension cells); PF: 1°C/min to −40°C; LN; TH: at 40°C	40% (suspension cells)	Binder and Zaerr (1980)
Aspens and hybrid aspen clones	Dormant buds	PF: slow freezing to −40°C; LN: 24 h; TH: rapid	Unspecified	Ahuja (1993)
P. alba x, P. grandidentata, P. glandulosa	Callus from cambial tissue	CP: DMSO (callus: DMSO, 1:5 w/v); PF: −25°C to −45°C (0.5–1°C/min); LN; TH: rapid	62%	Son et al. (1997)

PF: prefreezing temperature; LN: time in liquid nitrogen (when specified); TH: thawing procedure; CP: cryoprotective.

species, ranging from a minimum of −30 °C for *P. siebodi* (Sakai 1960) to a maximum of −15 °C for *P. maximowiczii*, *P. simonii* and *P. nigra* (Sakai 1965a,b). Twigs prefrozen at temperatures warmer than −15 °C never flushed and rooted, after they were thawed, potted in moist sand and kept in a greenhouse. In another study (Ahuja 1993), dormant buds from 13 aspen and hybrid aspen clones remained viable after slow-freezing to −40 °C, 24-h storage in liquid nitrogen (LN) and rapid thawing. Shoots differentiated from cryopreserved buds of all clones, although with marked differences among genotypes.

When dedifferentiated cells of poplar (i.e., callus and suspension cells originated from cambial tissue) were used for ultra-low temperature storage, best results were obtained with cells slowly cooled to −30 °C (*P.* × *eurameri-cana*, Sakai and Sugawara 1973), −40 °C (Binder and Zaerr 1980), or a temperature ranging from −25 to −45 °C (*P. alba* × *P. grandidentata* and *P. glandulosa*; Son et al. 1997), prior to immersion in LN. Pretreatments with cryoprotective chemicals (glycerol and DMSO, alone or in combination) and rapid thawing were generally required for cell survival. Successful recovery after 5 years storage was reported (Son et al. 1997).

Seeds of a hybrid aspen (*P. tremula* × *P. tremuloides*) showed no loss of viability following rapid cooling and storage at −196 °C for 6 days, provided that they were warmed in a water bath at 40 °C for about 2 min, and then kept at room temperature for 0.5 h before sowing (Ahuja 1986).

2.2 Protocol

In our studies, shoot tips from in vitro grown stock plants of white poplar (*Populus alba* L.), black poplar (*Populus nigra* L.) and grey poplar (*Populus canescens* Sm.) have been utilized to test the efficiency of a vitrification technique to preserve shoot-tip viability during direct cooling in LN. In order to develop a successful procedure of cryopreservation, white poplar was the first species tested. Experimental material was collected from stock plants monthly subcultured in a gelled Quoirin and Le Poivre medium, containing 0.09 M sucrose, 2.5 µM BA, 0.3 µM GA_3 and 0.3 µM NAA. During proliferation, stock plants were maintained at 22 ± 1 °C, under a 16-h photoperiod at a photon fluence rate of 60 µmol m^{-2} s^{-1}. Before bud excision, the plantlets were transferred to 5 °C and 8-h photoperiod for a 3-week cold-hardening period. Axillary buds (Fig. 3A), excised under a laminar-flow hood, were utilized to obtain shoot tips (1–2 mm long and 1–1.5 mm base diameter), consisting of the apical meristem and 4–5 leaf primordia.

Excised shoot tips were precultured for 2 days at 5 °C, under a 8-h photoperiod and a photon fluence rate of 30 µmol m^{-2} s^{-1}, on hormone-free MS medium, gelled with 0.7% agar and supplemented with 0.09, 0.3 or 0.7 M sucrose. Following this, shoot tips were loaded for 20 min at 25 °C with a cryoprotectant (2 M glycerol and 0.4 M sucrose) contained in 2-ml cryovials. The cryoprotectant was then replaced with the PVS2 vitrification solution (Sakai et al. 1990), consisting of 30% glycerol (w/v), 15% ethylene glycol (w/v), 15%

Fig. 3A–D. Shoot development and rooting from successfully cryopreserved shoot tips. **A** Axillary buds of white poplar (*arrows*) from which shoot tips were obtained (*bar* 0.25 cm). **B** Successfully vitrified and thawed buds remained green after plating on gelled MS medium, containing 1.5 μM BA and 0.5 μM GA$_3$, and resumed growth within 1–2 weeks without callusing (bar 0.2 cm). **C** Shoot elongation after transfer onto hormone-free MS medium containing 0.1% activated charcoal (*bar* 0.25 cm). **D** Rooting of elongated shoots was easily achieved on MS medium containing 3 μM IBA (*bar* 1.5 cm). (Lambardi et al. 2000a, modified)

DMSO (w/v) in MS medium containing 0.4 M sucrose (pH 5.8). In order to choose the longest possible exposure time of shoot tips to the vitrification solution, avoiding toxic effects, shoot tips were exposed to PVS2 for 20, 40, 60, 80, 100, and 120 min at 0 or 25 °C, and evaluated for survival. Time exposures of 20 min, at 25 °C, and 60 min, at 0 °C, were then selected for cooling trials. After treatment with the vitrification solution, the shoot tips were suspended in

0.6 ml fresh PVS2 solution in the cryovials and cooled to −196 °C by plunging directly into LN, where they were kept for at least 1 h.

In order to determine the best warming temperature, shoot tips of white and grey poplar were treated with PVS2 for 60 min at 0 °C, immersed in LN, and warmed by removing the cryovials from the LN, maintaining them for 5–10 s at air temperature, and then plunging them for 50 s into a water bath, at either 30, 35, 40, 45 or 50 °C. As best survival was obtained at 40 °C, this warming temperature was then used in all later experiments.

After thawing, PVS2 was drained from the cryovials and replaced with a liquid MS medium containing 1.2 M sucrose (Sakai et al. 1991) for a 20-min washing. The shoot tips were then plated on a gelled MS medium, lacking NH$_4$NO$_3$ (Kuriyama et al. 1990), and containing 0.09 M sucrose and different hormone combinations. Three weeks after plating on MS medium supplemented with growth regulators, the surviving shoot tips were transferred onto a gelled, hormone-free MS medium, containing 0.1% activated charcoal, for shoot development. Shoots longer than 1 cm were then rooted on a gelled MS medium, containing 3 µM indole-3-butyric acid (IBA).

The procedure selected for cryopreservation of white poplar shoot tips was then tested on shoot tips from in vitro plants of black poplar and grey poplar.

2.3 Results and Discussion

Toxicity of the Vitrification Solution. Prolonged exposures to a concentrated vitrification solution can injure experimental material, due to chemical toxicity or strong osmotic stress. In poplar, a typical symptom of toxicity of the PVS2 solution was a profuse blanching of uncooled shoot tips. However, this negative effect was markedly influenced by temperature and duration of treatments (Fig. 4). At 25 °C shoot formation strongly decreased after exposure to PVS2 for 40 min or more; when treatments were performed at 0 °C, more than 90% of the shoot tips were not damaged by incubation of up to 60 min, and formed normal shoots. Therefore, in order to obtain satisfactory protection from cooling, these two times/temperatures were applied in the subsequent cryopreservation experiments.

Effect of Preculture and Exposure to the PVS2 Solution. Successfully vitrified and thawed buds remained green after plating on gelled MS medium, containing 1.5 µM BA and 0.5 µM GA$_3$, and resumed growth within 1–2 weeks (Fig. 3B). The effects of both preculture and loading in PVS2 on bud survival after cooling to −196 °C and thawing at 40 °C are summarized in Fig. 5 (*top*). Higher survival was achieved with buds loaded with PVS2 for 60 min at 0 °C, in comparison with 20-min exposures at 25 °C.

Shoot-tip preculture in media containing high concentrations of sucrose (0.3–0.7 M) often improved survival after retrieval from LN (Niino et al. 1992, 1997; Matsumoto et al. 1994). In white poplar, after cooling, a steady decrease in shoot-tip survival was observed with increased sucrose concentrations in

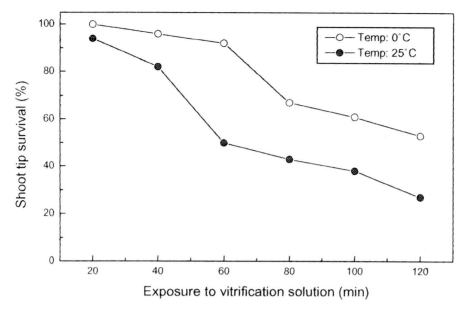

Fig. 4. Survival (%) of white poplar shoot tips after exposure to the vitrification solution (PVS2) at 0 or 25 °C. Shoot tips were treated with PVS2 for increasing time periods, after which they were immediately washed with liquid MS medium containing 1.2 M sucrose, plated on gelled MS medium and evaluated for survival after 1 week. The majority of dead shoot tips were profusely blanched. (Caccavale et al. 1998)

the preculture medium. Best protection from cooling in LN was obtained when a 2-day bud preculture on MS medium, supplemented with 0.09 M sucrose, was combined with a 60-min incubation at 0 °C, as shown by the percentage of shoot-tip survival (90%), which was the same as for unfrozen shoot tips.

Preculture and vitrification treatments can also affect the subsequent development of shoots from the surviving meristems (Fig. 5, *bottom*). Again, the combination 0.09 M sucrose/60-min PVS2 at 0 °C was the one that induced the highest percentage (62%) of healthy shoot formation (i.e., the percentages of shoot tips that developed without any defect or malformation per 100 survived shoot tips), although no statistical differences among treatments occurred.

Effect of Warming Temperature. As during cooling, the transition of the vitrification solution from a vitreous to a crystalline phase during warming can cause considerable damage to the tissues and compromise the regrowth of meristems after plating. It has often been reported that rapid warming in a water bath can avoid recrystallization, and ensure a satisfactory recovery of vitrified material. However, how fast this "rapid warming" should be is still controversial, as a wide range of water-bath temperatures have been suggested, ranging from 22 °C (Towill 1995) to 45 °C (Reed 1995). Cryopreserved

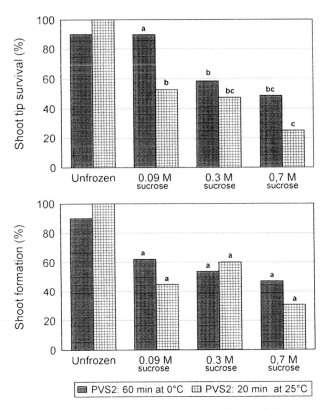

Fig. 5. Effect of sucrose concentration in the preculture medium and of exposure to the vitrification solution (PVS2) on shoot-tip survival (*top*) and shoot formation (*bottom*) after cryopreservation. Percentages on shoot-tip survival collected 3 weeks after warming at 40 °C and plating on a gelled MS medium, lacking NH_4NO_3, and containing 0.09 M sucrose, 1.5 µM BA and 0.5 µM GA_3; percentages on shoot formation, collected 4 weeks after transfer onto hormone-free MS medium containing 0.1% activated charcoal, refer to the number of shoot tips that developed into rootable shoots per 100 survived shoot tips. Percentages followed by different letters are significantly different at p ≤ 0.05

shoot tips of white poplar showed best regrowth (80%) when cryovials were maintained at room temperature for 5–10 s and then plunged for 50 s into a water bath at 40 °C, corresponding to a warming rate of 180 °C/min (calculated during the passage from −80 to −30 °C). Unlike white poplar, the effect of the warming temperature (in the range 30–50 °C) on shoot-tip survival of grey poplar appeared negligible (Fig. 6). However, for both species, shoot tips thawed at 40 °C produced a majority of healthy, well-developed shoots (data not shown), which proved that, in general, damage induced during rewarming, if it occurred, did not compromise meristem regrowth.

Shoot-Tip Regrowth, Elongation and Rooting. The notion of sublethal injuries, affecting almost all cryopreserved shoot tips, has highlighted the importance

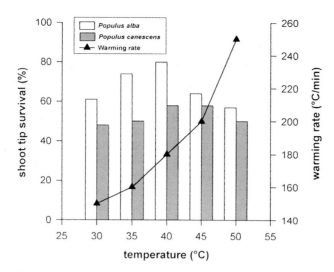

Fig. 6. Effect of bath temperature on warming rate and shoot-tip survival of *Populus alba* and *P. canescens*. Before thawing, shoot tips, after 2-day preculture at 5 °C on 0.09 M sucrose MS medium and 20-min loading with 2 M glycerol plus 0.4 M sucrose, were treated with PVS2 for 60 min at 0 °C and plunged into LN. Survival was evaluated 3 weeks after washing and plating on gelled MS medium, lacking NH_4NO_3, and containing 3 µM zeatin plus 1 µM GA_3. (Lambardi et al. 2000b, modified)

of a phase of repair after warming (Grout 1995). An appropriate hormone composition of plating medium can stimulate cell divisions of cryopreserved shoot tips and, consequently, allow regrowth. At the same time, callus proliferation prior to shoot development may cause genetic alterations and, therefore, it is always undesirable. Among the various cytokinins tested in combination with GA_3, BA at a concentration of 1.5 µM was the only treatment that never stimulated callus proliferation (Table 2). Shoot tips plated on hormone-free MS medium showed similar behaviour for both survival and callus production, but shoot formation was stimulated to a lesser extent. In comparison with 1.5 µM BA, the application of 3 µM zeatin markedly improved shoot-tip survival (66.7% vs. 89.5%) and speeded up shoot development, with a callus production lower than that induced by both thidiazuron and BA at a concentration of 3 µM.

After 3 weeks on cytokinin-containing MS medium, subculturing the developing shoots on hormone-free medium containing 0.1% activated charcoal gave faster elongation (Fig. 3C). Almost 100% rooting was achieved when shoots longer than 1 cm were transferred onto MS medium containing 3 µM IBA (Fig. 3D).

Application of the Vitrification Procedure to Black and Grey Poplar. With the step-by-step methodology described above for white poplar, the vitrification procedure was tested on another species of poplar (*P. nigra*), and a hybrid between poplar and aspen (*P. canescens* = *P. alba* × *P. tremula*). Significant

Table 2. Effect of different plating media on survival, shoot formation and callus proliferation of cryopreserved shoot tips; basal medium: MS, lacking NH_4NO_3 and containing 0.09 M sucrose and 0.7% agar. Percentages followed by different letters are significantly different at $P \leq 0.05$. (Lambardi et al. 2000a)

Plating medium (µM)[a]	Shoot-tip survival (%)[b]	Shoot formation (%)[c]	Callus proliferation[d]
BA 1.5 + GA₃ 0.5	66.7bc	89.7a	−
BA 3.0 + GA₃ 1.0	80.6abc	58.3b	+ +
TDZ 3.0 + GA₃ 1.0	84.6ab	48.6c	+ +
ZEA 3.0 + GA₃ 1.0	89.5a	52.6bc	+
No hormone	48.0c	25.0c	−

[a] Before plating, shoot tips, after 2-day preculture at 5°C on 0.09 M sucrose MS medium and 20-min loading with 2 M glycerol plus 0.4 M sucrose (1:1), were treated with PVS2 for 60 min at 0°C, plunged into LN, thawed at 40°C and washed for 20 min with liquid MS medium (TDZ, thidiazuron; ZEA, zeatin).
[b] Data collected 3 weeks after warming at 40°C and plating.
[c] Percentages refer to the number of shoot tips that developed into rootable shoots per 100 survived shoot tips. Data collected 4 weeks after transfer on hormone-free MS medium, containing 0.1% activated charcoal.
[d] + , Moderate callus proliferation; + + , high callus proliferation; −, no callus proliferation.

Table 3. Shoot-tip survival of white, grey and black poplar after cryopreservation, following the procedure determined for white poplar

Species	Shoot-tip survival (%)[a]
Populus alba	82.1a
Populus canescens	54.0b
Populus nigra	22.2c

[a] Percentages followed by different letters are significantly different at $P \leq 0.05$.

differences were observed among the three species as regards their response to cryopreservation (Table 3), as the percentages of shoot-tip survival ranged from a minimum of 22% (black poplar) to a maximum of 82% (white poplar).

Morphological and Histological Observations. Shoot tips that did not survive cooling to −196°C showed clear symptoms of blanching and browning. In order to reveal non-lethal damage produced by the cryopreservation procedure, histological observations were carried out on 2-µm thick sections of white poplar shoot tips that had been subjected to the cooling/thawing procedure. The shoot tips were embedded in historesin following the procedure of Yeung and Law (1987), after which they were sectioned, mounted on microscope slides, and stained with the periodic acid/Schiff's reagent and aniline blue black. Although diffuse breakages were observed, they were mainly localized in the tissues of leaflets (Fig. 7A). The vitrification procedure was effective in protecting the cells of the meristematic apex during rapid cooling to −196°C, as the apex and the surrounding area generally appeared unaffected

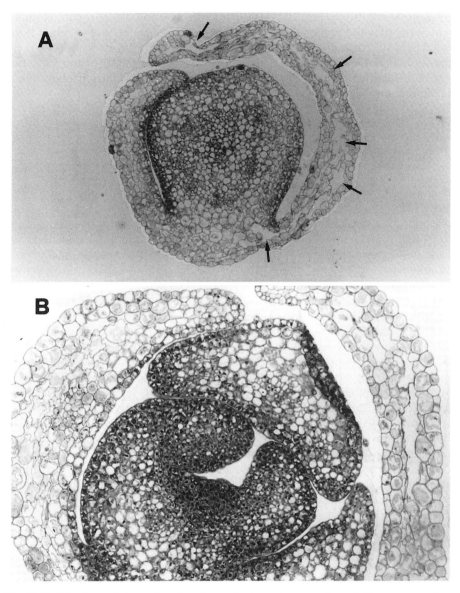

Fig. 7A,B. Histology of shoot tips after the cooling/warming procedure. **A** Transverse section about 1 mm below the meristematic apex (×120). When shoot tips are successfully vitrified, injuries (i.e., breakages) can be observed only in tissues of leaflets surrounding the meristematic apex (*arrows*). **B** Transverse section of the meristematic apex (×240). The small, densely packed cells of a properly vitrified meristematic apex appear undamaged and functional after thawing

by tissue disruptions (Fig. 7B). Moreover, after plating on regrowth medium, a majority of white poplar shoot tips exhibited an efficient phase of repair from breakages of leaflet tissues and originated normal-developing shoots. Among the surviving meristems, those that did not develop into healthy shoots showed symptoms of hyperhydricity or callused profusely, particularly if plated on a medium containing a high concentration of cytokinins.

3 Summary and Conclusions

It is well known that cryopreservation in LN, suggested for plants since the early 1970s, is a technique that has met with several obstacles to its large scale adoption. These are mainly due to the utilization of expensive machinery for the slow and controlled cooling of tissue that avoids lethal intracellular ice formation. Innovative preconditioning methodologies for tissue and organs from in vitro culture (e.g., vitrification and encapsulation-dehydration), before plunging the explants directly into LN, have been successfully tested in recent years with several species, and could open the door to a more convenient approach to the preservation of woody plant germplasm.

The present report shows that a one-step vitrification methodology can be successfully applied to the cryopreservation of Poplar species. An effective procedure for cryopreservation of shoot tips of *Populus alba* is summarized in Table 4, from which almost 60% of rooted shoots can be obtained from cryopreserved shoot tips. However, it is apparent from studies with other species that differences among optimal conditions (relative to cold-hardening, cryoprotectants, vitrification solution application, warming, plating media) can be ascribed to specific characteristics of the stock plants, such as genotype, physiological status, type of explant, etc. In corroboration of this, the procedure selected for *Populus alba* proved to be, at the same time, an excellent starting point for cryopreservation of *Populus canescens*, but unsatisfactory for

Table 4. Summary of the procedure of vitrification, cryopreservation and recovery, selected for white poplar (*Populus alba* L.) shoot tips

Phase	Treatment
1. Stock plant cold-hardening	5 °C, 8-h photoperiod, 3 weeks
2. Explant material	Shoot tips (1–2 mm long, 1–1.5 mm diameter)
3. Explant cold-hardening	0.09 M sucrose MS, 5°C, 8-h photop., 2 days
4. Cryoprotectant	2 M glycerol + 0.4 M sucrose, 20 min
5. Vitrification solution	PVS2, 60 min at 0 °C
6. Cryopreservation	Direct immersion in LN
7. Thawing	10 sec at room temperature + 50 sec at 40 °C
8. Washing	Liquid MS medium (1.2 M sucrose), 20 min
9. Plating	MS ($-NH_4NO_3$), 1.5 µM BA + 0.5 µM GA_3, 3 wk
10. Shoot elongation	MS, containing 0.1% activated charcoal
11. Shoot rooting	MS, containing 3 µM IBA

cryopreservation of *Populus nigra*. Research should move in the direction of making procedures as widely applicable as possible.

Acknowledgements. Stock material was kindly provided by the "Vitrocoop Laboratory – Centrale Ortofrutticola" (*P. alba*) and by the "Vivai Battistini Giuseppe" (*P. nigra* and *P. canescens*), Cesena, Italy. The author thanks Stefano Bisoffi and Lorenzo Vietto for the information on the clonal collections at the Poplar Research Institute, Casale Monferrato, Italy. Final thanks go to Angela Caccavale for technical assistance, and to Prof. Andrea Fabbri for his valuable comments on the manuscript.

References

Ahuja MR (1986) Storage of forest tree germplasm in liquid nitrogen (–196°C). Silvae Genet 35:249–251

Ahuja MR (1988) Differential growth response of aspen clones stored at sub-zero temperatures. In: Ahuja MR (ed) Somatic cell genetics of woody plants. Kluwer Academic, London, pp 173–180

Ahuja MR (1993) Regeneration and germplasm preservation in aspen-*Populus*. In: Ahuja MR (ed) Forestry sciences, vol 41. Micropropagation of woody plants. Kluwer, London, pp 187–194

Binder WD, Zaerr JB (1980) Freeze preservation of suspension cultured cells of a hardwood poplar. Cryobiology 17:624–625

Caccavale A, Lambardi M, Fabbri A (1998) Cryopreservation of woody plants by axillary bud vitrification: a first approach with poplar. Acta Hortic 457:79–83

Cagelli L (1997) Guidelines for seed and pollen storage. In: Turok J, Lefèvre F, de Vries S, Toth B (eds) *Populus nigra* network. Report of the 3rd Meeting. Sàrvàr, Hungary, 5–7 Oct 1996. IPGRI, Rome, pp 12–13

Chun YW (1993) Clonal propagation in non-aspen poplar hybrids. In: Ahuja MR (ed) Forestry sciences, vol 41. Micropropagation of woody plants. Kluwer, London, pp 209–222

Douglas GC (1989) Poplar (*Populus* spp.). In: Bajaj YPS (ed) Biotechnology in agriculture and forestry, vol 5. Trees II. Springer, Berlin Heidelberg New York, pp 300–323

FAO (ed) (1980) Poplars and willows in wood production and land use. FAO, For Ser 10

FAO (1996) Report of the 20th session of the international poplar commission and 38th session of its executive committee. Budapest, 1–4 Oct, http://www.agro.ucl.ac.be/efor/ipc/w3777/w3777e03.htm

Grout B (1995) Introduction to the in vitro preservation of plant cells, tissue and organs. In: Grout B (ed) Genetic preservation of plant cells in vitro. Springer, Berlin Heidelberg New York, pp 1–20

Hermann S (1976) Verfahren Konservierung und Erhaltung der Befruchtungsfähigkeit von Waldbaumpollen über mehrere Jahre. Silvae Genet 25:223–229

Kuriyama A, Watanabe K, Ueno S, Mitsuda M (1990) Inhibitory effect of ammonium ion on recovery of cryopreserved rice cells. Plant Sci 64:231–235

Lambardi M, Fabbri A, Caccavale A (2000a) Cryopreservation of white poplar (*Populus alba* L.) by vitrification of in vitro grown shoot tips. Plant Cell Rep 19:213–218

Lambardi M, De Carlo A, Benelli C, Bartolini G (2000b) Cryopreservation of woody species by vitrification of shoot tips and embryogenic tissue. Proceedings of the 4th International Symposium on In Vitro Culture and Horticultural Breeding. Tampere, Finland, 2–7 July 2000. Acta Hortic (in press)

Lubrano L (1992) Micropropagation of poplars (*Populus* spp.). In: Bajaj YPS (ed) Biotechnology in agriculture and forestry, vol 18. High-tech and micropropagation II. Springer, Berlin Heidelberg New York, pp 151–178

Matsumoto T, Sakai A, Yamada K (1994) Cryopreservation of in vitro grown apical meristems of wasabi (*Wasabia japonica*) by vitrification and subsequent high plant regeneration. Plant Cell Rep 13:442–446

Murashige T, Skoog F (1962) A revised medium for rapid growth and bioassay with tobacco tissue cultures. Physiol Plant 15:473–497

Niino T, Sakai A, Yakuwa H, Nojiri K (1992) Cryopreservation of in vitro grown shoot tips of apple and pear by vitrification. Plant Cell Tissue Org Cult 28:261–266

Niino T, Tashiro K, Suzuki M, Ohuchi S, Magoshi J, Akihama T (1997) Cryopreservation of in vitro grown shoot tips of cherry and sweet cherry by one-step vitrification. Sci Hortic 70:155–163

Quoirin M, Le Poivre P (1977) Etudes de milieux adaptes aux cultures in vitro de *Prunus*. Acta Hortic 78:437–442

Reed BM (1995) Cryopreservation of in vitro grown gooseberry and currant meristems. Cryo Lett 16:131–136

Sakai A (1960) Survival of the twig of woody plants at −196 °C. Nature 185:393–394

Sakai A (1965a) Determining the degree of frost-hardiness in highly hardy plants. Nature 206: 1064–1065

Sakai A (1965b) Survival of plant tissue at super-low temperatures III. Relation between effective prefreezing temperatures and the degree of frost hardiness. Plant Physiol 40:882–887

Sakai A (1995) Cryopreservation of germplasm of woody plants. In: Bajaj YPS (ed) Biotechnology in agriculture and forestry, vol 32. Cryopreservation of plant germplasm I. Springer, Berlin Heidelberg New York, pp 53–69

Sakai A (2000) Development of cryopreservation techniques. In: Engelmann F, Takagi H (eds) Cryopreservation of tropical plant germplasm. IPGRI, Rome, pp 1–7

Sakai A, Sugawara Y (1973) Survival of poplar callus at super-low temperatures after cold acclimation. Plant Cell Physiol 14:1201–1204

Sakai A, Kobayashi S, Oiyama I (1990) Cryopreservation of nucellar cells of navel orange (*Citrus sinensis* Osb. var. *brasiliensis* Tanaka) by vitrification. Plant Cell Rep 9:30–33

Sakai A, Kobayashi S, Oiyama I (1991) Survival by vitrification of nucellar cells of navel orange (*Citrus sinensis* var. brasiliensis) cooled to −196 °C. J Plant Physiol 137:463–470

Son SH, Chun YW, Hall RB (1991) Cold storage of in vitro cultures of hybrid poplar shoots (*Populus alba* L. × *P. grandidentata* Michx.). Plant Cell Tissue Org Cult 27:161–168

Son SH, Park YG, Chun YW, Hall RB (1997) Germplasm preservation of *Populus* through in vitro culture systems. In: Klopfenstein NB, Chun YW, Kim MS, Ahuja MR (eds) Micropropagation, genetic engineering, and molecular biology of *Populus*. Gen Tech Rep RM-GTR-297. USDA, Forest Service, Fort Collins, CO, pp 44–49

Towill LE (1995) Cryopreservation by vitrification. In: Grout B (ed) Genetic preservation of plant cells in vitro. Springer, Berlin Heidelberg New York, pp 99–111

Turok J, Lefèvre F, de Vries S, Alba N, Heinze B, Van Slycken J (eds) (1998) *Populus nigra* network. Report of the 4th meeting. Geraardsbergen, Belgium, 3–5 Oct 1997. IPGRI, Rome, 85 pp

Yeung EC, Law SK (1987) Serial sectioning techniques for a modified LKB historesin. Stain Technol 62:147–153

Zsuffa L, Sennerby-Forsse L, Weisberger H, Hall RB (1993) Strategies for clonal forestry with poplars, aspens, and willows. In: Ahuja MR, Libby WJ (eds) Clonal forestry II. Conservation and application. Springer, Berlin Heidelberg New York, pp 91–119

III.8 Cryopreservation of *Prunus*

Marie-Thérèse de Boucaud, Marthe Brison, Bertrand Helliot, and Valérie Hervé-Paulus

1 Introduction

1.1 Plant Distribution and Important Species

The genus *Prunus* which belongs to the family of Rosaceae is widely distributed mostly in the north temperate zone. Some species are found in the Indo-Malaysian Mountains and the Andes.

In Europe, North America and China, some native or subspontaneous species were selected, improved and became trees for fruit production. They comprise three subgenera: *Prunophora*, *Amygdalus* and *Cerasus* (Rehder 1947). The subgenus *Prunophora* contains plum (*euprunus* section) native to Anatolia and Persia and apricots (*armeniaca* section) native to central Asia or China, then imported into Persia, Armenia, and the rest of Europe. The subgenus *Amygdalus* comprises peach trees native to East Asia, introduced into Persia, Greece and to the rest of Europe and almond trees native to Asia and North Africa, introduced into Europe by the Romans. The subgenus *Cerasus* (*eucerasus* section) includes fruit species producing cherries. Its origin is complex. It may have been imported from Asia. Several species of *Prunus* and interspecific hybrids are used as rootstocks for fruit-trees belonging to the same species or different species to increase their geographic area of culture (Bernhard 1994). Other species are grown for their wood or for medicine (Dosba et al. 1994).

In Japan, many native species that were grafted for better genetic stability are cultivated as ornamental plants for their attractive blossoms or foliage.

1.2 Methods of Storage of *Prunus* Germplasm

Maintaining the diversity of *Prunus* germplasm is very important because of the economic importance of many of the species. It is necessary to maintain a gene reservoir for the breeding programs of cultured fruit and decorative

Laboratoire de Physiologie Cellulaire Végétale, Université Bordeaux 1, Avenue des Facultés, 33405 Talence Cedex, France

Biotechnology in Agriculture and Forestry, Vol. 50
L.E. Towill and Y.P.S. Bajaj (Eds.) Cryopreservation of Plant Germplasm II
© Springer-Verlag Berlin Heidelberg 2002

species. The many and varied species of exotic *Prunus*, from bushes to big trees, are a very interesting and valuable source of characteristics.

1.2.1 Seeds

Conservation of genetic resources of *Prunus* as seed collections is difficult. Although they tolerate desiccation, they do not store well at subzero temperatures. Therefore, the storage classification of *Prunus* seeds is termed "intermediate" (Ellis et al. 1991b) corresponding to the previous "suborthodox" category (Bonner 1986). In addition, they need chilling (4 °C) to break dormancy. In any case, they often lose their germination capacity in a relatively short time.

1.2.2 In Situ Conservation

In situ conservation, which involves conserving plants in their ecosystems, maintains adaptation to changes of the environment. However, loss of germplasm in natural populations may occur rapidly and so it is necessary to consider alternatives for the preservation.

1.2.3 Ex Situ Conservation and Database

This method is usually adopted for clonal maintenance in greenhouses, orchards and repositories. Interesting cultivars of *Prunus* have been selected and hybridized on the basis of various characteristics.

Many ex situ collections of *Prunus* in Europe are catalogued in a European database (Gass et al. 1996; Zanetto et al. 1997). In 1993, the International Plant Genetic Resources Institute (IPGRI) entrusted to INRA Villenave d'Ornon, France, the management of European Data of *Prunus* (EPDB, European *Prunus* database) which replaced the Nordic Gene Bank created in Sweden in 1983 (Dosba et al. 1994). In 1996, 12910 accessions were registered in 95 institutes of 25 European countries. This database is divided into five main files: peach-trees, plum-trees, cherry-trees, apricot-trees, and almond-trees. These "Europe wide passport data" (Gass et al. 1996; Zanetto et al. 1997) include the species names, synonyms, the name of the institute that maintains the accessions, and also descriptive data for the varieties (fruit and flower color).

1.2.4 In Vitro Plant Preservation

In vitro culture complements traditional field methods of conservation of genetic resources, particularly for species whose seeds are difficult to preserve. In vitro culture is a rapid means to propagate and distribute material for a collection in comparison with collection in field. In vitro culture also requires less space.

1.3 Need for Conservation Genetic Resources of the Genus *Prunus*

Conservation of genetic resources has two purposes: (1) preservation of genetic diversity, and (2) preservation of selected varieties for the economic value of their characteristics.

Maintaining genetic diversity of cultivated plants and their wild relatives is important for the continued improvement of crop species. There are several reasons for action to conserve the diversity of *Prunus* resources. These include the loss of important varietal collections, formerly grown by nursery gardeners, and the loss of many of the old varieties because they are short lived and disease often causes death in European conditions. There is more interest from amateur associations for preserving the apple group rather than the *Prunus* group.

Maintaining plants of vegetatively propagated and selected cultivars in field conditions may have disastrous consequences, e.g., loss of valuable material due to phytoplasmas and viruses that are very difficult to eradicate. *Prunus* spp. are particularly threatened by European Stone Fruit Yellows (ESFY) phytoplasmas, such as Apricot Chlorotic Leaf Roll (ACLR), which can provoke decline of entire orchards, and by Isometric Labile Ring Spot (ILAR) viruses such as the *Prunus* Necrotic Ring Spot Virus (PNRSV), Prune Dwarf Virus (PDV), Apple Chlorotic Leaf Spot Virus (ACLSV), Apple Mosaic Virus (APMV) and Peach Latent Mosaic Viroid (PLMVd). The most important disease originating in Eastern Europe, specifically in the Mediterranean area, is caused by Sharka disease (Plum Pox Potyvirus, PPV) which is an RNA virus transmitted by aphid vectors or by grafting infected material. Symptoms of the disease may appear on leaves or fruits and cause considerable loss. Infected material must be destroyed because no treatment has been found to cure them. In the last 15 years, breeding programs have been developed to limit the spread of this virus. These are aimed at screening PPV-resistant cultivars and introducing durable resistance into susceptible cultivars by crossing (Dosba et al. 1994). More recently, genetic transformations have been attempted. However, transgenic resistant plants are not commercially available and the safekeeping of threatened species is still dependent on the use of more traditional methods such as intensive micropropagation, chemotherapy, thermotherapy, meristem culture and micrografting, provided that the excised shoot tip including meristematic cells is smaller than 0.5mm (Colin and Verhoyen 1976).

No single strategy for *Prunus* genetic conservation is truly satisfying. In situ and ex situ conservation methods have problems. In vitro culture has a high risk of bacteria and fungal cell infection and is labor-intensive. In addition, use of in vitro culture can select variants. This disadvantage can be limited by slow growth of in vitro cultures, whereby the frequency of subculture can be reduced to every 12 months. However, the principal inconvenience is the possible occurrence of somaclonal variants, incompatible with the aim of genetic resource conservation.

For all these reasons, attempts have been made to develop cryopreservation procedures at ultra-low temperatures (–196 °C) that completely arrest cell metabolism. Cryopreservation as a long-term storage method for in vitro culture is becoming feasible for maintaining genetic diversity or selected cultivars. It allows duplication of field and in vitro collections with the possibility to check the sanitary state of the material. Cryopreservation of selected species requires the use of shoot tip culture and might be considered as a possible combined method of virus elimination and conservation of the shoot tips (Brison et al. 1997). This method can be used to eliminate other pathogens.

This chapter describes different techniques applied to embryonic axes, shoot tips and somatic embryos for cryopreservation of *Prunus* species and outlines the problems encountered, progress made to date and future prospects.

2 Cryopreservation

2.1 Cryopreservation of Embryonic Axes of *Prunus*

2.1.1 General Account

Cryopreservation of excised embryonic axes is a way to maintain genetic resource diversity of material if seeds do not store well. *Prunus* seeds are large seeds and difficult to conserve. Seed viability varies from year to year and can significantly decrease within 18–24 months following harvest. In addition, seeds show pronounced dormancy imposed by the endocarp, cotyledons and testa. Chilling treatments as long as 6 months may be required to break dormancy before testing germination.

In general, *Prunus* seeds withstand desiccation but not freezing. The problem of sensitivity to desiccation or freezing is complex. The different tissues of seeds, such as the embryonic axes and cotyledons, do not have the same critical moisture level (Finch-Savage 1992) and do not have the same response to freezing. Their response varies with development and germination stage and also with the time of shedding from the plant. In several cases, it has been noted that embryos survive desiccation and freezing better the fresher they are (Pence 1990; de Boucaud et al. 1991). However, because of the variety of responses to desiccation and freezing of recalcitrant and intermediate seeds, it appears that no single method is ideal for all species.

Two general approaches have been directed at the cryopreservation of these tissues. In some cases whole seeds, a part or only embryonic axes, in desiccated or partially desiccated states, are exposed to liquid nitrogen (LN). This is done when the tissue is dried sufficiently to avoid freezing damage. When this is not possible, tissue can be frozen while hydrated or partially hydrated, using cryoprotectants and controlled cooling to minimize freezing damage. In some cases, these two approaches have been combined.

Not much work has been done on embryonic axes of *Prunus*. Chaudhury and Chandel (1995) examined embryonic axes of almond (*Prunus amygdalus* Batsch). Almond seeds are important commercial products for the nutritious sweet kernel and almond oil. Almond seeds have the same storage characteristics as the other *Prunus* seeds, as described above. Chaudhury and Chandel (1995) showed that excised embryonic axes survived LN exposure after desiccation to between 10%–4% moisture content (% fresh weight), with viability ranging from 100%–40%. The highest survival (100%) was obtained for axes desiccated to about 7% moisture. Normal healthy shoots and roots emerged from 66% of surviving axes. Desiccation below 7% moisture decreased viability and rate of growth.

A study was undertaken by de Boucaud et al. (1996) on seeds from peach (*Prunus persica* L. Batsch), another economically important species that has storage characteristics like that of other *Prunus* species. This research used cv. GF305, which is a vigorous rootstock and a good indicator of fruit-tree virus, and confirmed the LN sensitivity of peach seeds. Two cryopreservation procedures for excised embryonic axes were used: either desiccation alone under a sterile laminar air flow or a 1 h dehydration period followed by application of a cryoprotective solution.

2.1.2 Materials and Cryopreservation Procedure

All experimental procedures applied to embryonic axes are shown in Fig. 1.

Seeds. Peach seeds of cv. GF 305 were provided by A Pierronnet, Unité de Recherche sur les Espèces Fruitières et la Vigne, Institut National de Recherches Agronomiques (INRA), Villenave d'Ornon, France. Seeds were from the previous harvest season and had been kept in their stones for 6 months at 4 °C. The moisture content at the time of the experiments was about 6%. After breaking the stones, controls with or without integuments were either directly cultured (CCs) or cultured after 3 or 5 h of dehydration at 23 °C (CsA, CsB), after which their moisture content was determined. For cryopreservation, samples with or without integuments were quickly plunged into LN and afterwards warmed in a 45 °C water-bath during 1 min. All seeds, before culturing, were sterilized in 14% calcium hypochlorite for 20 min, rinsed in sterile deionized water and left overnight in water in order to soften the integuments. A set of seeds (FsA) was peeled and cultured without integuments, another set (FsB) was cultured with their integuments (see left part of Fig. 1).

Embryonic Axes. All embryonic axes were excised from sterilized seeds left overnight in sterile water. Control samples (CC) were then cultured directly. Other samples (sets of 20) were subjected to one of the following treatments indicated on the right-hand side of Fig. 1.

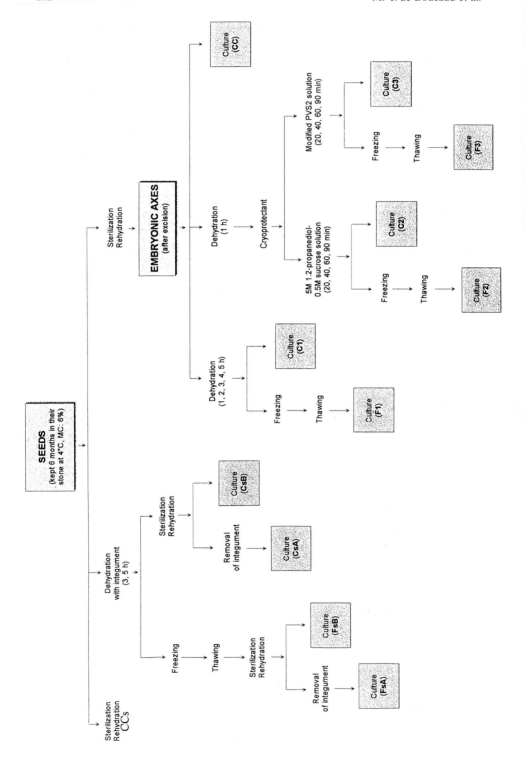

1. Dehydration on a Petri dish under sterile air flow for 1–5 h and culture on the medium described below (C1).
2. Dehydration in a Petri dish under sterile air flow for 1–5 h followed by rapid plunging into LN and thawing in water-bath at 45 °C and reculture (F1).
3. Dehydration in a Petri dish for 1 h followed by pretreatment for 20–90 min at room temperature in a cryoprotective solution, either 38% 1,2-propanediol/0.5 M sucrose (C2) or PVS2 solution (Sakai et al. 1990, modified by Brison et al. 1995) containing 12.5% w/v dimethyl sulfoxide (DMSO), 15% w/v ethylene glycol (EG), 25% w/v glycerol, 3% w/v polyethylene glycol (PEG) 8000 and 13.6% w/v sucrose (C3).
4. Dehydration for 1 h, treatment in a cryoprotective solution followed by rapid plunging of cryovials into LN (F2 and F3) and subsequent thawing in a water-bath during 1 min at 45 °C and culture as above.

All seeds (CCs, Cs, Fs) and all excised embryonic axes (CC, C1, C2, C3, F1, F2, F3) were cultured on Murashige and Skoog (MS) medium (Murashige and Skoog 1962) supplemented with the iron provided by sequestrene (R) (Protex) at 200 mg/l, and vitamins (1 mg/l thiamine HCl, nicotinic acid, pyridoxine HCl, Ca pantothenate, cysteine HCl, 0.1 mg/l biotin and 100 mg/l myoinositol). Growth regulators consisted of 1 mg/l 6-benzylaminopurine (BAP), 100 µg/l indole-3-butyric acid (IBA), and 100 µg/l µM gibberellic acid (GA3). The medium contained 30 g/l sucrose and 7 g/l agar (Arlès, Marseille, France). The pH was adjusted to 5.6 prior to autoclaving for 20 min at 115 °C under 1.7 bar pressure.

Results were expressed as percent survival and regrowth. The survival of embryonic axes and seeds was evaluated after 2 weeks of culture. The regrowth was evaluated after 2 months of culture as the number of shoots developing directly from embryonic axes (without callus formation) relative to the total number of axes cultured.

The moisture content of seeds was determined on a set of 20 seeds and expressed as percentage of dry weight. Samples were oven-dried at 70 °C until a constant weight was obtained.

The moisture content of embryonic axes was also expressed as a percentage of dry weight and was determined using two replicate samples of 100 embryonic axes. Embryonic axes were weighed under sterile conditions after every dehydration period (1, 2, 3, 4, 5 h) then cultured.

Thermal analyses of the two cryoprotectant solutions (1,2-propanediol/sucrose and modified PVS2) were performed with a differential scanning calorimeter (DSC-7, Perkin Elmer) using cooling and warming rates of 50 °C/min.

Fig. 1. Schematic representation of experiments performed with seeds and embryonic axes of peach. *CCs* Control embryonic axes cultured without dehydration; *CsA, CsB* control seeds; *FsA, FsB* cooled seeds; *CC* control embryonic axes cultured without dehydration; *C1, C, C3* control embryonic axes under different experimental conditions; *F1, F2, F3* cooled embryonic axes under different experimental conditions

Table 1. Moisture content and germination of control and cooled seeds of *Prunus persica* after various desiccation durations.

Desiccation (h)	Moisture content % dry weight	Germination (%)			
		CsA	CsB	FsA	FsB
0*	6	75	0	15	0
3	5	50	0	0	0
5	4.9	35	0	0	0

* CsA (0h) corresponds to directly cultured seeds (CCs in **Fig. 1**).
CsA: Control seeds without integument
CsB: Control seeds with integuments
FsA: Cooled seeds without integument
FsB: Cooled seeds with integuments

2.1.3 Results and Discussion

Dehydration and Cooling of Seeds. The survival of seeds and the decrease in their moisture content during dehydration are shown in Table 1.

During desiccation, the moisture content of the seeds with their integuments, which was already low, decreased from 6 to 4.9% over 5h. Seeds without integuments germinated without dehydration or with only slight dehydration (CsA), but germination decreased rapidly with decreasing moisture content. The presence of integuments completely inhibited germination (CsB). Only a low percentage (15%) of seeds cooled without integuments germinated.

These results confirm the intermediate storage characteristics of peach seeds, which can tolerate extensive desiccation but do not survive after exposure to LN. Other species also show sensitivity to low temperature but tolerance to desiccation. For example, seeds from some species of *Juglans*, identified as intermediate, can survive after careful drying down to 5% moisture (Gordon 1992) but they are still short-lived in storage because of their high lipid content. Seeds of other species such as *Carica papaya, Coffea arabica, Oreodoxa regia* and *Elaeis guineensis* (Ellis et al. 1991a–c) also show this characteristic. These intermediate species do not form a homogeneous group because they are not equally tolerant to desiccation. The difficulty encountered in cryopreservation of entire seeds was mainly overcome by cryopreservation of excised embryonic axes.

Cooling of Embryonic Axes. Embryonic axes excised directly from non-rehydrated seeds, frozen in LN and sterilized only before culturing, survived. However, they always produced rosette shoots that remained small and did not develop further (data not shown).

Therefore, we used embryonic axes excised from rehydrated seeds, i.e., soaked overnight, in further experiments. Results reported in Tables 2 and 3 compare the two cryopreservation methods examined, i.e., partial desiccation under laminar air flow (Table 2), and combination of a 1h dehydration period and loading with a cryoprotective solution, either 1,2-propanediol/sucrose or

Table 2. Survival and regrowth of embryonic axes of *Prunus persica* after desiccation and cryopreservation. Results are given as the ratio n/N and as percentages in parentheses. n-Number of surviving or regrowing embryonic axes; N-number of cultured embryonic axes

Desiccation (h)	Moisture content % dry weight	Survival*		Regrowth**	
		C1	F1	C1	F1
0***	80	19/20 (95)	0/20 (0)	14/17 (82)	2/12 (16)
1	35.4	16/20 (80)	16/20 (80)	18/21 (85)	20/21 (95)
2	30.1	14/20 (70)	15/20 (75)	18/21 (85)	25/26 (96)
3	11.9	15/20 (75)	16/20 (80)	20/26 (76)	22/26 (84)
4	8.9	18/20 (90)	16/20 (80)	19/20 (95)	15/20 (75)
5	7.4	9/20 (45)	10/20 (50)	9/16 (56)	19/25 (76)

* Survival rate after 15 days of culturing
** Regrowth rate after 2 months of culturing with formation of shoot
*** (0 h) corresponds to axes cultured directly (CC)
C1: Control samples
F1: Cooled samples

Table 3. Survival and regrowth of embryonic axes of *Prunus persica* after 1 h dehydration followed by application of a cryoprotective solution and cryopreservation. Results are given as the ratio n/N and as percentages of survival and regrowth in parentheses. n-number of surviving or regrowing embryonic axes; N = number of cultured embryonic axes

Duration of application of cryoprotectant (min)	1,2-Propanediol-sucrose solution				Modified PVS_2 solution			
	Survival*		Regrowth**		Survival*		Regrowth**	
	C2	F2	C2	F2	C3	F3	C3	F3
20	14/20 (70)	17/20 (55)	16/20 (80)	12/20 (60)	18/20 (90)	17/20 (85)	17/20 (85)	17/20 (85)
40	14/20 (70)	10/20 (80)	12/20 (60)	5/10 (50)	16/20 (80)	10/20 (50)	16/20 (85)	11/20 (55)
60	19/20 (95)	18/20 (90)	17/20 (85)	9/10 (90)	17/20 (85)	18/20 (90)	16/20 (80)	17/20 (85)
90	18/20 (90)	19/20 (95)	18/20 (90)	19/20 (95)	15/20 (75)	18/20 (90)	13/20 (90)	18/20 (85)

* Survival after 2 weeks culturing.
** Regrowth after 2 months culturing.
C2, C3: Control samples.
F2, F3: Cooled samples.

modified PVS2 (Table 3). Differences between survival and regrowth rates can be explained by the delay in germination and growth of some embryonic axes that occurred after the first observations performed 2 weeks after the experiments.

Survival and regrowth rates (about 70–90%) were similar for control and cooled axes after 1, 2, 3 or 4 h dehydration, corresponding to moisture contents of 35.4, 30.1, 11.9 and 8.9%, respectively (Table 2). Beyond this value, regrowth rates decreased for both control and cooled samples.

Table 3 shows survival and regrowth rates after 1 h dehydration, application of either 1,2-propanediol/sucrose (de Boucaud and Cambecedes 1988) or modified PVS2 solutions and after cryopreservation. Results were erratic but the highest levels of survival were obtained for loading periods of 60 and 90 min. Shoots regenerated from cooled (F2) and control (C2, CC) after 1 h of dehydration and 45 min of loading with the 1,2-propanediol/sucrose solution are shown in Fig. 2.

Although we were not able to determine the amount of cryoprotective solution penetrating cells, we studied the ability of both the solutions to vitrify using differential scanning calorimetry (DSC) with cooling and warming at 50 °C/min. (Fig. 3). For 1,2-propanediol/sucrose and modified PVS2 solutions no freezing exotherms were observed during cooling. A phase of glass (transition) occurred at about −115 and at −73 °C, respectively.

Fig. 2. Two-month old peach plant from cooled (*F2*), treated (*C2*) and control (*CC*) embryonic axes. Axes submitted to the cryoprotective, (1h dehydration and 45min loading 1,2-propanediol/sucrose solution) and cooling treatments show delayed regrowth in comparison with control (*CC*). *Bar* 1 cm

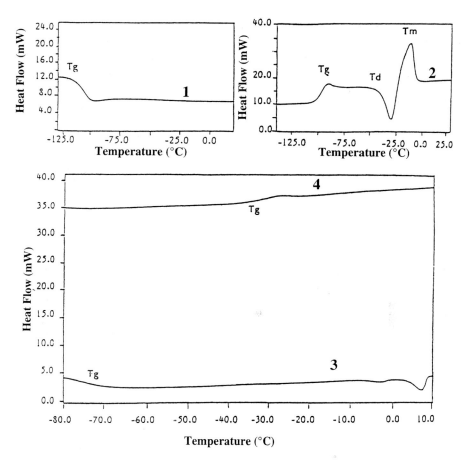

Fig. 3. Differential scanning calorimetry of cryoprotective solutions. The cooling and warming rate was 50 °C/min. Curves *1* and *3* are cooling profiles for 1,2-propanediol/sucrose and modified PVS2 solutions, respectively. Curves *2* and *4* are warming profiles of 1,2-propanediol/sucrose and modified PVS2, respectively. *Tg* Temperature of glass transition; *Td* temperature of devitrification; *Tm* melting temperature

During rewarming, the metastable 1,2-propanediol/sucrose solution devitrified at −49 °C with a maximum exotherm at −28 °C. The melting point appeared to be at −13 °C. This crystallization may be avoided by application of the higher warming rate (about 200 °C/min) in a water-bath at 45 °C, as employed for vials in our experiments. The modified PVS2 solution did not show devitrification with rewarming at 50 °C/min. This solution exhibited a glass transition at −37 °C. However, it is difficult to extrapolate these results to cryoprotectant-treated embryonic axes, since we do not know the amount of cryoprotectant that permeated. It was only possible to determine the moisture content of embryonic axes after various dehydration periods (Table 2).

Rapid cooling is the best means to obtain vitrification of material for a given moisture content according to the detailed study on inter-relationships between water content, rate of freezing to −196 °C, thermal properties of water and survival reported for excised embryonic axes of tea by Wesley-Smith et al. (1992). In our study, the two methods (desiccation alone or desiccation associated with cryoprotective treatment) both gave good levels of survival.

If we consider work carried out by others on recalcitrant or intermediate seeds, we observe that cryopreservation of embryonic axes was successfully achieved with both methods employed in our experiments. Pence (1990, 1992) studied the ability of embryonic axes of 18 species from seven tree genera of temperate regions to survive desiccation and dry freezing. She obtained either direct germination (*Aesculus*) or partial germination with callus formation (*Juglans, Carya, Castanea* and *Quercus*). Cryopreservation of hazelnut embryonic axes by desiccation also succeeded provided that hazelnut seeds were previously stratified (Reed et al. 1994). Cryopreservation using dehydration was also developed for embryonic axes of *Olea europaea* (Gonzalez-Rio et al. 1994), tea, cocoa and jackfruit (Chandel et al. 1995) and *Prunus amygdalus* (Chaudury and Chandel 1995). Embryonic axes excised from 48-h imbibed seeds of *Fraxinus excelsior* were cultured 48 h before being dehydrated down to various moisture contents and cryopreserved (Brearley et al. 1995).

Methods associating dehydration with the use of cryoprotectants were successfully developed for several genotypes of *Arachis hypogaea* (Runthala et al. 1993). Other experiments were also carried out on embryonic axes excised from imbibed *Pisum sativum* seeds (Berjak et al. 1995) in order to determine the dehydration period after which imbibed axes, which started germinating, lost their desiccation tolerance. A cryoprotective treatment allowed these imbibed germinating axes that had become desiccation-sensitive to withstand cooling.

The success of cryopreservation of embryonic axes by rapid cooling in LN depends on determining the optimal moisture content that minimizes both desiccation and freezing damage. It is necessary to adapt the method to requirements of the species. In cases where the moisture content cannot be decreased without irreversible damage, a cryoprotective treatment is necessary before cooling. In the case of peach embryonic axes, both treatments were very effective. Since these axes could withstand desiccation down to moisture contents as low as 8.9% enabling cryopreservation, the use of a cryoprotective treatment was unnecessary. The simplest method, which also avoids treatment with toxic chemicals is, therefore, recommended for cryopreservation of peach embryonic axes.

This protocol for cryopreservation of *Prunus persica* (L.) Batsch is applicable only to embryonic axes, since seeds were shown to be LN-sensitive. The use of excised embryonic axes is apparently the best way to overcome the difficulty of either storing or cryopreserving whole seeds of recalcitrant or intermediate species such as *Prunus*. Cryostorage of embryonic axes is easy and does not require the use of an expensive freezing apparatus. This allows considerable saving in space that compensates for the added costs for culture of axes after excision, cooling and warming. In comparison with classical conservation of seeds, it ensures stability of germplasm, requiring less periodic testing for viability and overcomes the problems of dormancy of *Prunus* seeds.

2.2 Cryopreservation of Shoot Tips of *Prunus*

2.2.1 General Account

Few publications report cryopreservation of *Prunus* using in vivo (winter buds) or in vitro (shoot tips) material. With winter leaf buds, Katano and Irie (1991) and Towill and Forsline (1999) obtained some growth after freezing. In the first work, Katano and Irie dissected buds of Japanese flowering cherry that had been cooled to −40 °C and then immersed in LN. In the second study, Towill and Forsline desiccated single-bud nodal sections of sour cherry before cooling at 1 °C/h to −30 °C and then plunging into LN. These methods, however, have some limitations. In both studies buds were harvested within a short period during maximum cold hardiness, and then underwent controlled desiccation in order to withstand freezing. Materials are not always available at the desired times and not all species are as desiccation-tolerant.

The results reported here with *Prunus* were obtained using apices excised from in vitro shoots. These are preferable sources of material for germplasm preservation. Indeed, shoots cultured under disease-free conditions are mass propagated and their growth can be easily controlled. Three different protocols have been successfully used for cryopreservation of this material.

Two-Step Cooling. Shoot tips, placed in cryotubes containing a cryoprotectant, are cooled slowly at 0.5 °C, 1 °C, or 2 °C/min to a given temperature and then plunged into LN. Slow cooling rates induce freeze dehydration of cells and therefore avoid damaging ice crystallization.

Vitrification. Shoot tips are exposed to a highly concentrated vitrification solution in cryotubes for a defined time before being plunged directly into LN. The solution concentration and the rapidity of cooling are such that glass formation occurs during freezing.

Encapsulation-Dehydration. Shoot tips are encapsulated in 3% alginate beads, incubated in liquid medium supplemented with different concentrations of sucrose, and then desiccated either in an air-flow cabinet or with the help of silica gel before being plunged into LN.

For each method, success depends on the *Prunus* species and also on several parameters of the protocol. Recently, two reports on the cryopreservation of *Prunus* species have been published. In the first one, Niino et al. (1997) cryopreserved shoot tips of two cherry cultivars (*P.* cv. *jamasakura* and *P.* cv. *lannesiana*) and five sweet cherry cultivars by one-step vitrification with vitrification solution exposure times of 90–120 min. They obtained 55–96% survival depending on the line used. In the second study, Shatnawi et al. (1999)

compared one-step vitrification and encapsulation-dehydration, applied to an almond rootstock (*Prunus dulcis*). With vitrification, under the described conditions, only 10% survival occurred whereas the encapsulation-dehydration technique gave 50%–62% survival. In both reports the importance of plantlet cold hardening and shoot tip preculture was emphasized for the successful cryopreservation.

Our work on cryopreservation of *Prunus* shoot tips began in 1992 (Brison et al. 1992). Since then, either to optimize the results or to simplify the procedure and to adapt it to the plant material, we have successively developed different techniques modified from the above methods. In this chapter, four cryogenic protocols are presented for *Prunus* species.

2.2.2 Methodology

Plant Material. The main experiments were carried out with two interspecific plum hybrids *Fereley-Jaspi* (from the cross *Prunus salicina* L. × *Prunus spinosa* L.) and *Ferlenain-Plumila* (from an open pollination of *Prunus besseyi* x unknown parent) selected and provided by INRA, Unité de Recherches sur les Espèces Fruitières et la Vigne, Villenave d'Ornon, France. The rootstock GF8$_1$, a hybrid of *Prunus marianna* selected for its resistance to wet soil, was also provided by INRA. Another *Prunus*, cv. *Ogden* (Bailey 1894; Moretti 1932), was provided by Dr. S. Spiegel, the Volcani Center, Virus Laboratory, Bet Dagan, Israel. All these belong to the species grouped as plums. One experiment was also performed on peach, *Prunus persica* L. Batsch cv. GF305, provided by INRA.

Culture Media. Prunus shoot cultures were maintained on modified MS basal medium supplemented with the iron provided by Séquestrène (Protex) 40 mg/l complemented by $FeSO_4$ Na_2 EDTA, vitamins:1 mg/l thiamine HCl, nicotinic acid, pyridoxine HCl, Ca pantothenate, cysteine HCl, 0.1 mg/l biotin and 100 mg/l myo-inositol. Growth regulators consisted of 1 mg/l 6-benzylaminopurine (BAP), 100 μg/l indole-3-butyric acid (IBA) and 100 μg/l gibberellic acid (GA3). 30 g/l sucrose and 7 g/l agar (Arlès, Marseille) were also incorporated. The pH was adjusted to 5.6 prior to autoclaving.

For the culture of *Prunus* cv. *Ogden*, the growth regulator concentrations were modified (0.75 mg/l BAP, 25 μg/l IBA, 50 μg GA3) and the culture medium was gelled with 1 g/l Phytagel (Sigma) and 3.5 g/l agar.

Shoot tip culture medium containing half-strength MS macronutrients with a quarter of ammonium concentration, full strength iron, MS micronutrients, was adjusted to pH 5.6 and gelled with a mixture of 3 g/l Difco Bacto agar and 1.25 g/l Phytagel. Vitamins and growth regulators (with varying concentrations of IBA, from 100 to 10 μg/l, according to plant species or to the nature of culture, e.g., preculture or standard culture) were always filter-sterilized and added to the autoclaved basal medium. All plant material was maintained in a 23 °C (16 h light), 18 °C (8 h dark) regime with a light intensity of $39 \mu E m^{-2} s^{-1}$ and was subcultured each month.

Cold Hardening and Preculture. One or 2 weeks after subculture, in vitro plantlets were placed at 4°C for 1–6 days to induce cold hardening. Shoot tips then were excised (often sorted by size) and precultured under the same conditions for 24h on shoot tip medium supplemented with 5% dimethyl sulfoxide (DMSO) + 1 or 2% proline for uncoated shoot tips or with different sucrose concentrations (from 0.25 to 1 M) for shoot tips embedded in alginate beads.

Cryoprotection, Cooling and Warming. The three methods described in the introduction were used, as well as a mixed method.

1. Two-step cooling: After preculture on the medium enriched with DMSO and proline, shoot tips were transferred with 1 ml cryoprotectant to cryotubes on ice. Modified PVS2 of Sakai et al. (1990) containing 25% w/v glycerol +15% w/v ethylene glycol +12.5% w/v DMSO +3% w/v polyethylene glycol 8000 + 0.4 M sucrose was gently added to the shoot tips. Cryotubes were then left at room temperature for 20–40 min depending on shoot tip size (0.5–2 mm) and 0.5 ml cryoprotectant was removed. Samples were cooled at 1°C/min to −40°C and then immersed in LN. Samples were rapidly warmed in water at 40°C, and the cryoprotectant drained and replaced with liquid medium containing 1.2 M sucrose. Rinsed shoot tips were plated on the standard medium enriched with 100 µg/l IBA and maintained in the dark for at least 48h.
2. Encapsulation-dehydration: Shoot tips were precultured for 24h at 4°C on standard medium complemented with 0.25 M sucrose. They were then encapsulated in alginate as described by Dereuddre et al. (1990) and cultured successively on solid medium enriched with 0.50, 0.75 and 1 M sucrose before being dried in a laminar air flow for 3–4h. After the beads had lost about 60% of their initial fresh weight, they were placed in cryotubes (about 10 per tube) which are immersed in LN. After rapid warming, beads were cultured for 2 days on media containing decreasing concentrations of sucrose and finally on standard medium. Shoot tips were removed from the alginate bead 24 or 48h later.
3. A mixed method combining the two preceding ones was also used in our experiments. This consisted of exposing the alginate-coated shoot tips to modified PVS2 and cooling them with the two-step method.
4. Vitrification: The first trials were performed with shoot tips prepared as for two-step cooling but cryotubes were plunged directly into LN. We improved this by using ultra-rapid cooling of shoot tips in droplets. After preculture on a medium containing DMSO and proline, shoot tips were incubated in modified PVS2 for 75–120 min and then transferred to droplets of the PVS2 placed on small pieces of aluminum foil. Each piece of foil was grasped with forceps and dropped into a polystyrene box filled with LN. Foils were then inserted vertically into the LN-filled cryotubes, which were closed and stored in LN. For warming, tubes were opened in LN and the small aluminum foils were grasped with forceps and quickly plunged into a solution containing 1.2 M sucrose. Rinsed shoot tips were plated on standard medium.

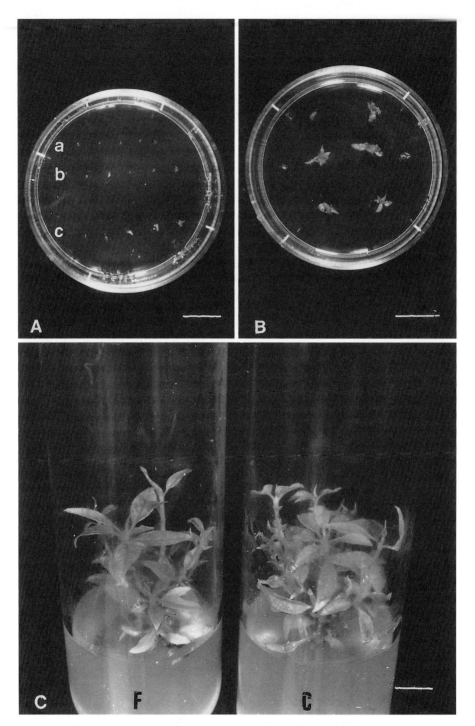

Fig. 4. A Effect of size of *Fereley* shoot tips on regrowth after cryopreservation 15 days after plating. Size of shoot tips before cryopreservation *a* 0.3–0.5 mm; *b* 0.5–1 mm; *c* 1 mm. **B** Cryoprotectant-treated control of *Fereley* 25 days after plating. **C** Plantlets developed from cryoprotectant-treated control (*C*) and cooled-warmed (*F*) shoot tips of *Ferlenain* 5 months after plating. No phenotype difference appears between plantlets *C* and *F*. *Bar* 1 cm

2.2.3 Results and Discussion

1. Using two-step cooling, the main results were obtained with the two *Prunus* interspecific hybrids, *Fereley* and *Ferlenain* (Brison et al. 1995). Regrowth occurred for 69.3% of shoot tips for *Fereley* and 74.5% for *Ferlenain* (Table 4). The cryoprotectant mixture did not significantly affect viability since 86% of cryoprotectant treated non-cooled shoot tips of both rootstocks grew compared with about 90% for untreated ones. Only shoot tips smaller than about 0.8mm for *Fereley* (Table 5, Fig. 4) and 0.5mm for *Ferlenain* (data not shown) did not show good growth after LN exposure.

   Different parameters were subsequently examined, such as the duration of exposure to cryoprotectant (15–45 min), and effect of cold hardening (0, 1, or 7 days) and preculture conditions (23 or 4 °C; Table 6). The highest survivals were obtained with 24-h hardening followed by 24-h preculture at 4 °C. The optimum exposure time was determined for each of the rootstocks. Apices were differently sensitive to the cryoprotectant based upon size.

   Other parameters were studied by Helliot and de Boucaud (1997) in a report on the cryopreservation of *Ferlenain* where they confirmed the positive influence of 24-h plantlet acclimation and 24-h shoot tip preculture at 4 °C. Likewise, in this work, 5% DMSO in preculture medium was the most effective concentration. They also pointed out the beneficial effect of proline as well as DMSO in the preculture medium (at a 20%

Table 4. Regrowth, 21 days after plating, of cryopreserved shoot tips of two *Prunus* interspecific hybrids

Hybrid	Regrowth (% total shoot tips)	
	Treated control[a]	Cooled[b]
Fereley	86[w]	69.3[x]
Ferlenain	86.8[w]	74.5[w]

[a] Cryoprotectant treated control: Mean of five replicates. Total number of shoot tips used: 84 for *Fereley*, 82 for *Ferlenain*.
[b] Treated shoot tips cooled in LN_2 and warmed. Mean of seven replicates (167 shoot tips used) for *Fereley* and five replicates (171 shoot tips used) for *Ferlenain*.
[w,x] Means followed by the same letter are not significantly different at $p = 1\%$, according to the *F*-test and after arc sin transformation.

Table 5. Effect of shoot tip size on growth of *Fereley* after LN exposure, 21 days after plating

Shoot tip size (mm)	0.3–0.5	0.5–1	1	1–1.5	1.5–2
Regrowth (% total shoot tips)	6.5[x]	55.8[w]	58[w]	67.6[w]	70[w]

Approximately 16 shoot tips were tested for each of two replicates.
[w,x] Means followed by the same letter are not significantly different at $p = 5\%$. The two groups (marked w and x) are significantly different from each other at $p = 0.03\%$ according to the *F*-test, after arc sin transformation.

Table 6. Effect of hardening and preculture conditions on growth of cryopreserved shoot tips from two *Prunus* interspecific hybrids, 21 days after plating

| | | Regrowth (% total shoot tips) | |
		Fereley	*Ferlenain*
Control		89[w]	86[w]
Cooled samples	A	75[w]	78[w]
	B	77[w]	86[w]
	C	47[x]	68[w]

Regrowth was evaluated as percentage of shoot tips producing shoots. A, B and C represent different culture conditions of the shoot tips before cryopreservation as follows. (A) Seven day hardening. Plantlets were hardened 6 days before shoot tip excision. Shoot tips were then precultured 24 h in hardening conditions. (B) one-days hardening. Excised shoot tips were precultured only 24 h in hardening conditions. (C) no hardening. Excised shoot tips precultured at 23 °C. Approximately 25 shoot tips were tested for each of three replicates.
[w,x] Figures followed by the same letter are not significantly different at $p = 5\%$ according to F-test after arc sin transformation. Those marked w and x are significantly different from each other at $p = 1\%$. When the two values (*Fereley* and *Ferlenain*) are added, line C differs significantly at $p = 0.5\%$.

concentration) and in PVS2 (at a 5% concentration). With these conditions, the best cooling rate was 5 °C/min.

2. With the encapsulation-dehydration experiments we aimed (a) to avoid DMSO application because it might induce genetic changes, and (b) to avoid use of a cooling apparatus. This method was successfully applied to *Fereley*, *Ferlenain* and *Prunus marianna* cv. GF8$_1$ with regrowth of 47, 55 and 40%, respectively (Table 7). These results are similar to those recently reported with almond by Shatnawi et al. (1999). Contrary to our data from two-step cooling, some shoot tips developed into calluses that rarely gave rise to shoots. Callus formation occurred to the same extent for dehydrated shoot tips as for dehydrated, cooled ones. The number of calluses obtained after cooling depended on the rootstock and ranged from 11.5% for *Fereley* to 23% for GF8$_1$. This represents the difference between survival and regrowth values given in Table 4. Callus formation is certainly due to the severe dehydration the shoot tips experience. Indeed, the water content of beads after successive cultures on media with increasing sucrose concentration was about 76% fresh weight. After dehydration, the final average water content of the beads was about 16%.

3. For the cryopreservation of peach (*Prunus persica* L. Batcsh), Paulus et al. (1992) used another protocol that combined encapsulation-dehydration with two-step cooling. Shoot tips in alginate beads were maintained for 30 min in a PVS2 mixture where sucrose was replaced by 15% mannitol and then were introduced into straws and cooled in two steps. Regrowth of

Table 7. Percentage of regrowth of control, treated control and percentage of survival and regrowth of different *Prunus* species shoot tips cryopreserved by different procedures, 30 days after plating

Prunus species	Cryopreservation procedure	Control (%) regrowth	Treated control (%) regrowth	Cryopreserved (%) survival[a]	regrowth[b]
Plums					
P. Fereley	Two-step cooling	86% (84)	–	59% (257)	69%(167)
	Alginate beads	88% (97)	84% (133)		47%
P. Ferlenain	Two-step cooling	87% (82)	–		74% (171)
	Alginate beads	–	57% (23)	72% (102)	55%
P. Marianna (GF 8_1)	Alginate beads	95% (48)	75% (44)	63% (143)	40%
P. cv Ogden	Droplets	–	37% (75)	–	52% (111)
P. cv Torinel	Droplets	97% (176)	57% (176)	–	47% (348)
P. domestica	Droplets	98% (194)	45% (178)	–	32% (379)
Almonds					
P. cv Chasanov	Droplets	100% (56)	50% (52)	50% (95)	40%
P. cv Samisch	Droplets	97% (86)	79%(57)	50% (52)	58%

[a] Surviuval: all surviving shoot tips including calluses.
[b] Regrowth: all shoot tips that grew into plantlets.
In parentheses: total number of plated shoot tips.
Control: shoot tips without any treatment.
Treated control: cryoprotectant or alginate-treated shoot tips.
Cryopreserved: cryoprotectant or alginate treated, cooled and warmed shoot tips.

treated controls was 50% while that of the cryopreserved material was 47%. This was a very significant result because of the difficulty of handling *Prunus persica* shoot tips. Unfortunately, growing cultures developed into small shoots that rarely proliferated, and had to be grafted for effective shoot recovery.

4. In the first trials with vitrification, shoot tips were placed in PVS2 in cryotubes and then directly immersed in LN. It is noteworthy that the modified PVS2, which vitrifies with cooling as confirmed by differential scanning calorimetry (see Fig. 3), did not cryopreserve any shoot tips of *Fereley* and *Ferlenain,* even with rapid cooling at an estimated rate of 170 °C/min (Brison et al. 1995). These observations were confirmed by Shatnawi et al. (1999) in their report on almond apices where they reported growth of only 10%. However, this vitrification method has been successfully applied to cherries and sweet cherries by Niino et al. (1997) using 45-day cold hardening, and to other woody plants (Niino and Sakai 1992; Niino et al. 1992, 1997; Kuranuki and Sakai 1995).

5. A simplified version of the method using vitrification of shoot tips in droplets of cryoprotectant as devised by Schäfer-Menuhr et al. (1994) and Schumacher (pers. comm.) for the cryopreservation of potato varieties was successfully applied to the Japanese plum cv. *Ogden* whose culture and cryopreservation had been difficult. Our first results were very encouraging since the main shoot tips survived freezing and regrowth of cryopreserved tips was even higher than that of the control (Table 8). Best regrowth was obtained for 1.5 mm long shoot tips with 74% surviving. If we did not take into account the response of very small shoot tips, the average regrowth reached 60%. We have to note, particularly in this case, that success was closely related to the rapid and good growth of the plantlets from which shoot tips were excised. The success of this method encouraged us to apply it immediately to other *Prunus* species such as almond and another plum. Two almonds cultivars, cv. *Chasanov* and cv. *Samisch,* and a plum cultivar, cv. *Torinel,* were thus easily cryopreserved (Table 7).

There is thus a wide range of reliable methods for effective cryopreservation of plum and peach shoot tips. All of them could certainly be improved by maintaining the plantlet cultures in suitable conditions (as noted above for plum cv. *Ogden*) and by studying carefully the best cold hardening time required for every protocol. The method of choice then depends on the plant material and on the apparatus available. Ultra-rapid vitrification with a concentrated cryoprotectant mixture, which is both cheap, easy, and time saving, seems the method to try first.

Table 8. Effect of shoot tip size on regrowth of *Prunus* cv *Ogden* after treatment and after LN exposure, 30 days after plating

Shoot tip size	(mm)	0.5–1	1	1.5	2
Regrowth	Treated control	11% *(18)*	47% *(19)*	52% *(21)*	41% *(17)*
Total shoot tips	Cooled	19% *(21)*	46% *(35)*	74% *(35)*	60% *(20)*

In parentheses: number of treated shoot tips in three replicates.

2.3 Cryopreservation of Somatic Embryos

Somatic embryos produced in large number from immature cotyledons or embryonic axes can be a powerful tool for conserving genetic diversity. To date few studies have succeeded in obtaining somatic embryos from lines of *Prunus* except *Prunus avium* L. (Garin et al. 1997; Reidiboym-Talleux et al. 1999).

Prunus avium is an economically important species for the wood industry. It would be rewarding to use embryogenic lines for large-scale multiplication on condition that they can be easily converted to plants but this is not always possible. We tried to conserve globular embryos without calluses from cultures showing relatively good proliferation.

2.3.1 Plant Material

Somatic embryos of wild cherry (*Prunus avium* L.) were provided by Dr. G. Grenier de March, Institut Supérieur Agricole de Beauvais (ISAB, France). The embryogenic culture line was derived from immature cotyledons of zygotic embryos of *Prunus avium*. This strain was maintained in the dark at 23 °C (8 h)–19 °C (16 h) and subcultured every 2–3 weeks on MS medium as modified by Garin et al. (1997).

2.3.2 Cryopreservation Procedure

We examined a procedure to directly plunge samples into LN after dehydration, thus avoiding the need for a programmable freezer. The different steps of the procedure are as follows:

1. Hardening: Cold hardening of somatic embryos was carried out for 3–4 days at 4 °C. One set used the standard condition, another set used cryoprotective treatment.
2. Cryoprotective treatment: 3–4 mm somatic embryo masses were transferred to a multiplication medium containing increasing concentrations of sucrose (from 0.25 to 1 M) and maintained at 4 °C for 3 days. Before cooling, they were placed in a laminar air-flow for about 4 h to accelerate dehydration to 50%–60% fresh weight. Dehydrated material was divided into two sets. The first one was put in culture and the other was cooled in LN.
3. Cooling: Masses were directly transferred to cryotubes already opened and filled with LN. Cryotubes were then closed and material was conserved in LN until warming.
4. Warming: Embryo masses were warmed rapidly in 1.2 M sucrose solution at 40 °C for 10 min. They were then placed on media containing decreasing sucrose concentrations (1–0.5–0.25 M each for 24 h) and after placed in standard growth conditions. Both controls (CC and C) and cooled-warmed samples (F) were all cultured under the same conditions.

Table 9. Summary of cryopreservation studies on *Prunus*

Plant material	Cryogenic protocol	Viability	References
Embryonic axes			
Prunus persica	desiccation + LN		de Boucaud et al. (1995)
	cryoprotectant + LN	66%	
Prunus amygdalus	desiccation to 7% moisture content + LN		Chaudhury and Chandel (1995)
Dormant buds			
Cherry	cooling to –40°C + LN		Katano and Irie (1990)
Sour cherry	cooling 1 min to –40°C + LN		Towill and Forsline (1996)
In vitro shoot tips			
Peach			
Prunus persica	encapsulation/ dehydration + two-step cooling	47%	Paulus et al (1992)
Plum rootstocks			
Prunus			
cv *Fereley*	two-step cooling	69%	Brison et al (1995)
cv *Ferlenain*	two-step cooling	74%	
cv *Ferlenain*	two-step cooling	67%	Helliot and de Boucaud (1997)
cv *Fereley*	encapsulation	47%	Brison et al.
cv *Ferlenain*	dehydration for the	55%	unpubl.
cv *Marianna* (GF8$_1$)	three cv	40%	
cv *Ogden*	vitrification in droplets (ultra-rapid cooling)	52%	Brison unpubl.
cv *Torinel*	droplets	47%	de Boucaud
P. domestica	droplets	32%	unpubl.
Cherries			
Prunus Jamasakura	vitrification	96%	Niino et al. (1997)
Prunus Lannesiana	vitrification	55–70%	
Prunus Avium	vitrification	60–80%	
Almonds			
Prunus dulcis	vitrification	10%	Shatnavi et al. (1999)
	encapsulation dehydration	55%–62%	
P. cv. *Chazanov*	droplets	40%	de Boucaud
P. cv. *Samisch*	droplets	58%	unpubl.

2.3.3 Results and Discussion

The experiments described above are preliminary and have to be improved. This *Prunus* somatic embryogenic line grows slowly. Dehydration and cooling further delayed regrowth. Regrowth from treated samples occurred as a callus after 2 months of culture. Calluses have the advantage of being quite distinct from embryogenic masses and allow the evaluation of regrowth. However, difficulties were encountered in the production of new embryos. The only means of quantitatively estimating this consisted of measuring the increase in weight of every embryo mass at each subculture.

Callus formation suggests cell damage and the extent should depend on the physiological state of the strain and culture conditions. However, we observed that globular embryos appeared de novo on callus, principally on dehydrated samples (C) only. Successful cryopreservation would consist of abundant production of globular embryos similar to the control (CC). Further research is ongoing in this direction. Once this is efficiently achieved, somatic embryo cryopreservation would be another way of conserving *Prunus* germplasm. A summary of cryopreservation studies on *Prunus* is given in Table 9.

3 Conclusions

Prunus species conserved in situ, ex situ and in vitro are threatened by different pathogens. Old varieties are disappearing and the germination capacity of seeds is short and limited by problems of dormancy. Thus, it is useful to develop methods which allow in the long term: (1) the conservation of germplasm diversity such as by embryonic axes for enhanced longevity; (2) conservation of rare selected clones, which are usually vegetatively propagated; (3) cryopreservation of disease tested, meristem-derived cultures and maintenance of pathogen-free stock; (4) conservation of somatic embryos with very high propagation potential considered as a useful substitute for seeds; and (5) international exchange of germplasm which could be provided from a bank responsible for the storage, maintenance and distribution of *Prunus*. This bank could be at the same place as the European database for the genus.

Cryopreservation of embryonic axes, shoot tips and somatic embryos is an interesting strategy to duplicate in field and in vitro collections even if the cryoprotective and freezing procedure has to be adapted to each species. A single method applied in all cases would be desirable, but is not feasible. The present thrust of research is to minimize use of expensive equipment and to simplify the procedure to make cryopreservation accessible to the budget of laboratories in developing countries.

Acknowledgements. The authors are grateful to Dr A.Pierronnet (Unité de Recherches sur les Espèces Fruitières et la Vigne, INRA, Villenave d'Ornon, France) for providing cv. GF 305 seeds and *Prunus* in vitro material and to Dr S. Spiegel (Department of Virology, The Volcani Center, Bet Dagan, Israel), for in vitro plum cv. *Ogden* and almonds cv. *Samisch* and cv. *Chasanov*. We thank also Dr G. Grenier de March (Institut Supérieur Agricole de Beauvais, France) for supplying somatic embryos of wild cherry.

References

Bailey LH (1894) The Japanese plums in North America. Stn Bull NY St Agric Exp Stn 62:36

Berjak P, Mycock DJ, Watt P, Wesley-Smith J, Hope B (1995) Cryostorage of pea (*Pisum sativum*). In: Bajaj YPS (ed) Biotechnology in agriculture and forestry, vol 32. Cryopreservation of plant germplasm I. Springer, Berlin Heidelberg New York, pp 292–307

Bernhard R (1994) Introduction. CR Acad Agric Fr 80(5):44

Bonner F (1986) Technologies to maintain plant diversity, vol 2, part D. Office of Technology Assessment, Washington DC, pp 630–672

Brearley J, Henshaw GG, Davey C, Taylor NJ, Blakesley D (1995) Cryopreservation of *Fraxinus excelsior* L. zygotic embryos. Cryo Lett 16:215–218

Brison M, Paulus V, de Boucaud MT, Dosba F (1992) Cryopreservation of walnut and plum shoot tips. Cryobiology 29:738

Brison M, de Boucaud MT, Dosba F (1995) Cryopreservation of in vitro grown shoot tips of two interspecific *Prunus* rootstocks. Plant Sci 105:235–242

Brison M, de Boucaud MT, Pierronnet A, Dosba F (1997) Effect of cryopreservation on the sanitary state of a cv. *Prunus* rootstock experimentally contaminated with Plum Pox Potyvirus. Plant Sci 123:189–196

Chandel KPS, Chaudhury R, Radhamani J, Malik SK (1995) Desiccation and freezing sensitivity in recalcitrant seeds of tea, cocoa and jackfruit. Ann Bot 76:443–450

Chaudhury R, Chandel KPS (1995) Cryopreservation of embryonic axes of almond (*Prunus amygdalus* Batsch) seeds. Cryo Lett 16:51–56

Colin J, Verhoyen M (1976) Micrografts with meristematic tissues, a possible technique to eliminate viruses from *Prunus* trees. Acta Hortic 67:97–102

de Boucaud MT, Cambecedes J (1988) The use of 1,2-propanediol for cryopreservation for recalcitrant seeds: the model case of *Zea mays* imbibed seeds. Cryo Lett 9:94–101

de Boucaud MT, Brison M, Ledoux C, Germain E, Lutz A (1991) Cryopreservation of embryonic axes of recalcitrant seed: *Juglans regia* L. cv. Franquette. Cryo Lett 12:163–166

de Boucaud MT, Helliot B, Brison M (1996) Desiccation and cryopreservation of embryonic axes of peach. Cryo Lett 17:379–390

Dereuddre J, Scottez C, Arnaud Y, Duron M (1990) Resistance of alginate-coated axillary shoot tips of pear tree (*Pyrus communis* L. Beurre Hardy) in vitro plantlets to dehydration and subsequent freezing in liquid nitrogen: effect of previous hardening. CR Acad Sci Paris 310 (Ser III):317–323

Dosba F, Bernhard R, Zanetto A (1994) Importance des ressources génétiques des *Prunus*. CR Acad Agric 80(5):45–57

Ellis RH, Hong TD, Roberts EH (1991a) Effect of storage temperature and moisture on germination of papaya seeds. Seed Sci Res 1:69–72

Ellis RH, Hong TD, Roberts EH (1991b) An intermediate category of seed storage behavior? II. Effect of provenance, immaturity and imbibition on desiccation tolerance in coffea. J Exp Bot 42:653–657

Ellis RH, Hong TD, Roberts EH, Soetisna V (1991c) Seed storage behaviour in *Elaeis guineensis*. Seed Sci Res 1:99–104

Finch-Savage WE (1992) Embryo water status and survival in the recalcitrant species *Quercus robur*: evidence for a critical moisture content. J Exp Bot 43:663–669

Garin E, Grenier E, Grenier de March G (1997) Somatic embryogenesis in wild cherry (*Prunus avium*). Plant Cell Tissue Organ Cult 48:83–91

Gass T, Tobutt KR, Zanetto A (1996) Report of working group on *Prunus*. 5th meeting, Menemeni-Izmir Turquie, 1–3 Feb 1996, IPGRI, pp 1–70

Gonzalez-Rio F, Gurriaran MJ, Gonzales-Benito E, Revilla MA (1994) Desiccation and cryopreservation of olive (*Olea europaea* L.). Cryo Lett 15:337–342

Gordon AG (1992) Seed manual for forest trees. Forestry Commission Bulletin 83. Her Majesty's Stationery Office, London, pp 98–104

Helliot B, de Boucaud MT (1997) Effect of various parameters on the survival of cryopreserved *Prunus* Ferlenain in vitro plantlet shoot tips. Cryo Lett 18:133–142

Katano M, Irie R (1991) Shoot tip culture of Japanese flowering cherry (*Prunus yedoensis* Matsum.) and a possible cryopreservation of shoot tip in liquid nitrogen. Proc Fac Kyusyu Univ 10:17–27

Kuranuki Y, Sakai A (1995) Cryopreservation of in vitro-grown shoot tips of tea *(Camellia sinensis)* by vitrification. Cryo Lett 16:345–352

Moretti A (1932) Self-sterility and self-fertility in plums. Ital Agric 69:961–983

Murashige T, Skoog F (1962) A revised medium for rapid growth and bioassays with tobacco tissue cultures. Physiol Plant 15:473–497

Niino T, Sakai A (1992) Cryopreservation of alginate-coated in vitro-grown shoot tips of apple, pear and mulberry. Plant Sci 87:199–206

Niino T, Sakai A, Enomoto S, Magoshi J, Kato S (1992) Cryopreservation of in vitro grown shoot tips of mulberry by vitrification. Cryo Lett 13:303–312

Niino T, Tashiro K, Suzuki M, Ohuchi S, Magoshi J, Akihama T (1997) Cryopreservation of in vitro grown shoot tips of cherry and sweet cherry by one-step vitrification. Sci Hortic 70:155–163

Paulus V, Brison M, de Boucaud MT, Dosba F (1992) Preliminary results of cryopreservation of peach meristems. Book of poster abstracts, XIIIème Congr Eucarpia Angers, France

Pence VC (1990) Cryostorage of embryos axes of several large-seeded temperate tree species. Cryobiology 27:212–218

Pence VC (1992) Desiccation and the survival of *Aesculus*, *Castanea* and *Quercus* embryo axes through cryopreservation. Cryobiology 29:391–399

Reed BM, Normah MN, Yu X (1994) Stratification is necessary for successful cryopreservation of axes from stored hazelnut seed. Cryo Lett 14:377–384

Rehder A (1947) Manual of cultivated trees and shrubs. Macmillan, New York, pp 452–481

Reidiboym-Talleux L, Diemer F, Sourdioux M, Chapelain K, Grenier-de March G (1999) Improvement of somatic embryogenesis in wild cherry (*Prunus avium*). Effect of maltose and ABA supplementation. Plant Cell Tissue Organ Cult 55:199–209

Runthala P, Jana MK, Mohanan K (1993) Cryopreservation of ground nut (*Arachis hypogaea* L.) embryonic axes for germplasm conservation. Cryo Lett 14:337–340

Sakai A, Kobayashi S, Oiyama I (1990) Cryopreservation of nucellar cells of navel orange (*Citrus sinensis* Obs. var. *brasiliensis* Tanaka) by vitrification. Plant Cell Rep 9:30–33

Schäfer-Menuhr A, Schumacher HM, Mix Wagner G (1994) Langzeitlagerung alter Karoffelsorten durch Kryokonservierung der Meristeme in flüssigen Stickstoff. Landbauforschung Völkenrode 44:301–313

Shatnawi MA, Engelmann F, Frattarelli A, Damiano C (1999) Cryopreservation of apices of in vitro plantlets of almond. (*Prunus dulcis* Mill.). Cryo Lett 20:13–20

Towill LE, Forsline PL (1999) Cryopreservation of sour cherry (*Prunus ceraceus* L.) using a dormant vegetative bud method. Cryo Lett 20:215–222

Wesley-Smith J, Vertucci CW, Berjak P, Pammenter NW, Crane J (1992) Cryopreservation of desiccation sensitive axes of *Camellia sinensis* in relation to dehydration, freezing and the thermal properties of tissue water. J Plant Physiol 140:596–604

Zanetto A, Dosba F, Tobutt K, Maggoni L (1997) Report of an extraordinary meeting of the EECP/GR *Prunus* working group and the 2nd coordination meeting of the EU Project Genres 61. IPGRI, 13–15 Nov 1997, Zaragosa, Spain, pp 1–50

III.9 Cryopreservation of *Quercus* (Oak) Species

M. Elena González-Benito[1] and Carmen Martín[2]

1 Introduction

1.1 Plant Distribution and Important Species

The genus *Quercus* belongs to the Fagaceae family and includes 600 species distributed along the temperate Northern Hemisphere, South to the Malaysia and Colombian summits (Mabberley 1987). Many species are cultured as ornamental and timber trees, either for construction, furniture or floorboards. Other uses include: acorns as swine-food or to obtain tanning agents, and the bark of some species for dye or cork. In some cases, galls are also employed for tanning, and leaves to feed silkworms (Mabberley 1987; Table 1).

Many species contain subspecies, varieties or races, and hybrids are quite frequent among subspecies and species. Besides their social and economic importance, oaks form forests that have a protective role in the ecosystems. However, their distribution is being reduced by human pressure, which sometimes leads to habitat fragmentation (McNeely et al. 1995).

1.2 Germplasm Conservation

Measurements for the in situ conservation of oak forests have been implemented in some countries, e.g., Czech Republic (Sinderlar 1991), Germany (Kleinschmit 1993), Spain (Ayala de Sotilla 1975; Gil 1995), and the USA (Packard 1993). Various genetic studies have also been carried out (Isagi and Suhandono 1997), which should enhance use of the genetic resources. Plant breeding programs are being carried out in some species with more direct economic importance, such as *Q. suber* (cork oak) (Varela and Eriksson 1995).

Seeds of *Quercus* species are considered recalcitrant (homoiohydrous) (Roberts 1973; King and Roberts 1980; Roberts and King 1980). Depending on species acorns can be stored from anywhere between 2 months and 3.5 years (King and Roberts 1980). In general, after 5–8 months in storage, around

[1] Departamento de Biología Vegetal, Escuela Universitaria de Ingeniería Técnica Agrícola, Universidad Politécnica de Madrid, Ciudad Universitaria, 28040 Madrid, Spain
[2] Departamento de Biología Vegetal, Escuela Técnica Superior de Ingenieros Agrónomos, Universidad Politécnica de Madrid, Ciudad Universitaria, 28040 Madrid, Spain

Biotechnology in Agriculture and Forestry, Vol. 50
L.E. Towill and Y.P.S. Bajaj (Eds.) Cryopreservation of Plant Germplasm II
© Springer-Verlag Berlin Heidelberg 2002

Table 1. Some uses of *Quercus* species (Mabberley, 1987)

Species	Distribution	Use
Q. alba L. (white or Quebec oak)	East North America	Timber and fuel
Q. cerris L. (Turkey or manna oak)	South Europe, West Asia	Timber, sweetmeats, . . .
Q. coccifera L. (Kermes oak)	West Mediterranean Basin	Feeding of cochineal insects
Q. coccinea Muenchh. (scarlet oak)	East North America	Timber
Q. gambelii Nutt. (shin oak)	South North America	Edible acorns
Q. garryana Douglas (Oregon oak)	West North America	Timber, fuel
Q. ilex L. (holm or holly oak)	Mediterranean Basin	Edible acorns, galls in tanning
Q. ilicifolia Wangenh. (bear oak)	East North America	Shrub
Q. imbricaria Michaux (shingle oak)	East North America	Timber
Q. infectoria Oliver (gall oak)	East Mediterranean Basin	Galls for tanning and medicinal use
Q. lobata Née (California oak)	California	Edible acorns
Q. macrocarpa Michaux (bur oak)	East North America	Timber
Q. macrolepis Kotschy	South Europe, West Asia	Cupules for leather tanning
Q. marilandica Münchh. (jack oak)	East North America	Charcoal
Q. mongolica Fisher ex Turcz. (Japanese oak)	Japan, Sakhalin	Timber
Q. nigra L. (possum oak)	South East USA	Edible acorns
Q. obtusiloba Michaux (iron or post oak)	South East USA	Timber, edible acorns
Q. petraea (Matt.) Liebl. (sessile oak)	Europe, West Asia	Timber
Q. prinus L. (chestnut oak)	South East USA	Tanbark, timber, edible acorns
Q. robur L. (pedunculate oak)	Europe, Mediterranean Basin	Timber, acorns eaten by pigs, coffee substitute
Q. rubra L. (red oak)	East North America	Timber
Q. suber L. (cork oak)	Southern Europe, North Africa	Cork
Q. velutina Lam. (quercitron)	South USA	Tanbark
Q. virginiana Miller (live oak)	North America	Timber

60–75% germination can be expected. However, in some cases, germination was significantly less (e.g. *Quercus petraea* Lieblen, to 20%). Moisture contents lower than 20% are deleterious for the seed. Conservation is usually carried out in sealed polyethylene bags or in moist sand or peat (King and Roberts 1980; Roberts and King 1980). In practice, storage is not usually longer than 6 months, from harvesting in late autumn-winter to sowing in spring (Catalán Bachiller 1991). Temperatures of 0–2 °C, high relative humidity (approximately 90%) and aeration are recommended for acorn conservation (Catalán Bachiller 1991).

When preservation of certain genotypes is required, vegetative propagation is necessary. Conventional and in vitro propagation techniques have not been studied for many oak species (see review by Bellarosa 1989). These species are vegetatively propagated with difficulty, especially from mature trees. Probably the oak species most studied is *Q. suber*. In this species in vitro conservation has been developed and plantlets can be stored for 1 year without subculturing (Romano and Loucao 1994). Shoot cultures were maintained at 5 °C in the dark without losing multiplication and rooting capacity when cultured under standard conditions.

1.3 Need for Conservation and Cryopreservation

Because of the difficulties in storing recalcitrant seeds for long periods of time without high viability loss due to desiccation, germination during storage or fungal infection, cryopreservation has been recommended as a viable alternative (Chin 1988; Pence 1995). Oak seeds are usually large; therefore, cryopreservation of embryo axes is required. This procedure requires the use of in vitro techniques for subsequent recovery. For the conservation of selected genotypes, cryopreservation of micropropagated shoot apices or somatic embryos could be feasible when the in vitro techniques are successfully developed.

2 Cryopreservation

2.1 General Account

Cryopreservation studies of different plant organs and structures have been reported for a few oak species.

Pollen. Q. petraea and *Q. robur* pollen showed the same germination percentage after cryopreservation, by direct immersion in liquid nitrogen (LN), compared to control pollen (stored at −18 °C) (Jörgensen 1990).

Seeds. There has not been any success with seed cryopreservation of *Q. petraea*, *Q. robur* and *Q. rubra* using slow or rapid cooling (Jörgensen 1990; Ahuja 1991).

Immature Embryos. Jörgensen (1990) reported 90%–100% survival of *Q. petraea* immature embryos after slow cooling to −40 °C prior to immersion in LN. Embryos had been previously treated for 1 h at 0 °C with a solution of 10% dimethyl sulfoxide (DMSO) and 5% sucrose. Warming was carried out at 40 °C.

Embryo Axes. Due to the large size of *Quercus* seeds, most cryopreservation work has been carried out with embryo axes (Table 2). The use of isolated axes provides more uniform desiccation. However, in vitro techniques must be used for the development of these axes. Therefore, appropriate procedures need to

Table 2. Studies carried out on oak embryo axes cryopreservation (approximate data as some are taken from figures)

Species	Cryoprotection	MC[1]	Cooling	Warming	Recovery conditions[2]	Recovery percentage[3] Control[4]	Recovery percentage[3] Cooled	Reference
Q. alba	Drying[5]	ND	Rapid	Ambient	MS + 2 mg/l BAP + 2mg/l IAPhe, 27°C	100% C?	25% C	Pence (1990)
Q. faginea	Drying	22%	Rapid	40°C[6]	WPM + 1.5 mg/l BAP, 25°C	65% S and/or R	60% S and/or R	González-Benito and Pérez (1992)
Q. faginea	15%DMSO	–	Rapid	40°C	WPM + 1.5 mg/l BAP, 25°C	95% S and/or R	40% C	González-Benito and Pérez (1992)
Q. falcata	Drying	37%	Rapid	Ambient	MS + 2 mg/l BAP + 2mg/l IAPhe, 27°C	100% E	70% C	Pence (1992)
Q. macrocarpa	Drying	36%	Rapid	Ambient	MS + 2 mg/l BAP + 2mg/l IAPhe, 27°C	Callus	Callus	Pence (1992)
Q. macrocarpa	Drying	ND	Rapid	Ambient	MS + 2 mg/l BAP + 2mg/l IAPhe, 27°C	100% C?	20% C	Pence (1990)
Q. marilandica	Drying	ND	Rapid	Ambient	MS + 2 mg/l BAP + 2mg/l IAPhe, 27°C	70% R?	0%	Pence (1990)
Q. muhlenbergii	Drying	ND	Rapid	Ambient	MS + 2 mg/l BAP + 2mg/l IAPhe, 27°C	100% C?	40% C	Pence (1990)
Q. nigra	Drying	20%	Rapid	Ambient	MS + 2 mg/l BAP + 2mg/l IAPhe, 27°C	100% E	50% C	Pence (1992)
Q. palustris	Drying	20%	Rapid	Ambient	MS + 2 mg/l BAP + 2mg/l IAPhe, 27°C	50% C	10% C	Pence (1992)
Q. robur	Drying	25% or 39%	1°C/min to –38°C or rapid	40°C	WPM + 0.01 mg/l NAA + 0.3mg/l BAP	85% S	0%	Poulsen (1992)
Q. robur	Sucrose and glycerol, followed by encapsulation-dehydration	24%	1°C/min to –20°C	40°C	QL + 1mg/l BAP + 0.25 mg/l zeatin, 26°C	81% S and /or R	12% S and/ or R	Chmielarz (1997)
Q. rubra	Drying	20%	Rapid	Ambient	MS + 2 mg/l BAP + 2mg/l IAPhe, 27°C	100% R, S or E	80% R, S or E	Pence (1992)
Q. rubra	Drying	36% to 63%	Rapid	45°C	MS + 2mg/l BAP + 1 mg/l NAA	ND	0%	Sun (1999)

[1] MC = moisture content, fresh weight basis, of axes with best response after cooling. ND = not determined

[2] BAP = 6-benzyladenine, NAA = 1-naphthaleneacetic acid, IAPhe = indoleacetylphenylalanine, QL = Quoirin and Lepoivre (1977)

[3] C = callus; E = enlargement (hypocotyl elongation or swelling); S = shoot elongation; R = radicle elongation. Symbols followed by question marks mean that it is not clear if in control axes this was the type of response or was only stated for frozen axes

[4] Development of non-desiccated non-cooled embryos

[5] Drying = drying in a laminar flow bench

[6] In a water bath

be developed for in vitro embryo axis culture prior to the establishment of the cryopreservation protocol.

In most cases cryoprotection was achieved by drying axes in the air flow of a laminar flow bench ("Drying" in Table 2), which could be considered rapid drying (Pence 1995). Of those studies, the two cases where development of shoot and/or root was observed (Q. faginea and Q. rubra), moisture content had been reduced to 20%–22%. However, in other species, axes with these moisture contents only led to callus formation after retrieval from LN. This indicated that some cell survival was obtained after cooling and warming but that damage was too extensive to allow organised growth. It is uncertain whether altering culture conditions could produce more normal growth of the axes.

Direct immersion in LN ("Rapid cooling" in Table 2) was used in most of the studies. Slow cooling of Q. robur embryo axes has been attempted by decreasing temperature at a rate of 1 °C/min to −20 or −38 °C, followed by immersion in LN (Poulsen 1992; Chmielarz 1997). However, in these two studies, the authors followed a different approach. Poulsen (1992) used desiccation as a cryoprotective treatment prior to cooling, whereas Chmielarz (1997) imbibed axes stepwise in a series of sucrose and glycerol solutions prior to encapsulation in alginate beads. The encapsulated axes were then desiccated in a laminar flow bench followed by maintenance over silica gel for 20 h. Those treatments gave a moisture content of 24%. Axes were subjected to slow cooling, stored in LN for 24 h and warmed in a water bath (40 °C). After 3 weeks in culture, 12% of them showed either shoot and/or root development. Although survival was as high as 96%, most axes (78%) only produced callus.

Up to now, the best results obtained in the cryopreservation of oak species have been obtained with Q. faginea (González-Benito and Pérez 1992). This study will be described in more detail in the following sections.

2.2 Cryopreservation of Embryonic Axes of *Quercus faginea*

2.2.1 Method

Green acorns, or those just turning brown, were collected, washed in tap water and soap and kept moist at 5 °C until used (never longer than 10 days). Pericarp was removed and half of the acorn (containing the embryo axes) was soaked in a 10% commercial bleach solution (0.5% active chlorine) for 20 min, followed by three rinses in sterile, distilled water. Axes were excised in sterile conditions. After removal of excess water with filter paper, axes were placed on open Petri dishes and dried in the air flow of a laminar flow bench for different periods (1, 3, 5, and 8 h). Moisture contents were determined after drying axes in an oven at 105 ± 2 °C for 24 h, and were calculated as a percentage of fresh weight.

Axes were placed in polyethylene cryovials and plunged into LN. After 1 h, vials were warmed by immersion in sterile water at 40 °C. Axes were cultured on WPM medium (Lloyd and McCown 1981) salts plus MS modified vitamins (1 mg/l thiamine instead of 0.1 mg/l) (Murashige and Skoog 1962) supplemented with 1.5 mg/l BAP (6-benzyladenine). Samples were incubated at 25 °C in darkness for 1 week, and afterwards in a 16 h photoperiod with an irradiance of 50 μmol m^{-2} s^{-1}.

2.2.2 Results

In the non-desiccated axes radicle growth started 2 weeks after culture. Desiccated embryonic axes germinated maximally after 4 weeks (Table 3). Axes survival (root, shoot or callus growth) after cooling was similar to non-cooled ones when they were desiccated for 3 or 5 h (21 and 19% moisture content, respectively). Maximum recovery percentage (embryos showing shoot and/or root development) of cooled axes was obtained after 3 h desiccation. Callus induction (Table 4) took place in embryo axes desiccated for more than 1 h and in the cooled ones. Maximum callus induction was observed with 1 h desiccation in cooled axes and in 8 h dried, uncooled axes.

Table 3. Recovery percentage (development of root or/and shoot 4 weeks after initiation) of embryo axes of *Q. faginea* after drying to different moisture contents and cooling by direct immersion in liquid nitrogen

Desiccation time (h)	Moisture Content (%)*	Recovery (%)	
		Control	Cooled in LN
0	64 ± 3	100	–
1	39 ± 6	80	10
3	21 ± 5	65	60
5	19 ± 5	25	35
8	16 ± 3	44	20

–: treatment not carried out.
* Fresh weight basis.

Table 4. Percentage of embryo axes of *Q. faginea* which showed callus induction 4 weeks after initiation. (Means with the same letter not significantly different with Duncan's test, $p < 0.05$)

Desiccation Time (h)	% Callus induction	
	Control	Cooled in LN
0	0 c	–
1	0 c	36 a
3	11 bc	15 bc
5	8 bc	8 bc
8	23 ab	16 ab

–: treatment not carried out.

2.2.3 Discussion

These results are comparable to those observed with recalcitrant seeds from other species such as *Hevea brasiliensis, Araucaria hunsteinii* and *Juglans regia*. With *Q. faginea* the highest value for axis growth after cooling was with 3h desiccation (21% moisture content). For *H. brasiliensis* and *A. hunstenii* this water content was 16 and 20.4% (fresh wt. basis), respectively (Normah et al. 1986; Pritchard and Prendergast 1986). In *J. regia*, embryo axes with 5% moisture content after harvesting were more tolerant to LN than those with 20% moisture content (Boucaud et al. 1991). In *Q. faginea*, below 21% moist, desiccation damage occurred. Therefore, desiccation to approximately 21% (under our conditions, 3h desiccation in the laminar flow bench) was the best treatment for the cryopreservation of *Q. faginea* axes.

2.3 Development of Cryopreservation Protocols for *Quercus* Species

A similar procedure to the one described above for *Q. faginea* was followed with axes of *Q. suber* (Nuñez Moreno 1997). However, due to the considerable contamination, data were not analysed and only some general observations were made after 2 weeks in culture (before fungal growth was too extensive). Most control axes and those desiccated for 1 or 3h showed radicle elongation. Axes not desiccated or desiccated for 1h and cooled did not show any growth and turned brown. This was similar for axes dried for 5h, cooled or not. However, axes desiccated for 3h and cooled showed some white tissues. Axes moisture contents were 60% for undried axes, and after desiccation were 40, 20 and 18% for 1, 3, and 5h, respectively.

The encapsulation-dehydration and in vitro culture of embryo axes of *Q. suber* and *Q. ilex* has been examined as a first step to using these procedures in cryopreservation (González-Benito et al. 1999). Embryo axes of both species were aseptically excised, encapsulated in alginate beads, cultured in 0.75 M sucrose liquid medium, desiccated for different periods in a flow bench and cultured on basal WPM medium. Moisture content of encapsulated axes dropped from 74–71% to 25–21% after 6h desiccation, respectively, for the two species. At the lower moisture contents germination dropped to 20% in both species. Germination and shoot elongation of encapsulated embryos (non-desiccated or desiccated for 4h, 31 and 25% moisture content, respectively, for both species) were studied after culture on WPM medium supplemented with different concentrations of BAP and IBA (indole-3-butyric acid). A high percentage of germination (89%) and development of shoots (67%) from desiccated axes of *Q. suber* occurred on a medium with 0.1 mg/l BAP (González-Benito et al. 1999). The in vitro growth of *Q. ilex* axes was low in all media tested, especially shoot elongation of desiccated axes.

Studies carried out on the cryopreservation of encapsulated axes of these two species have so far been unsuccessful (data not published).

Studies on the cryopreservation of *Q. robur* axes have shown that an extreme surface-sterilisation treatment and prolonged periods of dehydration, although not lethal by themselves, resulted in cellular damage that gave no survival after cooling in LN (Berjak et al. 1999). Comparing different protocols, the authors found that fast drying and cooling gave best survival. Although results were still preliminary, warming axes in a Ca^{2+}/Mg^{2+} solution seemed to improve organised growth, promoting both shoot and root elongation.

Sun (1999) found that slow-to-intermediate cooling ($\leq 10\,°C/min$) produced dehydration damage (measured as electrolyte leakage) in a study on the state and phase transitions in *Q. rubra* axes and cotyledonary tissues. The critical water content (content below which electrolyte leakage increased rapidly) was 0.30 g/g dry weight (23% fresh weight basis). Under rapid cooling ($>10\,°C/min$) freeze-induced dehydration damage could be avoided if the initial water content was higher than 0.50 g/g dry weight (approx. 33% fresh weight basis). However, at that water content, the vitrified cellular matrix was very unstable upon warming ($10\,°C/min$). Based on the value of the critical water content, cryopreservation was attempted with axes dehydrated to a water content between 0.55 and 1.70 g/g dry weight (33%–63% fresh wt. basis). Axes were placed inside cryovials and cooled in LN. No survival was observed after 20 day culture (Table 2).

2.4 Discussion

Embryo axis cryopreservation could be a useful tool for *Quercus* species germplasm preservation. The species with highest economical importance is probably *Q. suber*; however, for genetic resource conservation, it is also important to preserve related species.

In oak, as with other species having recalcitrant seeds, research is in a preliminary stage. In some cases the number of axes per treatment was low, appropriate in vitro culture protocols were not thoroughly studied and, in most studies, callus was the predominant form of growth.

Since acorns do not store well, it is difficult to carry out experiments with axes in a similar physiological state. Important variations in moisture content and physiological state among different provenances occur with time in recalcitrant seeds (Engelmann 1997). Seeds are usually stored moist at low temperatures and, generally, they start the germination processes (Berjak and Pammenter 1997). Under those conditions, the frequent development of fungi makes in vitro establishment difficult. Appropriate fungicides must be used.

In an attempt to promote in vitro growth, cryopreservation of *Q. faginea* axes was carried out leaving some part of the cotyledons on the axis (González-Benito and Pérez 1992). However, in this case, no survival after freezing was obtained even for the longest desiccation period (8 h, 25% moisture content) or with the use of DMSO (15% for 1 h at room temperature).

The cryopreservation of oak axes and their subsequent in vitro cultures often produce callus, instead of organised growth (Pence 1990, 1992). However, few studies exist on the effect of culture medium on embryo or embryo axis development. Poulsen (1992) observed that axes of *Q. robur* developed better on WPM than on MS medium.

As mentioned before, the cryoprotective method most frequently employed has been desiccation in the laminar flow bench. However, flash-drying (very rapid drying) has been effective for the cryopreservation of axes from recalcitrant seeds for some species (Wesley-Smith et al. 1992), including *Q. robur* (Berjak et al. 1999). There has been some criticism about the reproducibility of drying embryo axes in the airflow of a laminar flow bench (Poulsen 1992).

Other cryopreservation techniques have been tested, in some cases with success. Chmielarz (1997) reported a 12% recovery as shoot or root development with a protocol based on preculture-encapsulation-dehydration. Techniques based on encapsulation have been successful for the cryopreservation of shoot apices from many species (Engelmann 1997), and they could be explored. However, when used with recalcitrant seed axes, they have the inconvenience of providing a slower desiccation rate that is detrimental for this type of explant. Other techniques not thoroughly explored are those based on vitrification.

Differential scanning calorimetry elucidates the effect of warming and cooling rates and provides information on ice crystal formation (Dereuddre 1992; Martínez et al. 1998). In *Q. rubra* seed tissues, devitrification and melting transitions were absent in rapid cooling when water content was lower than 0.2 g/g dry weight (Sun 1999). However, that water content was below the critical value for dehydration damage. Based on the state and phase transition studies, Sun (1999) suggests that partial dehydration plus very rapid cooling is a possible protocol for recalcitrant seed cryopreservation.

Warming rate is another factor. In most of the published works warming took place at ambient temperature. Embryo axes with relatively high water content (20% 39%, see Table 2) may have had small ice crystals recrystallizing into larger ones during warming. Consequently, fast warming is recommended.

3 Summary and Conclusions

Cryopreservation of *Quercus* is in an early stage. Some success has been achieved with pollen, immature embryos and embryo axes. However, further work is required to develop practical protocols for the different species. Cryopreservation would also be useful for selected genotypes propagated through in vitro shoot culture or somatic embryogenesis (Manzanera and Pardos 1990; Bueno et al. 1992).

Acknowledgements. The work on *Q. suber* and *Q. ilex* cryopreservation was supported by Instituo Nacional de Investigación y Tecnología Agraria y Alimentaria (INIA) project No. FOA 1997–1650.

References

Ahuja MR (1991) Application of biotechnology to preservation of forest tree germplasm. In: Ahuja MR (ed) Woody plant biotechnology. Plenum Press, New York, pp 307–313

Ayala de Sotilla E (1975) *Quercus pyrenaica* Willd y la conservación del medio ambiente. Procedimientos para determinar la posibilidad de su conservación y mejora. Montes 31:139–142

Bellarosa R (1989) Oak (*Quercus* spp.). In: Bellarosa R (ed) Biotechnology in agriculture and forestry, vol 5. Trees II. Springer, Berlin Heidelberg New York, pp 387–401

Berjak P, Pammenter NW (1997) Progress in the understanding and manipulation of desiccation-sensitive (recalcitrant) seeds. In: Ellis RH, Black M, Murdoch AJ, Hong TD (eds) Basic and applied aspects of seed biology. Kluwer, Dordrecht, pp 689–703

Berjak P, Walker M, Watt MP, Mycock DJ (1999) Experimental parameters underlying failure or success in plant germplasm cryopreservation: a case study on zygotic axes of *Quercus robur* L. Cryo Lett 20:251–262

Boucaud MT, Brison M, Ledoux C, Germain E, Lutz A (1991) Cryopreservation of embryonic axes of recalcitrant seed: *Juglans regia* L. cv. Franquette. Cryo Lett 12:163–166

Bueno MA, Astorga R, Manzanera JA (1992) Plant regeneration through somatic embryogenesis in *Quercus suber*. Physiol Plant 85:30–34

Catalán Bachiller G (1991) *Quercus*. In: Catalán Bachiller G (ed) Semillas de árboles y arbustos forestales. MAPA-ICONA. Colección Técnica, Madrid, pp 318–324

Chin HF (1988) Recalcitrant seeds – a status report. IBPGR, Rome

Chmielarz P (1997) Preservation of *Quercus robur* L. embryonic axes in liquid nitrogen. In: Ellis RH, Black M, Murdoch AJ et al. (eds) Basic and applied aspects of seed biology. Kluwer, Dordrecht, pp 765–769

Dereuddre J (1992) Cryopreservation of in vitro cultures of plant cells and organs by vitrification and dehydration. In: Datte Y, Dumas A (eds) Reproductive biology and plant breeding. Springer, Berlin Heidelberg New York, pp 291–300

Engelmann F (1997) Importance of desiccation for the cryopreservation of recalcitrant seed and vegetatively propagated species. Plant Genetic Resour Newslett 112:9–18

Gil L (1995) Present state of *Quercus suber* in Spain: proposals for the conservation of marginal populations. In: Frison E, Varela MC, Turok J (compilers) *Quercus suber* Network. Report of the first two meetings, 1–3 Dec 1994 and 26–27 Feb 1995. IPGRI, Rome, pp 14–20

González-Benito ME, Pérez C (1992) Cryopreservation of *Quercus faginea* embryonic axes. Cryobiology 29:685–690

González-Benito ME, Herradón E, Martín C (1999) The development of a protocol for the encapsulation-desiccation and in vitro culture of *Quercus suber* L. and *Q. ilex* L. Silvae Genet 48:25–28

Isagi Y, Suhandono S (1997) PCR primers amplifying microsatellite loci of *Quercus myrsinifolia* Blume and their conservation between oak species. Mol Ecol 6:897–899

Jörgensen J (1990) Conservation of valuable gene resources by cryopreservation in some forest tree species. J Plant Physiol 136:373–376

King MW, Roberts EH (1980) Maintenance of recalcitrant seeds in storage. In: Chin HF, Roberts EH (eds) Recalcitrant crop seeds. Tropical Press SDN BDH, Kuala Lumpur, pp 53–89

Kleinschmit J (1993) Strategien zur Erhaltung forstlicher Genressourcen erlaeutert am Beispiel von Eiche, Fichte und Douglasie (*Quercus robur, Quercus petraea, Picea abies, Pseudotsuga menziesii*) [Strategies for conservation of forest genetic resources, demonstrated by example of oaks, Norway spruce and Douglas-fir (*Quercus robur, Quercus petraea, Picea abies, Pseudotsuga menziesii*)]. In: Meier-Dinkel A, Schuete G, Su KT, Kleinschmit J (eds) Beitraege zur In vitro Vermehrung und Wurzelentwicklung von Stiel- und Traubeneiche sowie zur Erhaltung forstlicher Genressourcen (*Quercus robur* L. und *Quercus petraea* (Matt.) Liebl.). Sauerlaender, Frankfurt am Main, pp 179–212

Lloyd G, McCown B (1981) Commercially-feasible micropropagation of mountain laurel, *Kalmia latifolia*, by the use of shoot-tip culture. Comb Proc Int Plant Prop Soc 30:421–427

Mabberley DJ (1987) The plant book. Cambridge Univ Press, Cambridge

Manzanera JA, Pardos JA (1990) Micropropagation of juvenile and adult *Quercus suber* L. Plant Cell Tissue Organ Cult 21:1–8

Martínez D, Revilla MA, Espina A, Jaimez A, García JR (1998) Survival cryopreservation of hop shoot tips monitored by differential scanning calorimetry. Thermochim Acta 317:91–94

McNeely JA, Gadgil M, Leveque C, Padoch C, Redford K (1995) Human influences on biodiversity. In: Heywood VH (ed) Global diversity assessment. Cambridge Univ Press, Cambridge, pp 771–821

Murashige T, Skoog F (1962) A revised medium for rapid bioassays with tobacco tissue cultures. Physiol Plant 15:473–497

Normah MN, Chin HF, Hor YL (1986) Desiccation and cryopreservation of embryonic axes of *Hevea brasiliensis* Muell.-Arg. Pertanika 9:299–303

Nuñez Moreno Y (1997) Optimización de protocolos de crioconservación mediante encapsulación-desecación aplicados a distintas especies y propágulos. Final Year Project, Escuela Universitaria de Ingeniería Técnica Agrícola, Universidad Politécnica de Madrid

Packard S (1993) Restoring oak ecosystems. Restoration Manage Notes 11:5–16

Pence VC (1990) Cryostorage of embryo axes of several large-seeded temperate tree species. Cryobiology 27:212–218

Pence VC (1992) Desiccation and the survival of *Aesculus, Castanea* and *Quercus* embryo axes through cryopreservation. Cryobiology 29:391–399

Pence VC (1995) Cryopreservation of recalcitrant seeds. In: Bajaj YPS (ed) Biotechnology in agriculture and forestry, vol 32. Cryopreservation of plant germplasm I. Springer, Berlin Heidelberg New York, pp 29–50

Poulsen KM (1992) Sensitivity to desiccation and low temperatures (–196 °C) of embryo axes from acorns of the pedunculate oak (*Quercus robur* L.). Cryo Lett 13:75–82

Pritchard HW, Prendergast FG (1986) Effects of desiccation and cryopreservation on the in vitro viability of embryos of the recalcitrant seed species *Auracaria hunsteinii* K. Schum. J Exp Bot 37:1388–1397

Quoirin M, Le Poivre P (1977) Etudes de milieux adaptes aux cultures in vitro de *Prunus*. Acta Hortic 78:437–442

Roberts EH (1973) Predicting the storage life of seeds. Seed Sci Technol 1:499–514

Roberts EH, King MW (1980) Storage of recalcitrant seeds. In: Withers LA, Williams JT (eds) Crop genetic resources. The conservation of difficult material. International Union of Biological Sciences, Series B42, pp 39–48

Romano A, Loucao MAM (1994) Conservaçao in vitro de germeplasma de sobreiro (*Quercus suber* L.) [In vitro conservation of cork oak (*Quercus suber*) germplasm]. Rev Biol 15:29–42

Sindelar J (1991) Nastin opatreni k zachrane a reprodukci genovych zdroju drevin listnatych v Cesske Republice. I. Uvodni poznamky. Druhy rodu *Quercus* L. [Outline of measures for the conservation and reproduction of genetic resources of broadleaved trees in the Czech Republic. I. Introductory remarks. *Quercus* species]. Zpravy Lesnickeho Vyzkumu 36:1–7

Sun WQ (1999) State and phase transition behaviors of *Quercus rubra* seed axes and cotyledonary tissues: relevance to the desiccation sensitivity and cryopreservation of recalcitrant seeds. Cryobiology 38:372–385

Varela MC, Eriksson G (1995) Multipurpose gene conservation in *Quercus suber* – a Portuguese example. Silvae Genet 44:1–37

Wesley-Smith J, Vertucci CW, Berjak P, Pammenter NW, Crane J (1992) Cryopreservation of desiccation-sensitive axes of *Camellia sinensis* in relation to dehydration, freezing rate and the thermal properties of tissue water. J Plant Physiol 140:596–604

III.10 Cryopreservation of *Ribes*

Barbara M. Reed and Kim E. Hummer

1 Introduction

The genus *Ribes* L., the currants and gooseberries, includes more than 150 described species of shrubs which are native throughout northern Europe, Asia, North America, and in mountainous areas of South America and northwest Africa (Brennan 1996). Only about 10 or 12 of these species comprise the primary gene pool from which domesticated currants and gooseberries were developed. The discussion in this chapter will focus on the background and cryopreservation of these economically important species.

Total world *Ribes* acreage (Table 1) has been stable over the past several decades although the breakup of the former Soviet Union greatly increased fruit availability. Black currants, the major crop, are primarily grown for the juice market. They are also valued for production of jams, jellies, liqueurs, such as creme de cassis, for the conversion of white wines to rosé, and as flavorants and colorants for dairy products. Black currant juice has intense flavor, color, high ascorbic acid, and other antioxidant levels which are now becoming recognized for their nutraceutical properties. Poland, the Russian Federation, the United Kingdom and the Scandinavian countries lead the world production in black currants.

Red currants are valued for the fresh market and for the production of preserves and juice. The main red currant producers are Poland, Germany, Holland, Belgium, France, and Hungary. Gooseberries, which are eaten fresh or processed into pies and jams, are primarily grown in Poland, Germany, and Hungary (Brennan 1996). Several species of currants have ornamental qualities for plant habit, flowering, or fall foliage.

Ribes production is negligible in North America because of white pine blister rust, *Cronartium ribicola* Fisch., a disease introduced from Asia, which kills susceptible five-needled white pines (Hummer 2000). *Ribes* and *Pinus* L. subgenus *Strobus* are alternate hosts for this disease. Since the early 1900s, *Ribes* culture has been restricted in parts of the United States in an attempt to curtail white pine blister rust.

USDA-ARS National Clonal Germplasm Repository, 33447 Peoria Rd., Corvallis, OR 97333-2521, USA

Biotechnology in Agriculture and Forestry, Vol. 50
L.E. Towill and Y.P.S. Bajaj (Eds.) Cryopreservation of Plant Germplasm II
© Springer-Verlag Berlin Heidelberg 2002

Table 1. World production (MT) of *Ribes,* currants and gooseberries, 1998

Country	Black Currants	Gooseberries
Australia	665	
Austria	15,646	1,634
Azerbaijan	200	
Belgium-Luxembourg	2,500	
Bulgaria	100	
Czech Republic	21,000	8,532
Denmark	3.550	
Estonia	1,000	
Finland	2,592	
France	10,586	
Germany	121,200	70,000
Hungary	12,000	5,457
Ireland	1,000	
Italy	250	
Moldova, Republic of	1,000	1,500
Netherlands	1,600	
New Zealand	1,800	25
Norway	18,000	3,200
Poland	165,000	36,014
Romania	1,000	
Russian Federation	180,000	32,000
Slovakia	3,604	1,409
Switzerland	327	34
Ukraine	18,000	
United Kingdom	19,400	2,500
World	601,519	162,305

FAO, 1999.

1.1 *Ribes* Distribution and Important Species

1.1.1 Taxonomy

Ribes was originally placed in the Saxifragaceae (Vetenant 1799; Engler and
Prantl 1891), but more recent taxonomic treatments classify the genus in the
family Grossulariaceae because of wholly inferior ovary, totally syncarpous
gynoceum, and fleshy fruit (Lamarck and De Candolle 1805; Cronquist 1981;
Sinnott 1985).

Early classifications also recognized two genera, *Ribes* and *Grossularia*
(Coville and Britton 1908; Berger 1924; Komarov 1971). Numerous infrageneric classifications are proposed for these two genera. Prevalent mono-
graphs recognize a single genus, *Ribes* (de Janczewski 1907; Sinnott 1985).
Crossability between gooseberry and currant species supports the concept
of a single genus (Keep 1962). De Janczewski (1907) subdivided the genus
into six subgenera: *Coreosma,* the black currants; *Ribes* (+ Ribesia), the red
currants; *Grossularia,* the gooseberries; *Grossularioides,* the spiny currants;
Parilla, the Andean currants; and *Berisia,* the European alpine currants.

The centers of diversity for *Coreosma*, *Ribes*, and *Berisia* include Northern Europe, Scandinavia and the Russian Federation (Jennings et al. 1987); and for *Grossularia* in the Pacific Northwest of North America (Rehder 1986). In addition, several species of black currants with sessile yellow glands are native to South America.

1.1.2 Cytology and Evolution

The basic chromosome number of *Ribes* is x + 8 (Zielinski 1953) and all species and cultivars are diploid. The chromosome complement and karyotype are highly uniform (Sinnott 1985) and the chromosomes are 1.5 to 2.5 μm (Darlington 1929). Mitotic and meiotic processes are also highly uniform (Zielinski 1953).

The principal evolutionary pressure in the genus appears to be geographical adaptation (Sinnott 1985). Messinger et al. (1999) examined subgeneric taxa of *Ribes* for restriction site variation in two cpDNA regions. While several infrageneric lineages were strongly supported, *Grossularioides* spp. were unexpectedly united with those from *Grossularia*. *Coreosma* spp. exhibited high divergence and were not monophyletic in the analysis. Messinger et al. (1999) consider two possible, not mutually exclusive, evolutionary scenarios for *Ribes*: (1) a long period of stasis is interrupted by sudden radiation of species; and (2) gene flow due to hybridization as a force for diversification.

1.1.3 Important Taxa

Ribes is cultivated for edible fruit, ornamental plant habit, and bloom. The main economically important crop groupings include the black currants, red and white currants, gooseberries, hybrid berries, and ornamentals. The important species within each of these crop groupings are discussed below.

1.1.3.1 Black Currants

Coreosma, the subgenus for black-fruited currants, has sessile resinous glands. The species of most economic importance is *R. nigrum* L., which is native through northern Europe and central and northern Asia to the Himalayas and includes subsp. *europaeum,* subsp. *scandinavicum,* and *R. nigrum* var. *sibiricum* Wolf (Brennan 1996).

Ribes nigrum is an unarmed, strongly aromatic shrub, growing as tall as 2 m (Rehder 1986). The leaves are lobed, up to 10 cm per side, glabrous above, slightly pubescent with numerous sessile, aromatic glands beneath; the racemes droop and have 4 to 10 flowers. The flowers have reddish – or brownish-green – campanulate hypanthia and recurved sepals. The whitish petals are about two-thirds as long as the sepals. The fruits are globose, up to 10 mm diameter, and are generally shiny black when ripe, although green- and yellow-fruited forms exist (Liberty Hyde Bailey Hortorium 1976). This species was domesticated within Northern Europe by 1600 and is described in early herbals (Brennan 1996). Recent breeding efforts have doubled fruit size compared to wild fruits. Breeders cross *R. nigrum* with *R. ussuriense* Jancz., *R. dikuscha* Fisch. and *R. nigrum* var. *sibiricum* for disease resistance; with *R. bracteosum* Dougl. for longer racemes resulting in higher yield (Brennan 1996). *Ribes hudsonianum* Rich., the northern North American black currant, and *R. americanum* Mill., the American black currant, have desirable traits which may be useful for broadening the gene pool. *Ribes nigrum* cultivars have a range of descriptive characters (Table 2).

1.1.3.2 Red Currants

Ribes, the red currant subgenus, has crystalline glands on young growth (Brennan 1996). Several species have economic importance: selections of *R. sativum* Syme (+ *R. vulgare* Jancz.) were initially made from native stands in northwestern Europe. *Ribes petraeum* Wulf., a montane species, was also selected from the wild, while most of the cultivated red currants were derived from *R. rubrum* L., a Scandinavian species which is native as far north as 70°N latitude (Brennan 1996). *R. rubrum* is an unarmed shrub that grows to 2m (Rehder 1986). The shoots are glabrous or have glandular hairs. Stems are covered with a smooth pale yellow bark. The leaves are deeply cordate, 3- to 5-lobed, 6×7cm in diameter. Flowers, which occur on long racemes, are greenish tinged with purple. The hypanthium is almost flat and the petals are very small. The fruit is globose, 6–10mm in diameter, red and glabrous.

Ribes triste Pallas is the North American red currant, with similar fruit quality to European red currants, but not developed for cultivation. Two additional species, *Ribes spicatum* Robs. from Norway and *R. multiflorum* Kit. from England, are used in red currant genetic improvement programs. White and pink currants are a color form of the red species.

1.1.3.3 Gooseberries

Gooseberry species have nodal spines. The European gooseberry, *R. uva-crispa* L. (+ *R. grossularia* L.), native in the United Kingdom eastward through northern Europe, the Caucasus, and North Africa, is most frequently selected for cultivar development (Brennan 1996). *R. uva-crispa* is a spiny shrub that grows as high as 1.5m tall (Liberty Hyde Bailey Hortorium 1976). Stems have two to three spines at the nodes. Leaves are as large as 5 × 6.5cm, sparsely pubescent or glabrous. Flowers occur in axillary clusters of one to three (much fewer than those on currant racemes), are pale green, sometimes pinkish, and have a hemispherical hypanthium, reflexed sepals, and short white petals. The fruit, which can be hispid, is globose to ovoid, about 10mm in diameter, green, yellow, or purplish-red.

American species, such as *R. divaricatum* Dougl., *R. hirtellum* Michx and *R. oxyacanthoides* L., are used extensively in breeding with the European gooseberry. In England in the late 18th to 19th centuries, groups of amateur growers formed organizations with the purpose of increasing gooseberry fruit

Table 2. Descriptive characteristics of selected *Ribes nigrum* L. black currant cultivars

Cultivar	Country of origin	Vigor	Habit	Yield	Vitamin C	Flowering season	Cold hardy	Spring frost	Rust
Baldwin	United Kingdom	2	3	2	3	2	1	1	1
Beloruskaya sladkaya	Belarus	2	3	2	3	2	3	3	2
Ben Alder	United Kingdom	2	3	2	2	3	1	2	1
Ben Lomond	United Kingdom	2	3	2	3	3	1	2	2
Ben Tirran	United Kingdom	3	3	2	3	3	1	2	2
Blackdown	United Kingdom	3	3	3	2	2	1	1	1
Boskoop Giant	The Netherlands	3	2	2	3	1	2	2	1
Brodtorp	Finland	2	1	2	2	1	3	3	1
Consort	Canada	2	3	2	2	1	2	2	3
Crandall (*R. odoratum*)	United States	3	2	3	1	3	2	2	3
Crusader	Canada	2	2	2	2	1	2	2	3
Golubka	Russian Federation	2	2	2	3	2	3	3	1
Minaj Smyriov	Belarus	2	2	2	2	2	3	3	2
Noir de Bourgogne	France	3	2	3	2	1	1	2	1
Ojebyn	Sweden	2	2	2	1	2	3	3	1
Pilot A. Mamkin	Russian Federation	2	2	2	2	2	3	3	2
StorKlas	Sweden	2	1	2	1	1	3	3	1
Silvergieters Zwarte	Germany	3	3	2	nd	1	2	2	1
Titania	Sweden	3	2	3	2	1	3	3	3
Wellington XXX	United Kingdom	3	3	1	2	2	1	1	1

Vigor[1]: 1 = least, 3 = most; Habit[1] 1 = spreading, 3 = erect; Yield[1] 1 = least, 3 = high; Flowering season[1]: 1 = early, 3 = late; Cold hardy[2]: 1 = least, 3 = most; Spring frost resistance[2]: 1 = least, 3 = most; vitamin C in fruit[2]: 1 = low, 3 = high; White pine blister rust, *Cronartium ribicola* Fisch., (rust) resistance[1]: 1 = least, 3 = most.

[1] Data collected at the USDA-ARS National Clonal Germplasm Repository, Corvallis, Oregon.
[2] Adapted from Brennan (1996) and Tuinyla and Lukosevicius (1996).

size and improving quality. These groups were quite successful and the crop thrived until the introduction of American powdery mildew, *Sphaerotheca mors-uvae* (Schw.) Berk. The gooseberry cultivars of the time were quite susceptible and acreage was greatly reduced (Brennan 1996).

1.1.3.4 Hybrids

Ribes x *nidigrolaria* Bauer is a hybrid cross of black currants with gooseberries (Brennan 1996). These man-made hybrids, commonly referred to as jostaberries, are very vigorous, do not have the acrid odor of black currants, have no or reduced spines, and are disease-resistant. These cultivars are grown more by homeowners rather than commercial growers. Their disease resistance and large size fruit are popular with organic farmers.

1.1.3.5 Ornamentals

Ornamental and flowering *Ribes* species have a broad range of colors. The American species, *R. aureum* Pursh and *R. odoratum* Wendl., have fragrant yellow flowers with tubular hypanthia that bloom in spring (Rehder 1986). The fruits are black but do not have the "black currant odor" characteristic of *R. nigrum*. Another American species, *R. sanguineum* Pursh, has a range of flower color variants from white to dark red-purple, and is used in landscape plantings for spring bloom and wildlife habitat. Unfortunately, this species tends to be susceptible to white pine blister rust (Hummer and Finn 1999). Some of the American gooseberry species, such as *R. speciosum* Pursh., *R. lobbii* Gray, and *R. menziesii* Pursh, have very attractive fuchia-like flowers and are planted for their ornamental landscape attributes (Brennan 1996).

1.1.3.6 Endangered Species

The genus *Ribes* is fairly robust. Most species are broadly distributed and are not in danger of extinction. However, the World Conservation Monitoring Center 1997 *Red List of Threatened Plants* (www.wcmc.org) includes 18 *Ribes* species. *Ribes kolymense* (Trautv.) Komarov ex Pojark is extinct from the former Soviet states; three American and one Sardinian species are endangered; six American are vulnerable; two from the Pacific Northwest, another Sardinian and a Chilean species are rare; three Russian species are indeterminate (Table 3). *Ribes ussuriense*, one of the Russian species listed as indeterminate, contains the dominant gene, *Cr*, for immunity from white pine blister rust (Brennan 1996). Genes from this species have allowed the cultivation of black currants in white pine blister rust restricted zones of the United States.

The Endangered Species Act of the United States (Department of the Interior, Fish and Wildlife Service, 50 CFR, Part 17) lists the Miccosukee gooseberry, *R. echinellum* (Cov.) Rehder (Table 3). This spiny-fruited gooseberry species, whose native habitat occurs along the shoreline of Lake Miccosukee near Monticello, Florida, and in limited locations in South Carolina

Table 3. Threatened *Ribes* species

Species	Locality	Source	Designation
R. amarum Munz var. *hoffmannii* Munz	California, United States	WCMC	V
R. armenum Pojark.	Armenia	WCMC	I
R. binominatum Heller	California and Oregon, United States	WCMC	V
R. canthariforme Wiggins	California, United States	WCMC	V
R. divaricatum (Heller) Jepson var. *Parishii* (Heller) Jepson	California, United States	WCMC, ONHP	V
R. cereum var. *colubrinum* Hitchc.	Oregon, United States	ONHP	R
R. echinellum (Coville) Rehd.	Florida, Georgia, South Carolina, United States	WCMC, ESA	E
R. erythrocarpum Coville & Lieb.	Oregon, United States	WCMC	V
R. integrifolium Philippi	Chile	WCMC	R
R. klamathense (Cov.) Fedde	Oregon, United States	ONHP	I
R. kolymense (Trautv.)Komarov ex Pojark	former Soviet States	WCMC	EX
R. malvifolium Pojark.	Tajikistan, Uzbekistan	WCMC	I
R. menziesii Pursh var. *thacherianum* Jepson	California, United States	WCMC	E
R. niveum Lindl.	Idaho, Nevada, Oregon, and Washington, United States	WCMC	R
R. oxyacanthoides L. subsp. *irriguum* (Dougl.) Sinnot	British Columbia, Canada; Idaho, Montana, Oregon, and Washington, United States	WCMC	R
R. sandalioticum (Arrigoni) Arrigoni	Sardinia	WCMC	R
R. sardoum Martelli	Sardinia	WCMC	E
R. sericeum Eastw.	California, United States	WCMC	V
R. tularense (Coville) Fedde	California, United States	WCMC	V
R. ussuriense Jancz.	Russian Federation	WCMC	I

WCMC = World Conservation Monitoring Center, Red List of Threatened Plants, www.wcmw.org; ESA Endangered Species Act of the United States, Department of the Interior, Fish and wildlife Service, 50 CFR Part 17; ONHP = Oregon Natural Heritage Program (www.abi.org/nhp). Designations are: EX = extinct, E = endangered, R = rare, V = vulnerable, I = indeterminate.

and Georgia, is threatened by encroaching human development. Several accessions of this species are maintained ex situ at the National Clonal Germplasm Repository at Corvallis, Oregon. The Oregon National Heritage Program lists *R. cereum* var. *colubrinum* Hitchc. as rare and *R. divaricatum* and *R. klamathense* (Cov.) Fedde as indeterminate (Table 3).

1.2 Various Methods for the Storage of *Ribes* Germplasm

Ribes germplasm collections are maintained as plants in fields or screened houses to maintain clonal identities. Alternatively, they may be held as in vitro cultures or cryopreserved as shoot tips in liquid nitrogen. Species collections may be held as seed stored at -20, -80, or $-196\,°C$.

1.2.1 Propagation and Culture of Ribes Plants

Currants root readily from dormant stem cuttings taken in the fall, or softwood taken in the spring. Most black currants root well from any type of cutting, but red currants do not root as easily (Brennan 1996). Nurseries in milder climates, such as the Pacific Northwest of the United States, cut 15-cm-long stems of black currant cultivars in late October and cover the lower third of the stem with soil. These remain in the field throughout the winter while roots form. Rooted cuttings are dug and shipped in the early spring.

Gooseberries, particularly cultivars derived from European species, root less readily than currants. Generally, a basal application of auxin is needed for successful rooting of hardwood cuttings. In cases where cuttings will not root, mound layering or grafting must be attempted. Budding or whip-and-tongue grafting, with similar procedures to those used for temperate fruit trees, can be performed in the dormant season. Clones of *R. aureum* or *R. odoratum* have been selected for rootstocks (Hamat et al. 1989).

1.2.2 Cold Storage of In Vitro Cultures

Cold storage of in vitro cultures of *Ribes* germplasm was first reported by Gunning and Lagerstedt (1985) who found that plant condition declined within a few months for *Ribes* accessions stored at $5\,°C$ but plantlets remained in excellent condition with storage at $-1\,°C$. Brennan et al. (1990) stored three *Ribes* cultivars in vitro at $6\,°C$ for 3 months with good survival and morphogenic potential. Reed and Chang (1997) reported data for 80 *Ribes* genotypes in dark storage at $4\,°C$ with a mean storage time of 1.4 years and 40 accessions at $-1\,°C$ in the dark with a mean storage time of 2.76 years. The addition of a 12-h photoperiod improved the $4\,°C$ storage mean to 2.4 years making it nearly equivalent to the $-1\,°C$ storage. In vitro cultures of over 100 *Ribes* genotypes are preserved in the collection at NCGR-Corvallis (Reed and Chang 1997).

1.2.3 Cryopreservation

Early cryopreservation studies of *Ribes* species and cultivars involved evaluation of freezing survival of dormant plant buds in mid-winter. Sakai and Nishiyama (1978) obtained excellent results when testing dormant buds of 'Oregon Champion' gooseberry and 'London Market' currant for survival following exposure to liquid nitrogen (LN). Dormant buds were frozen at 5 °C/day to −40 °C, plunged in LN, and thawed slowly with 100% survival. Reed and Yu (1995) cryopreserved shoot tips from in vitro grown *Ribes aureum*, *R. diacantha* Pall., and *R. rubrum* plants using a variety of techniques. Controlled cooling and vitrification were successful for two of three genotypes while encapsulation-dehydration worked well for all three. Benson et al. (1996) studied *R. nigrum* cultivars cryopreserved by the same three methods as Reed and Yu (1995). Some living plants were recovered with all three techniques, but encapsulation-dehydration was the most successful, followed by vitrification. Controlled cooling resulted in low viability in these tests. These studies also found that cold acclimatization and dimethyl sulfoxide (DMSO) pretreatments did not influence the ice nucleation and melt characteristics of the apices as determined by differential scanning calorimetry (DSC). Improved recovery of vitrified currant (*R. aureum* and *R. ciliatum* Humb. & Bonpl.) shoot tips and callus was obtained following a 2-h pretreatment in sucrose, proline, abscisic acid responsive proteins (RABP) or bovine serum albumin (BSA) (Luo and Reed 1997). A 1% BSA pretreatment was suggested as the most economical and available of the materials tested and increased shoot tip regrowth of some *Ribes* that had proved difficult to cryopreserve from 40% to nearly 70% following vitrification. Two *Ribes nigrum* cultivars, Ojebyn and Ben Lomond, cryopreserved by the encapsulation-dehydration technique, were successfully conditioned on a 0.75 M sucrose medium and produced recovery equivalent (90–100% regrowth of meristems) to those cold acclimated for 1 week (Dumet et al. 2000). Both vitrification and encapsulation-dehydration methods were shown to be suitable for storage of *Ribes* germplasm in international genebanks, although some differences in results were noted between the two laboratories involved in the study (Reed et al. 2000).

2 Methodology/Protocol for Cold Storage and Cryopreservation

2.1 Initiation and Multiplication of In Vitro Cultures

Explants from potted screenhouse-grown plants were disinfected in 10% commercial bleach (sodium hypochlorite 0.5%) with 5 drops of Tween 20 per 500 ml, shaken on a rotary shaker for 10 min, and rinsed 3 × in sterile, deionized water. Single node sections were transferred to 16 × 100 mm tubes with

10 ml of liquid Murashige and Skoog (MS; Murashige and Skoog 1962) medium (pH 6.9) with no growth regulators for contaminant detection (Reed et al. 1995). Uncontaminated explants were transferred into Magenta GA7 (Magenta, Chicago IL) boxes with 40 ml of NCGR-*Ribes* (RIB) medium and subcultured at 3-week intervals.

Micropropagated plantlets were multiplied on NCGR-RIB, which contains MS mineral salts and vitamins, but with only 30% of the normal ammonium and potassium nitrate concentrations, and (per liter): 50 mg ascorbic acid, 20 g glucose, 0.1 mg N6-benzyladenine (Sigma Chemical, St. Louis, MO), 0.2 mg GA3, 3.5 g agar (Bitec, Difco, Detroit, MI) and 1.45 g Gelrite (Kelco, San Diego, CA) at pH 5.7. Plants were grown at 25 °C with 16-h days ($25 \mu mol\, m^{-2} s^{-1}$).

2.2 Cold Storage of *Ribes* In Vitro Cultures

2.2.1 General Growth Conditions

Micropropagated plantlets were multiplied in Magenta GA7 boxes on NCGR-RIB as noted above. Plantlets were divided, transferred to fresh medium at 3-week intervals, and indexed for bacterial contaminants before storage (Reed and Tanprasert 1995; Reed et al. 1995).

2.2.2 Cold Storage of in Vitro Plantlets

Ten plantlets (2.5 3 cm) of each genotype were transferred to individual chambers of 5-chamber Star-pak bags (Gardner Enterprises, Willis, TX). Each chamber contained 10 ml hormone-free NCGR-RIB with 3.5 g agar and 1.75 Gelrite (equivalent to 8 g agar). Bags were sealed twice with an impulse sealer, labeled, returned to the growth room for 1 week, and then cold acclimatized (CA) for 1 week (8-h days at 22 °C and 16-h nights at −1 °C). Storage was at −1 °C in the dark or 4 °C with or without a 12-h photoperiod ($3 \mu mol\, m^{-2} s^{-1}$). Each shoot was evaluated at 3- to 4-month intervals. Shoots were rated on a vigor scale of 0 to 5, based on plant appearance: 5 + dark green leaves and stems, no etiolation; 4 + green leaves and stems, little etiolation; 3 + shoot tips and upper leaves green, some etiolation, 2 + shoot tips green, leaves and stems mostly brown, base may be dark brown, should be removed for subculture; 1 + plant mostly brown, only extreme shoot tip green, much of base dark brown; 0 + plant totally brown, no visible green on shoot tip, dead. A completely randomized design was used for the cold storage experiments. Data were analyzed for ANOVA using the factor program on MSTATC (1988).

2.3 Cryopreservation of *Ribes* Shoot Tips

2.3.1 General Growth Conditions

Micropropagated plantlets were multiplied and shoot tips recovered on NCGR-RIB under the growth conditions described above. All cultures were CA for one week (as noted above) before 0.8 mm shoot tips were excised. Additional CA may be useful for some genotypes as was found for pears and blackberries that exhibited low regrowth following 1-week CA and cryopreservation but high regrowth with 4–10 weeks CA (Chang and Reed 2000).

2.3.2 Protocol Optimization Experiments

2.3.2.1 Controlled Cooling

The method used was developed for pear (Reed 1990) and further modified for *Ribes* (Reed and Yu 1995). Shoot tips excised from CA plantlets were CA in the incubator for 2 days on NCGR-RIB with 5% DMSO and additional Gelrite (0.3 g/l) in the medium, then transferred to 0.25 ml liquid NCGR-RIB in 1.2-ml plastic cryotubes and 1 ml of the cryoprotectant PGD [10% each polyethylene glycol (MW 8000), glucose and DMSO in NCGR-RIB liquid medium] was added over 30 min. A 30-min equilibration at 4 °C was followed by cooling at 0.1 or 0.3 °C/min to −40 °C and plunging into LN. Samples were thawed for 1 min in 45 °C water then transferred to 22 °C water for 2 min, rinsed in liquid NCGR-RIB and plated on NCGR-RIB for recovery.

2.3.2.2 Encapsulation-Dehydration

This method was developed for pear (Dereuddre et al. 1990) and further adapted for *Ribes* (Reed, unpubl.). CA shoot tips were dissected onto agar plates then encased in alginate beads (3% low viscosity alginic acid with 0.75 M sucrose) for 18-h pretreatment in liquid NCGR-RIB with 0.75 M sucrose. Following pretreatment, beads were separated on sterile Petri dishes and air dried in the laminar flow hood for 4 h, placed in cryotubes and plunged into LN. Vials were rewarmed at room temperature for 15 min. Vials were filled with liquid NCGR-RIB for 10 min to rehydrate the beads, then drained, and the encapsulated shoot tips were planted on NCGR-RIB recovery medium.

2.3.2.3 Vitrification

The vitrification technique for white clover (Yamada et al. 1991) was modified by Reed and Yu (1995) and further modified by Luo and Reed (1997). CA shoot tips were pretreated for 2 days in a CA incubator (see general growth conditions) on NCGR-RIB with 5% DMSO and additional Gelrite (0.3 g/l). Shoot tips were removed from the pretreatment plates and placed in a cryotube with 1 ml of 1% bovine serum albumin (BSA) for 2 h at room temperature. The BSA solution was drawn off and PVS2 cryoprotectant (30% glycerol, 15% ethylene glycol and 15% DMSO in liquid NCGR-RIB with 0.4 M sucrose) was added to cryotubes on ice and stirred. After 20 min in PVS2, vials were submerged in LN. Samples were rewarmed for 1 min in 45 °C water, then transferred to 22 °C water for 2 min, and rinsed in liquid NCGR-RIB with 1.2 M sucrose before plating on NCGR-RIB recovery medium. One set of experiments compared the 5% DMSO pretreatment medium with 1.2 M sorbitol pretreatment medium but did not include the BSA presoak.

2.3.3 Sucrose Pretreatment as a Substitute for Cold Acclimatization

Ribes nigrum cvs. Ben Lomond and Ojebyn meristems were cryopreserved by encapsulation-dehydration (E-D) following the standard 1 week of cold acclimatization as listed above or 1 cm shoots with leaves removed were grown on NCGR-RIB with 0.75 M sucrose under standard light and temperature conditions (Dumet et al. 2000).

2.3.4 Pretreatment Effects on Vitrification Protocols

Pretreatments with 1–4 h immersion in 1, 5, 10 and 15% proline, 1 or 2% abscisic acid responsive proteins (RABP), 1% BSA, or 0.4 M sucrose in the standard NCGR-RIB were tested to determine their effectiveness on the standard vitrification protocol. After cold acclimatization and 48-h pregrowth on 5% DMSO medium, meristems were immersed in one of the solutions for 0–4 h, then it was removed, and vitrification solution was added followed by LN exposure, thawing, and regrowth. *Ribes ciliatum* Humb. & Bonpl. and *Ribes aureum* Pursh meristems and callus were used in this study (Luo and Reed 1997).

2.3.5 Thermal Analysis of Cryopreservation Protocols

Differential scanning calorimetry (DSC) was performed using a Perkin Elmer DSC7 and a TAC 7 PC (Benson et al. 1996). DSC profiles were constructed with a scanning rate of ± 10 °C/min. Zinc and indium standards were used for calibration. Samples of *Ribes nigrum* cvs. Ben Tron and Ben More were sealed in aluminum pans and scans were performed from +5 °C to −150 °C. Data from each cryopreservation technique included three cryovials. Recovery data was taken at 6 weeks. Cooling and warming profiles were constructed for each sample (3–6 replicates per technique).

2.3.6 Germplasm Storage Studies

A comparison of cryopreservation methods by separate laboratories using the same genotypes was initiated at NCGR in Corvallis, Oregon, USA, and the University of Abertay-Dundee, Dundee, Scotland. R*ibes nigrum* cv. Ojebyn and *Ribes rubrum* cv. Red Lake were cryopreserved using the three methods mentioned above (Reed et al. 2000).

2.4 Results

2.4.1 Cold Storage of Ribes In Vitro Cultures

Like many aspects of in vitro culture, plantlet response to cold storage is genotype dependent. Earlier work in this laboratory found 2.76 ± 0.7 years as the mean storage length for 40 *Ribes* accessions stored at $-1\,°C$ in darkness for 15–24 months (Reed and Chang 1997). Storage at $4\,°C$ with a 12-h photoperiod produced results similar to the $-1\,°C$ darkness ratings (Table 4). Dark $4\,°C$ storage required repropagation more than 1 year earlier (1.4 ± 0.8 year) than plantlets under the other two storage conditions. Ratings of plantlets stored at $-1\,°C$ in the dark were similar to those at $4\,°C$ with a 12-h photoperiod. Individual genotypes varied greatly as indicated by the standard

Table 4. Storage duration of representative *Ribes* genotypes at the National Clonal Germplasm Repository, Corvallis, which remained viable (rated 2 or above) at $4\,°C$ with a 12 hr dim photoperiod

Genotype	Fruit type	Rating[a]	Months stored
Oregon	Gooseberry	2	21
Tsema	Black Currant	2	21
Kosmiczenskaja	Black Currant	2	21
R. curvatum	Gooseberry	2	14
R. curvatum	Gooseberry	3	14
Rolan	Red Currant	2	14
Noire de Bourgogne	Black Currant	2	14
White Cherry	White Currant	2	14
Malling Redstart	Red Currant	2	13
Raby Castle	Red Currant	3	13
R. viscosissimum	Black Currant	3	13
Kerry	Black Currant	3	13
D. Young	Gooseberry	3	12
Malling Jet	Black Currant	2	12
Minj Smyriov	Black Currant	2	12
Baldwin	Black Currant	2	12

[a] Reed, 1998 unpublished data. Shoots were rated on a vigor scale of 0 to 5, based on plant appearance: 5 = dark green leaves and stems, no etiolation; 4 = green leaves and stems, little etiolation; 3 = shoot tips and upper leaves green, some etiolation, 2 = shoot tips green, leaves and stems mostly brown, base may be dark brown, should be removed for subculture; 1 = plant mostly brown, only extreme shoot tip green, much of base dark brown; 0 = plant totally brown, no visible green on shoot tip, dead.

deviations of 9 to 10 months. The storage range for individual genotypes varied from 8 months to 3 years.

2.4.2 Cryopreservation of Ribes Shoot Tips

2.4.2.1 Protocol Optimization

Ribes cryopreservation protocols were optimized in several ways (Reed and Yu 1995; Reed et al. 2000). In the controlled cooling protocol regrowth of meristems exposed to cooling rates of 0.3 °C/min and 0.5 °C/min were not significantly different for three genotypes tested (Fig. 1). Pregrowth before vitrification in PVS2 on either sorbitol or DMSO media produced significant differences for two of three genotypes. DMSO was significantly better for two of the three and the third showed no difference (Fig. 1). Length of drying of alginate beads in the E-D method was significant for two of three genotypes with 3-h drying generally better than 2 h (Fig. 1). The success of the three cryopreservation methods tested on *Ribes* in vitro grown shoot tips varied greatly with the genotype of the accession (Fig. 1, Table 5). Overall, the encapsulation-dehydration method gave the highest survival for the genotypes tested. Vitrification techniques produced moderate to high recoveries for over half of the genotypes tested, and the addition of a 2-h presoak in BSA prior to PVS2 addition further improved recovery. Five genotypes were stored using the presoak-vitrification technique and are the beginning of a long-term *Ribes* base-germplasm collection held at the USDA-ARS National Seed Storage Laboratory, Fort Collins, Colorado. Control shoot tips (exposed to cryoprotectant but not LN) were recovered in high percentages (80–100%) for slow freeze and encapsulation-dehydration techniques. Vitrification solutions were more toxic to the control shoot tips and reduced the percentage of regrowth (60–80%). Shoot tips produced from surviving meristems developed into normal plantlets (Fig. 2).

Table 5. Regrowth of *Ribes* meristems following 1-hr exposure to LN. Methods used were controlled cooling at 0.3 °C/min (CC), vitrification in PVS2 (Vit), and encapsulation-dehydration with a 3 hr dehydration (E-D)

Genotype	Control	CC	Vit	E-D
R. rubrum cv. Cherry	80–100%[a]	30%	10%	60%
R. diacantha	80–100%[a]	70%	70%	60%
R. aureum	80–100%[a]	55%	60%	90%
R. nigrum cv. Ben More	70–80%[b]	10%	20%	80%
R. nigrum cv. Ben Tron	100%[b]	–	60%	80%
R. aureum cv. Bronze	80%[c]	–	44%; 69%[d]	–
R. ciliatum	80%[c]	–	42%; 68%[d]	–

[a] Reed and Yu, 1995.
[b] Benson et al., 1996.
[c] Luo and Reed, 1997.
[d] standard method; with BSA presoak.

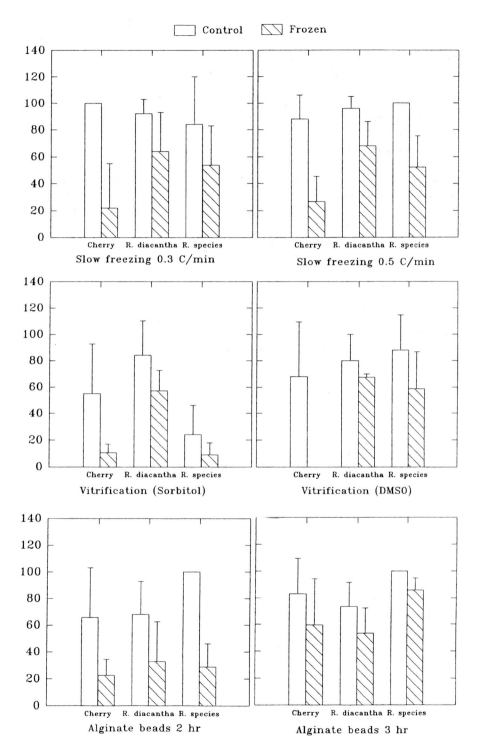

Fig. 1. Regrowth of control and frozen meristems of three *Ribes* genotypes following cryoprotection or cryoprotection and freezing with six different techniques (mean ± SD). Controlled cooling (*slow freezing*) at 0.3 or 0.5 °C/min; vitrification with a 1.2 M sorbitol pretreatment (*sorbitol*); or 5% DMSO pretreatment (*DMSO*); Encapsulation-dehydration with 2- or 3-h dehydration (alginate beads). *Ribes rubrum* cv. Cherry, *R. diacantha* and *R. aureum* (species). (Data from Reed and Yu 1995)

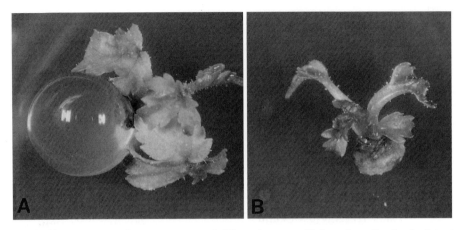

Fig. 2. Shoots regrown from cryopreserved *Ribes nigrum* cv. Ojebyn shoot tips by **A** alginate encapsulation-dehydration or **B** vitrification in PVS2 cryoprotectant. (Photographs courtesy of Rex Brennan, Scottish Crop Research Institute)

Table 6. Effect of 2-hr pretreatments on the survival of vitrified *Ribes aureum* and *R. ciliatum* meristems and callus 4 weeks after warming: 0.4 M sucrose in liquid medium (0.4 M sucrose in NCGR-RIB) alone or with 5% proline, 1% abscisic acid responsive protein (RABP), or 1% bovine serum albumin (BSA)

Genotype	Control	0.4 M Sucrose	5% Proline	1% RABP	1% BSA
Meristems					
R. aureum	44.0 ± 1.0^c	52.7 ± 1.0^b	64.7 ± 1.3^a	66.0 ± 1.3^a	68.7 ± 1.5^a
R. ciliatum	42.1 ± 0.0^c	53.8 ± 1.7^b	67.4 ± 2.4^a	76.3 ± 7.0^a	67.6 ± 2.5^a
Callus					
R. aureum	39.8 ± 0.5^c	45.8 ± 0.4^b	50.6 ± 0.8^a	51.0 ± 0.6^a	50.6 ± 0.5^a
R. ciliatum	39.9 ± 2.1^c	48.7 ± 1.2^b	61.0 ± 0.8^a	64.3 ± 1.9^a	64.2 ± 2.2^a

Percentage survival data were transformed by arcsine square root and expressed as means ± SEM%. Means separation by Duncan's multiple range test (p ≤ 0.05) (for meristems $n = 60$; for callus n = 45). Values in a row with different letters are significantly different. (Luo and Reed 1997).

2.4.2.2 Pretreatment Effects on Vitrification Protocols

Vitrified meristems and callus both exhibited improved recovery following vitrification when pretreated for 2h with sucrose, proline, RABP, or BSA. Immersion for 2h was better than longer or shorter intervals. Immersion for 2h in 0.4M sucrose solution significantly improved regrowth of callus and meristems of both species (Table 6). Pretreatment of meristems and callus with 5% or 10% proline, 1% RABP, or 1% BSA in 0.4M sucrose NCGR-RIB all had significantly better results than the sucrose medium alone (Table 6). Shoots grew directly from meristems without a callus phase. Pretreated meristems resumed growth 3 days after warming and reached maximum regrowth by 1week, compared with 2 weeks for controls (Luo and Reed 1997).

2.4.2.3 Sucrose Pretreatment as a Substitute for Cold Acclimatization

Both R. *nigrum* cultivars tested had high regrowth following encapsulation-dehydration (93–100%) and LN exposure (90–100%). There were no significant differences between cultivars or between the cold acclimatized meristems and those conditioned for 1 week on NCGR-RIB with 0.75 M sucrose (Dumet et al. 2000).

2.4.2.4 Thermal Analysis

Differential scanning calorimetry profiles showed highly reproducible ice nucleation and melting phenomena (Benson et al. 1996). PVS2 solution produced nucleation and melt profiles; however, profiles of vitrified meristems with the PVS2 solution removed showed no evidence of nucleation on cooling or warming indicating that the glass associated with the tissues is stable on warming. Analysis of dehydration treatments demonstrated reproducible ice nucleation and melt events associated with a major thermal peak. The stability of the glassy state on warming appeared marginal even after 4 h of dehydration before LN exposure (data not shown).

2.4.2.5 Germplasm Storage Studies

Some differences were evident between the two laboratories, so more precise utilization of protocols may be necessary to standardize the procedures. Encapsulation-dehydration produced excellent results for both genotypes at both locations (<90% regrowth). Vitrification was moderately successful at both locations for both genotypes (25–68% regrowth). Controlled freezing produced little regrowth from either genotype, perhaps due to differences in controlled rate freezers. Differences in familiarity with the process between the two laboratories may also have influenced the results (Reed et al. 2000).

2.5 Discussion

2.5.1 In Vitro Cold Storage

Cold storage at −1 °C in darkness or 4 °C with a 12-h photoperiod was successfully used as an intermediate-length germplasm conservation method. Most genotypes stored under these conditions were held for over 2 years without repropagation and recovered quickly when placed in growth room conditions on fresh culture medium. The cold hardiness of most *Ribes* genotypes made them excellent candidates for this form of intermediate storage. Cold storage of many temperate crops is best at −1 to 4 °C (Reed and Chang 1997). Differences in storage time among accessions may be due to the condition of the plant when stored, size of propagule, preparation of medium, or placement of the plant in the medium. Storage consistency depends on the

technical skills of laboratory workers. *Ribes* shoots in cold storage are particularly sensitive to size variation; small shoots store for shorter periods than do
intermediate or large shoots or plantlets. Genotypes with a short, bushy growth
habit exhibit a shorter storage duration than those that are more elongated
(data not shown). In addition to their adaptability to medium-term storage, in
vitro *Ribes* shoots are ideal for germplasm exchange. Shoots can be removed
from cold storage and shipped to requestors as needed. In vitro cultures are
often preferred for international shipments because many, although not all,
disease and insect threats are eliminated when the plants are initiated into
culture.

2.5.2 Cryopreservation

Seeds cannot be used as a base (long term) storage collection to preserve
specific *Ribes* genotypes, and alternative clonal methods must be employed.
Researchers can now recover 60–100% of cryopreserved shoot tips, indicating
that cryopreservation is an acceptable form of long-term storage for clonal
Ribes germplasm. Preliminary screening indicates that most genotypes can be
successfully cryopreserved using cold acclimation and the vitrification method
with presoak or the encapsulation-dehydration methods. Benson et al. (1996)
showed that devitrification of PVS2 can occur on warming of vials following
exposure to liquid nitrogen, thus reducing the survival of shoot tips. The addition of a 2-h presoak in proline or BSA to the vitrification protocol greatly
improved survival of *Ribes* shoot tips (Luo and Reed 1997). Immersion for
2 h in sucrose improved recovery of meristems and callus; however, extended
immersion was detrimental. Extended immersion may cause overdehydration
of the cells resulting in injury unrelated to the effects of low temperature.
Proline, BSA and RABP 2-h immersion greatly improved recovery of both
callus and meristems. These substances may have provided additional stability to the glassy state of the cryopreserved tissues. An unanticipated result of
the pretreatment protocols was the rapid recovery of meristems in almost all
pretreatment groups. Cryoprotectant injury to *Ribes* meristems is common
with highly concentrated vitrification solutions such as PVS2 but was overcome with these pretreatments (Luo and Reed 1997). For encapsulated shoot
tips adequate desiccation was needed to maintain the stability of the glassy
state (Benson et al. 1996). A sucrose-preconditioning step can substitute
for the cold-acclimatization period prior to the encapsulation-dehydration
process for some *Ribes* genotypes (Dumet et al. 2000). The increased tolerance to LN could be due to massive absorption of sucrose by the cells, or the
result of osmotic stress. Standardization of techniques between laboratories is
needed if cryopreservation is to be instituted on a large scale. The first attempt
at this standardization indicated that transfer of techniques between laboratories can be successful. These studies indicate that some protocols are more
easily transferred than others, and success with a protocol can be linked to
specific portions of each process. When transferring protocols to other labo-

ratories, training of personnel, standardization of growth rooms and equipment, and clearly written step-by-step protocols for each technique are key points to consider (Reed et al. 2000). Long-term *Ribes* germplasm storage has been initiated at NCGR, Corvallis, Oregon (PVS2 vitrified meristems) and at the University of Abertay-Dundee in Scotland (encapsulated meristems) (Reed et al. 2000).

3 Summary and Conclusions

Active-clonal field collections of *Ribes* germplasm can now be secured with secondary (active-backup) in vitro collections and cryopreserved (base) collections. Recent advances in the cold storage of *Ribes* in vitro cultures and cryopreservation make both techniques appropriate for germplasm conservation. Cold storage at −1 °C in darkness or 4 °C with a 12-h photoperiod is successful for medium-term storage of 2–3 years and allows availability for germplasm exchange. Cryopreserved collections of *Ribes* germplasm in liquid nitrogen were recently initiated at the National Clonal Germplasm Repository and the University of Abertay-Dundee. These in vitro techniques, when used in concert with active field collections, constitute a reliable system for conserving *Ribes* genetic resources.

References

Benson EE, Reed BM, Brennan RM, Clacher KA, Ross DA (1996) Use of thermal analysis in the evaluation of cryopreservation protocols for *Ribes nigrum* L. germplasm. Cryo Lett 17: 347–362

Berger A (1924) A taxonomic review of currants and gooseberries. Bull NY State Agric Exp Sta 109

Brennan RM (1996) Currants and gooseberries. In: Janick J, Moore JN (eds) Fruit breeding, vol II. Vine and small fruit crops. Wiley, New York, chap 3, pp 191–295

Brennan RM, Millam S, Davidson D, Wilshin A (1990) Establishment of an in vitro *Ribes* germplasm collection and preliminary investigations into long-term low temperature germplasm storage. Acta Hortic 280:109–112

Chang Y, Reed BM (2000) Cold acclimatization improves the cryopreservation of in vitro-grown *Pyrus* and *Rubus* meristems. In: Engelmann F, Takagi H (eds) Cryopreservation of tropical plant germplasm. Current research progress and application. Japan International Research Center for Agricultural Sciences, Tsukuba, Japan/International Plant Genetic Resources Institute, Rome, Italy, pp 382–384

Coville FV, Britton NL (1908) Grossulariaceae. North Am Flora 22:193–225

Cronquist A (1981) An integrated system of classification of flowering plants. Columbia Univ Press, New York

Darlington CD (1929) A comparative study of the chromosome complement in *Ribes*. Genetica 11:267–269

de Janczewski E (1907) Monograph of the currants *Ribes* L. (in French). Mem Soc Phys Hist Nat Geneve 35:199–517

Dereuddre J, Scottez C, Arnaud Y, Duron M (1990) Effects of cold hardening on cryopreservation of axillary pear (*Pyrus communis* L. cv. Beurre Hardy) shoot tips of in vitro plantlets. CR Acad Sci Paris 310:265–272

Dumet D, Chang Y, Reed BM, Benson EE (2000) Replacement of cold acclimatization with high sucrose pretreatment in black currant cryopreservation. In: Engelmann F, Takagi H (eds) Cryopreservation of tropical plant germplasm. Current research progress and application. Japan International Research Center for Agricultural Sciences, Tsukuba, Japan/International Plant Genetic Resources Institute, Rome, Italy, pp 385–387

Engler A, Prantl K (1891) Ribesioideae. Naturl Pflanzenfam 3:97–142

FAO (1999) Food Agriculture Organization of the United Nations, World Wide Web Statistical Database. www.fao.org

Gunning J, Lagerstedt HB (1985) Long-term storage techniques for in vitro plant germplasm. Proc Int Plant Prop Soc 35:199–205

Hamat L, Porpaczy A, Himelrick DG, Galletta GJ (1989) Currant and gooseberry management. In: Galletta GJ, Himelrick DG (eds) Small fruit crop management. Prentice Hall, New York, pp 245–272

Hummer K (2000) History of the origin and dispersal of white pine blister rust. HortTechnology 10:515–517

Hummer K, Finn C (1999) Third year update: *Ribes* susceptibility to white pine blister rust. Acta Hortic 505:403–408

Jennings DL, Anderson MM, Brennan RM (1987) Raspberry and blackcurrant breeding. In: Abbot AJ, Atkin RK (eds) Improving vegetatively propagated crops. Academic Press, London, pp 135–147

Keep E (1962) Interspecific hybridization in *Ribes*. Genetica 33:1–23

Komarov VL (ed) (1971) Flora of the [former] USSR, vol IX. In: Ribesioideae Engl (translated from the Russian by the Israel Program for Scientific Translation, Jerusalem). Keter, London, pp 175–208

Lamarck JB, De Candolle AP (1805) Flore francaise. Desray, Paris

Liberty Hyde Bailey Hortorium (1976) Hortus third. Macmillan, New York, pp 969–971

Luo J, Reed BM (1997) Abscisic acid-responsive protein, bovine serum albumin, and proline pretreatments improve recovery of in vitro currant shoot tips and callus cryopreserved by vitrification. Cryobiology 34:240–250

Messinger W, Liston A, Hummer K (1999) *Ribes* phylogeny as indicated by restriction-site polymorphisms of PCR-amplified chloroplast DNA. Plant Syst Evol 217:185–195

MSTATC (1988) MSTATC – a software program for the design, management, and analysis of agronomic research experiments. Michigan State University, East Lansing, Michigan

Murashige T, Skoog F (1962) A revised medium for rapid growth and bioassays with tobacco tissue cultures. Physiol Plant 15:473–497

Reed BM (1990) Survival of in vitro-grown shoot tips of *Pyrus* following cryopreservation. HortScience 25:111–113

Reed BM, Chang Y (1997) Medium- and long-term storage of in vitro cultures of temperate fruit and nut crops. In: Razdan MK, Cocking EC (eds) Conservation of plant genetic resources in vitro, vol 1. Science Publishers, Enfield, New Hampshire

Reed BM, Tanprasert P (1995) Detection and control of bacterial contaminants of plant tissue cultures. A review of recent literature. Plant Tissue Culture Biotech 1:137–142

Reed BM, Yu X (1995) Cryopreservation of in vitro-grown gooseberry and currant shoot tips. Cryo Lett 16:131–136

Reed BM, Buckley PM, DeWilde TN (1995) Detection and eradication of endophytic bacteria from micropropagated mint plants. In Vitro Cell Dev Biol 31P:53–57

Reed BM, Brennan RM, Benson EE (2000) Cryopreservation: an in vitro method for conserving *Ribes* germplasm in international genebanks. In: Engelmann F, Takagi H (eds) Cryopreservation of tropical plant germplasm. Current research progress and application. Japan International Research Center for Agricultural Sciences, Tsukuba, Japan/International Plant Genetic Resources Institute, Rome, Italy, pp 470–472

Rehder A (1986) Manual of cultivated trees and shrubs, 2nd rev and enlarged edn. Dioscorides Press Portland, Oregon, pp 293–311

Sakai A, Nishiyama Y (1978) Cryopreservation of winter vegetative buds of hardy fruit trees in liquid nitrogen. HortScience 13:225–227

Sinnott QP (1985) A revision of *Ribes* L. subg. *Grossularia* (Mill.) per. Sect. Grossularia (Mill.) Nutt. (Grossulariaceae) in North America. Rhodora 87:189–286

Tuinyla V, Lukosevicius A (1996) Pomology of Lithuania. Lithuanian Science and Encyclopedia Publisher; Vilnius Lietuva, Lithuania

Vetenant (1799) Tableau du regne vegetal. Drisonnier, Paris

Yamada T, Sakai A, Matsumura T, Higuchi S (1991) Cryopreservation of shoot tips of white clover (*Trifolium repens* L.) by vitrification. Plant Sci 78:81–87

Zielinski QB (1953) Chromosome numbers and meiotic studies in *Ribes*. Bot Gaz 114:265–274

III.11 Cryopreservation of *Rosa* (Rose)

P.T. LYNCH

1 Introduction

1.1 The Distribution and Importance of Roses

Roses are the most important ornamental crop plant, with worldwide cultivation as a garden and amenity plant, for cut flowers and as a source of aromatic oils. It has been estimated that over 200 million rose bushes are planted annually, which represents a retail market of $US 720 million (Short and Roberts 1991). In the UK approximately 30 million plants and 0.5 million cut flowers are produced annually.

The genus *Rosa* consists of over 100 species (Rehder 1960), which can be divided into four subgenera, namely *Platyrhodon*, *Hesperhodos*, *Hulthemia* (also known as *Simplicifolia*) and *Eurosa* (Rehder 1949). The modern rose is within *Eurosa*, whilst the remaining subgenera have contributed little to the history of the cultivated rose. Only eight *Rosa* species have contributed in any significant way to the gene pool of modern roses (Krüssman 1981). As a result of the numerous indigenous species found in Asia, it is likely that the ancestors of modern roses originated from this region, *Rosa indica* and *Rosa chinensis* being the most significant. Molecular analysis has confirmed the significance of Asian species in the ancestry of the modern rose (Piola et al. 2001).

1.2 The Need for and Approaches to Conservation

1.2.1 Clonal Conservation

Rose breeding from its origins in the 18th century (Krüssman 1981) has led to the production of numerous cultivars and even the development of new classes of roses, for example, the Hybrid Teas, Floribundas and, more recently, the English Roses (Austin 1988). The development of so much 'novel'

Division of Biological Sciences, School of Environmental and Applied Sciences, University of Derby, Kedleston Road, Derby DE22 1GB, UK

Biotechnology in Agriculture and Forestry, Vol. 50
L.E. Towill and Y.P.S. Bajaj (Eds.) Cryopreservation of Plant Germplasm II
© Springer-Verlag Berlin Heidelberg 2002

germplasm has been dependent on sexual crosses and the selection of mutants, although, in the future, genetic manipulation may be significant to enhance features such as disease resistance (Marchant et al. 1998). The desirable characteristics of rose cultivars have changed with time. For example, in Victorian times, large shapely flowers were desired for the show bench, whereas today ground-cover roses are favoured as labour-saving, cost-effective plants. Rose breeders do not always just follow fashion, but have on occasions created it, for example, the brightly coloured Floribunda roses of the 1950s and 1960s (Beales 1992). Hence, as a result of changes in horticultural fashion, cultivars can fall out of favour and unique germplasm may be lost.

Seed banking is the usual method of choice for plant germplasm conservation (Chin 1994). However, this method is inappropriate for roses, because elite varieties do not come true from seed. Furthermore, germination of rose seeds is problematic, with germination rates of less than 20% not uncommon (Marchant et al. 1994). Therefore, to support their breeding programmes, rose hybridisers keep collections of roses under glasshouse or field conditions. However, the number and types of accessions maintained in this way tend to be determined by commercial pressures and normally reflect the characteristics of the roses a company produces. Many rose cultivars are also maintained in gardens worldwide, often in association with rose and horticultural societies, for example, the Royal National Rose Society has a 27-acre rose garden near St Albans, UK. Collections of roses are also supported by other organisations, for example, in the UK by the National Council for the Conservation of Plants and Gardens. However, plant germplasm maintained in field or glasshouse collections is vulnerable to loss due to pests, diseases and environmental factors and is expensive to maintain in terms of labour and land use (Benson et al. 1998).

Successful in vitro propagation has been reported for many rose species and cultivars (Douglas et al. 1989; Skirvin et al. 1990; Leyhe and Horn 1994; Voyiatzi et al. 1995; Yan et al. 1996). Micropropagation is a highly effective means for the rapid propagation of disease-free, uniform rose plants (Tweedle et al. 1984) and is used for the commercial propagation of elite rose varieties (Martin 1985). It is possible to maintain, for short- to medium-term storage, vegetatively propagated germplasm in tissue culture, either in the actively growing state or under in vitro slow-growth regimes (Ford-Lloyd and Jackson 1991; Engelman 1997; Reed 1999). However, despite the potential advantages of using micropropagation, as compared with field and glasshouse maintenance, as a means of short- to medium-term storage of germplasm, this approach has not been significantly adopted for roses. This may in part be due not only to the costs, in vitro conservation can more expensive than in vivo approaches (Epperson et al. 1997), but also the potentially adverse effects of somaclonal variation (Arene et al. 1993). These shortcomings can be reduced by linking in vitro culture to cryopreservation procedures (Stacey et al. 1999). Indeed it is becoming routine to apply cryopreservation to some crop species that are propagated in vitro (Bajaj 1995). Although the development of such an approach for the conservation of roses is at an early stage, successful cryopreservation of in vitro grown rose shoot tips has been reported

(Lynch et al. 1996) based on the encapsulation/dehydration cryopreservation approach of Fabre and Dereuddre (1990).

1.2.2 Pollen Storage

Despite maintaining field and glasshouse collections of roses to support ongoing breeding programmes, it is also desirable for plant breeders to store viable pollen, to enable breeding accessions which flower at different times and/or are spatially separated, to be routinely crossed (Luza and Polito 1988). The storage of viable pollen is also a means of ensuring that pollen is available when stigmatic receptivity is maximal (Lee et al. 1985) and of safeguarding against shortages of pollen due to the effects of pests, diseases or adverse environmental conditions during a particular season. Long-term storage of pollen may also provide a useful strategy for the preservation of germplasm alongside clonal strategies (Towill 1985; Brennan and Millam 1999). The ability to store pollen may also, in the future, be desirable for haploid culture of roses (Wissemann et al. 1996).

There are several reports concerned with the storage of rose pollen. Khosh-Khui et al. (1976) noted that the viability of rose pollen grains can not be maintained at $0\,^{\circ}C$ and a relative humidity of 50% for periods in excess of 9 weeks. However, Visser et al. (1977) reported that when stored at $+1\,^{\circ}C$ and a relative humidity of 20% or less, rose pollen viability was maintained for up to 40 weeks, but seed set using this pollen declined the longer it had been stored. In a comparative study, Marchant et al. (1993) showed that the viability of pollen from two English rose varieties stored at +4 or $-20\,^{\circ}C$ significantly declined over an 8-week period, the rate of decline being greater for the pollen stored at $+4\,^{\circ}C$. In contrast, cryopreservation has been shown to permit the effective long-term storage (up to 1 year) of rose pollen without significant loss of viability or the ability to facilitate hybridisation (Marchant et al. 1993; Rajasekharan and Ganeshan 1994).

This chapter describes the approaches used for the cryopreservation of in vitro culture-derived rose shoot tips and rose pollen.

2 Materials and Methods

2.1 Shoot Tip Cryopreservation (Lynch et al. 1996)

Shoot cultures of Rosa *multiflora* were maintained on semi-solid Murashige and Skoog medium (Murashige and Skoog 1962) containing $30\,g\,l^{-1}$ sucrose, $8.0\,g\,l^{-1}$ agar (Difco Bacto agar, Difico Laboratories, Detroit, Michigan, USA) and supplemented with $0.1\,mg\,l^{-1}$ 6-benzylaminopurine, $0.1\,mg\,l^{-1}$ gibberellic acid and $0.004\,mg\,l^{-1}$ naphthaleneacetic acid (Lloyd et al. 1988), designated RSC1, at $18\,^{\circ}C$ with a 16h photoperiod and light intensity of $3\text{--}7\,\mu mol\,m^{-2}\,s^{-1}$

PAR. Apical shoot tips were excised under filter-sterilised anti-oxidant solution (0.2 M phosphate buffer, pH 5.7, containing $50 g l^{-1}$ ascorbic acid and $15 g l^{-1}$ sodium borate). Excised shoot tips (Fig. 1A) were suspended in liquid calcium-free RSC1 medium containing $30 g l^{-1}$ sodium alginate. The shoot tips were drawn up into sterile $10\text{-}cm^3$ disposable pipettes and dripped into liquid RSC1 medium containing $14.7 g l^{-1}$ $CaCl_2 \cdot 2H_2O$; thus, each shoot tip was encapsulated in a bead of calcium alginate (Fig. 1B). The beads were left to 'cure' for 15 min in the calcium-enriched RSC1 medium, and then collected on a 250-μm nylon mesh and transferred to liquid RSC1 preculture medium containing 0.5 M sucrose (10 ml of medium per 5-cm Petri dish containing 5 beads). The beads were incubated at 18 °C for 24 h in the dark with shaking (40 rpm). After preculture the beads were collected on a 250-μm nylon mesh and individually rapidly surface-dried by rolling them on filter paper (Whatman No 1). Beads (in groups of 10) were transferred to a 5.5-cm filter paper (Whatman No 1) in the lids of 5-cm Petri dishes, their fresh weight determined, and the dishes were then placed into 250-ml powder jars (SLS, Nottingham, UK) containing 80 g of silica gel (the jars were sterilised and the silica gel dried by incubation at 160 °C for 6 h). Jars were sealed with a screw lid and Nescofilm (Nippon Shoji Kaisha Ltd., Osaka, Japan) and maintained under standard culture conditions. Beads were exposed to silica gel for 2 h under the same incubation conditions as during the sucrose preculture treatment. After this the beads were transferred into 2-ml polypropylene vials (Starstedt Ltd.,

Fig. 1. Cryopreservation of shoot tips of *Rosa multiflora*. **A** Excised shoot tip. **B** Non-frozen shoot encapsulated in alginate bead. **C** Shoot regrowing, 50 days after thawing, note callus development. **D** Shoot regrowing, 80 days after thawing. *Bars* represent 5 mm

Boston Road, Leicester, UK; 10 beads per vial), then mounted on aluminium canes and plunged into liquid nitrogen.

Beads were warmed after a minimum of 24 h storage in liquid nitrogen, by plunging into sterile water at 40 °C. The shoot tips were excised from the alginate beads and placed on semi-solid RSC1 medium (4 shoot tips per 5-cm Petri dish) and incubated for 10 days at 18 °C in the dark, after which they were maintained under the same light conditions as the original micro-propagated shoot cultures and transferred to fresh medium at 30-day inter-vals. Approximately 120 days after thawing, the shoots were excised from any callus tissue and subsequently maintained as micropropagated shoot cultures.

2.2 Pollen Cryopreservation

2.2.1 Method of Marchant et al. (1993)

English rose cultivars 'Heritage' and 'The Countryman' were used in this study. Anthers were collected at between 07:30 and 08:30, from bushes grown in a mesh tunnel house from semi-open flowers, i.e. when the flower bud had reached its final and true varietal colour, where several petals of the outer whorl were slightly detached from the rest of the bud (Gudin et al. 1991). Detached anthers were sealed in 2.0-ml polypropylene vials and transported to the laboratory. After approximately 1 h the vials were opened and main-tained at 25 ± 2 °C, with a relative humidity of 30% under a 12-h photoperiod for 48 h, which caused the anthers to dehisce. Subsequently, the vials were vortexed to separate the pollen from the anthers, after which the anthers were removed. Using a camel hair brush small samples of pollen were removed to determine the pollen viability based on the frequency of in vitro germination, as detailed in Sect. 2.2.3.1. The vials were resealed, mounted on aluminium canes and plunged into liquid nitrogen.

Vials of pollen were warmed after a minimum of 48 h storage in liquid nitrogen by plunging into sterile water at 45 °C for 2 min, after which the pollen viability was determined by in vitro germination (Sect. 2.2.3.1) or used to cross-pollinate rose flowers.

2.2.2 Method of Rajasekharan and Ganeshan (1994)

Hybrid tea rose varieties 'Bulls Red', 'Happiness' and 'Paradise' and the Flori-bunda variety 'Queen Elizabeth' were used in this study. Rose flowers, which were due to open within 24 h and had been bagged previously, were collected between 10:00 and 11:00 and transported to the laboratory. The flower petals were removed and the pollen collected by gently tapping the flower hips on to butter paper. To determine pollen viability, samples were taken for in vitro germination assessment as described in Section 2.2.3.2. The remaining pollen was transferred to gelatine capsules and enclosed in laminated aluminium pouches. The pouches were sealed and plunged into liquid nitrogen. The pollen was stored for 48 h or 1 year and thawed to ambient temperature prior to viability and pollination assessments.

2.2.3 Determination of Pollen Viability

2.2.3.1 In Vitro Germination (Marchant et al. 1993)

Samples of pollen were transferred using a camel hair brush to the surface of semi-solid pollen grain germination medium (40.0 mg l^{-1} H$_3$BO$_3$, 226.5 mg l^{-1} CaCl$_2$·6H$_2$O, 150 g l^{-1} sucrose and 4.0 g l^{-1} Sigma agarose Type 1; adjusted to pH 5.6 prior to autoclaving) in 5-cm Petri dishes (8.0 ml of medium per dish). The dishes were incubated at 23 °C under a 12-h photoperiod. After 24 h the samples were examined under a microscope. Germination was deemed to have occurred when the length of the emerging pollen tube was greater than the diameter of the pollen grain.

2.2.3.2 In Vitro Germination (Rajasekharan and Ganeshan 1994)

Pollen germination medium consisting of 100.0 mg l^{-1} H$_3$BO$_3$, 226.5 mg l^{-1} CaNO$_3$·4H$_2$O, 200 mg l^{-1} MgSO$_4$·7H$_2$O, 100 mg l^{-1} KNO$_3$ and 150 g l^{-1} sucrose with suspended pollen grains was incubated as hanging drops on slides in the dark at 25 °C. After 4 h the samples were examined under a microscope. Germination was deemed to have occurred when the length of the emerging pollen tube was greater than the diameter of the pollen grain.

3 Results

3.1 Shoot Tip Cryopreservation

Non-encapsulated shoot tips exhibited shoot elongation and development within 28 days, but encapsulated shoots took up to 90 days to grow away from the calcium alginate beads. Hence, to reduce the inhibitory effect of encapsulation, shoot tips were excised prior to culture. The optimum concentration of sucrose during preculture was 0.5 M (Table 1). Higher sucrose concentrations resulted in a significant ($p < 0.005$) reduction in the frequency of shoot growth. Greater callus formation was also noted at the base of shoot tips treated with higher sucrose concentrations. Periods of exposure to silica gel in excess of 2 h resulted in a significant ($P < 0.005$) reduction in shoot tip growth (Table 1).

Post-warming shoot tip growth, at a frequency of 25 ± 7%, was only observed after preculture with 0.5 M sucrose and 2 h exposure to silica gel, followed by rapid cooling. Growth of shoot tips was first observed approximately

Table 1. Effect of sucrose preculture and exposure to silica gel on the growth of encapsulated rose shoot tips

Preculture sucrose concentration (M)	Duration of exposure to silica gel (h)	% post-treatment shoot tip growth
0	0	100 ± 0.0
0.25	0	100 ± 0.0
0.5	0	90 ± 5.2
0.75	0	75 ± 7.0
1.0	0	15 ± 7.0
0.5	1	90 ± 1.5
0.5	2	87 ± 3.4
0.5	3	50 ± 14
0.5	6	0
0.5	12	0

Values are the mean of replicate experiments ± standard deviation.

40 days after warming. Callus growth was observed around the base of shoot tips after thawing (Fig. 1C). No shoot regeneration was observed from the callus. Although occasionally the growing shoot was overgrown by the callus, generally the shoot developed to a stage were it could be excised from the callus approximately 60–80 days after thawing (Fig. 1D).

3.2 Pollen Cryopreservation

The reports of Marchant et al. (1993) and Rajasekharan and Ganeshan (1994) showed that the viability of fresh pollen from different rose cultivars, as determined by in vitro pollen germination, varies significantly ($p < 0.05$), as shown in Table 2. The viability of pollen from the English rose cultivars was consistently lower. This may be due to varietal effects, but the differences in the conditions in which the source plants were grown, and in the methods used to determine in vitro pollen germination, may also have contributed. Voyiatzi (1995) showed that the composition of the culture medium used for in vitro culture of rose pollen affected the frequency of pollen germination. Interestingly, Voyiatzi (1995) also used pollen from the cultivar 'Queen Elizabeth' in his study, but reported values for % pollen germination considerably lower than Rajasekharan and Ganeshan (1994). Factors such as the stage of flower development and season have also been shown to influence rose pollen viability (Gudin et al. 1991; Marchant et al. 1993).

Importantly, Table 2 shows that, despite variability between cultivars, cryopreservation did not significantly ($p < 0.05$) affect pollen viability. In a series of reciprocal crosses (shown in Table 3), cryopreserved rose affected fertilisation. The comparability in the number of seeds produced per hip using cryopreserved pollen, as compared with fresh pollen, indicates that cryopreservation has no apparent adverse effect on the ability of pollen grains to produce competent pollen tubes that are capable of delivering to the ovule male gametes able to undergo double fertilisation. This is consistent with the

Table 2. Effect of cooling on rose pollen viability

Rose Cultivar	% Viability		Methods according to reference
	Pre-cool	Post-warm	
Heritage	27.6	24.6	1
The Country Man	16.2	14.3	1
Happiness	58.6	65.3	2
Bulls Red	67.8	62.7	2
Paradise	53.9	63.7	2
Queen Elizabeth	69.7	60.4	2

1. Marchant *et al.* (1993).
2. Rajasekharan and Ganeshan (1994).

Table 3. Effect of cryopreservation on the ability of rose pollen to effect fertilisation

Pollen source	Pollen recipient	Pollen cryopreserved	Mean number of seeds per hip	Method according to reference
The Countryman	Heritage	No	21.4	1
The Countryman	Heritage	Yes	21.6	1
Heritage	The Countryman	No	22.6	1
Heritage	The Countryman	Yes	21.3	1
Happiness	Bulls Red	No	9.7	2
Happiness	Bulls Red	Yes	2.5	2
Bulls Red	Happiness	No	10.1	2
Bulls Red	Happiness	Yes	14.8	2
Queen Elizabeth	Paradise	No	18.8	2
Queen Elizabeth	Paradise	Yes	18.8	2

1. Marchant *et al.* (1993).
2. Rajasekharan and Ganeshan (1994).

results of Parfitt and Almehdi (1984), who showed that pollen tube development in other rosaceous species was unaffected by cryopreservation.

4 Summary and Conclusions

Cryopreservation has, as yet, been little developed or exploited in roses. The initial study of Lynch et al. (1996) showed that it was possible to recover shoot tips from *R. multiflora* after cryopreservation. However, the method requires further development to increase the frequency of post-thaw regrowth and determine the applicability of the encapsulation/dehydration approach to a wider range of *Rosa* species and *R. hybrida* cultivars. The genetic stability of rose germplasm recovered from cryopreservation also needs to be determined. Once cryopreservation protocols are more established for roses, the exploitation of this technology could be an appropriate means of safeguarding valuable germplasm currently maintained in field and glasshouse collections. The development of efficient cryopreservation protocols may also be useful to support future rose biotechnology programmes, as has been shown to be important for other plant species (Benson et al. 1998). The ability to cryopreserve rose germplasm may also become more commercially significant. When a patent involving a biological culture is filed a sample must be lodged with a laboratory accredited with International Depository Authority status by the World Intellectual Property Organisation, Geneva (Anon. 1995). This system currently applies to the registration of new plant cultivars, but will be of increasing importance for genetically modified plants.

Although rose pollen cryopreservation has not been widely reported, the comparable approaches successfully taken by the two groups are both efficient and relatively simple. Importantly, both groups reported no significant adverse effect on the ability of pollen to effect fertilisation after cryopreservation. The equipment requirements for rose pollen cryopreservation are simple and relatively cheap, making it a viable option to rose hybridisers who wish to store pollen for long periods.

References

Anon (1995) Guide to the deposit of micro-organisms under the Budapest Treaty. World Intellectual Property Organisation, Geneva

Arene L, Pellegrino C, Gudin S (1993) A comparison of the somaclonal variation level of *Rosa hybrida* L. cv. Meirutral plants regenerated from callus or direct induction from different vegetative and embryogenic tissues. Euphytica 71:83–90

Austin D (1988) The heritage of the rose. Antique Collectors Club, London, pp 176–225

Bajaj YPS (ed) (1995) Cryopreservation of plant germplasm I. Biotechnology in agriculture and forestry, vol 32. Springer, Berlin Heidelberg New York

Beales P (1992) Roses. Harvil, London, pp 1–26

Benson EE, Lynch PT, Stacey GN (1998) Advances in plant cryopreservation technology: current applications in crop plant biotechnology. AgBiotech News Info 10:133N–141N

Brennan RM, Millam S (1999) Conservation of small fruit germplasm. In: Benson EE (ed) Plant conservation biotechnology. Taylor Francis, London, pp 154–164

Chin HF (1994) Seedbanks: conserving the past for the future. Seed Sci Technol 22:385–400

Douglas GC, Rutledge CB, Casey AD, Richardson DHS (1989) Micropropagation of Floribunda, ground cover and miniature roses. Plant Cell Tissue Org Cult 19:55–64

Engelman F (1997) In vitro conservation methods. In: Calow JA, Ford-Lloyd BV, Newbury HJ (eds) Biotechnology and plant genetic resources. CAB International, Wallingford, pp 119–161

Epperson JE, Pachico DH, Guevara CL (1997) A cost analysis of maintaining cassava plant genetic resources. Crop Sci 37:1641–1649

Fabre J, Dereuddre J (1990) Encapsulation-dehydration: a new approach to cryopreservation of *Solanum* shoot tips. Cryo Lett 11:413–426

Ford-Lloyd BV, Jackson MT (1991) Biotechnology and methods of conservation of plant genetic resources. J Biotech 17:247–256

Gudin S, Arene L, Bulard C (1991) Influence of season on rose quality. Sex Plant Reprod 4: 113–117

Khosh-Khui M, Bassrii A, Niknejad M (1976) Effects of temperature and humidity on pollen viability of six rose species. Can J Plant Sci 56:517–523

Krüssman G (1981) Roses. Timber Press/Portland Press, Oregon, pp 67–100

Lee CW, Thomas JA, Buchmann SC (1985) Factors affecting in vitro germination and storage of Jojoba pollen. J Am Soc Hortic Sci 110:671–676

Leyhe U, Horn W (1994) A contribution to micropropagation of *Rosa hybrida*. Gartenbauwissenschaft 59:85–88

Lloyd D, Roberts AV, Short KC (1988) The induction of in vitro adventitious shoots in *Rosa*. Euphytica 37:31–36

Luza JG, Polito VS (1988) Cryopreservation of English Walnut (*Juglans regia* L.) pollen. Euphytica 37:141–148

Lynch PT, Harris WC, Chartier-Hollis JM (1996) The cryopreservation of shoot tips of *Rosa multiflora*. Plant Growth Regul 20:43–45

Marchant R, Power JB, Davey MR, Chartier-Hollis JM, Lynch PT (1993) Cryopreservation of pollen from two rose cultivars. Euphytica 66:235–241

Marchant R, Power JB, Davey MR, Chartier-Hollis JM (1994) Embryo rescue for the production of F1 hybrids in English Rose. Euphytica 74:187–193

Marchant R, Davey MR, Lucas JA, Lamb CJ, Dixon RA, Power JB (1998) Expression of a chitinase transgene in rose (*Rosa hybrida* L) reduces development of blackspot disease (*Diplocarpon rosae* Wolf). Mol Breed 4:187–194

Martin C (1985) Plant breeding in vitro. Endeavour 9:81–86

Murashige T, Skoog F (1962) A revised medium for rapid growth and bioassays with tobacco cell cultures. Physiol Plant 15:473–497

Parfitt DE, Almehdi AA (1984) Liquid nitrogen storage of pollen from five *Prunus* species. HortScience 19:69–70

Piola MM, Chessel D, Jay M, Heizmann P (2001) The domestication of the modern rose: genetic structure and allelic composition of the rose complex. Theor Appl Genet 102:398–404

Rajasekharan PE, Ganeshan S (1994) Freeze preservation of rose pollen in liquid nitrogen: feasibility, viability and fertility status after long-term storage. J Hortic Sci 69:565–569

Reed BM (1999) In vitro conservation of temperate tree fruit and nut crops. In: Benson EE (ed) Plant conservation biotechnology. Taylor Francis, London, pp 139–154

Rehder A (1949) Bibliography of cultivated trees and shrubs hardy in the cooler temperate regions of the northern hemisphere. Arnold Arboretum of Harvard University, Jamaica Plain, MA, USA, pp 296–318

Rehder A (1960) Manual of cultivated trees and shrubs hardy in North America, 2nd edn. Springer, Berlin Heidelberg New York, pp 424–482

Skirvin RM, Chu MC, Young HJ (1990) Rose. In: Ammirato PV, Evans DA, Sharp WR, Bajaj YPS (eds) Handbook of plant cell culture, vol 5. Ornamental species. McGraw-Hill, New York, pp 716–743

Short KC, Roberts AV (1991) *Rosa* sp. (roses) in vitro culture, micropropagation and the production of secondary products. In: Bajaj YPS (ed) Biotechnology in agriculture and forestry, medicinal and aromatic plants III. Springer, Berlin Heidelberg New York, pp 376–397

Stacey GN, Lynch PT, Benson EE (1999) Plant gene banking: agriculture, biotechnology and conservation. Agro Food Industry Hi Tech 10:9–14

Towill LE (1985) Cryopreservation. In: Dodds JH (ed) In vitro methods for conservation of plant genetic resources. Chapman and Hall, London, pp 41–70

Tweedle D, Roberts AV, Short KC (1984) In vitro culture of roses. In: Novak FJ, Havel L, Dolezel J (eds) Plant tissue and cell culture: application to crop improvement. Czech Acad Sci, Prague, pp 529–530

Visser T, De Vries DP, Welles GWH, Scheurink JAM (1977) Hybrid tea-rose pollen. I. Germination and storage. Euphytica 26:721–728

Voyiatzi CI (1995) An assessment of the in vitro germination capacity of pollen grains of five tea-hybrid rose cultivars. Euphytica 83:199–204

Voyiatzi CI, Voliatzis DG, Tsiakmaki V (1995) In vitro shoot proliferation rates of the rose cv. (Hybrid-Tea) Dr Verhage, as affected by apical dominance regulation substances. Sci Hort 61:241–249

Wissemann V, Mollers C, Hellwig FH (1996) Microspore and anther culture in the genus *Rosa*, investigations and current status. J Appl Bot 70:218–220

Yan M, Byrne DH, Jing C (1996) Propagation of rose species in vitro. In Vitro Cell Dev Biol Plant 32:103–108

Section IV
Australian Species

IV.1 Cryostorage of Somatic Tissues of Endangered Australian Species

D.H. Touchell[1], S.R. Turner[1,2], E. Bunn[1], and K.W. Dixon[1]

1 Introduction

1.1 Background

The flora of southwest Western Australia is remarkably rich in species and is recognized as one of the top ten most botanically diverse regions in the world (Brown et al. 1998). There are over 8000 plant taxa within the region (32% of the Australian flora) with 70% of species endemic to the region (Hopper et al. 1996). In common with other biodiverse regions, threatening processes, such as land clearing for agricultural, mining and urban development, as well as the spread of plant diseases and pests, have resulted in major habitat reduction and disturbance. The number of rare and threatened species has increased from 238 to 327 over the past 10 years within Western Australia, with 1959 plant species currently classified as in need of conservation action (Hopper et al. 1990; Brown et al. 1998). With little or no foreseeable changes in the mitigating factors accompanying the increase in endangered plant species, conservation agencies must turn towards ex situ methods in order to compliment in situ conservation programs for the increasing number of threatened plant species.

Ex Situ Methods for Plant Conservation. The methods available to conserve threatened flora have traditionally focused on seed banks and active growth collections, primarily maintained in botanic gardens (i.e., container and in vitro collections). More recently, in vitro plant culture has become a highly favored means for ex situ conservation of rare and threatened plant taxa. Allied to the ability to conserve clonal germplasm in vitro is the emerging technology of cryostorage, which utilizes liquid nitrogen for maintenance of germplasm collections under conditions of high environmental stability. These methods are considered in more detail below, with reference to the integrated conservation program for rare and threatened native plants in operation at Kings Park and Botanic Garden.

[1] Kings Park and Botanic Garden, West Perth, WA 6005, Australia
[2] Curtin University of Technology, Bentley, WA 6845, Australia
Current address: D.H. Touchell, School of Forestry and Wood Products, Michigan Technological University, Houghton, Michigan 49931, USA

Biotechnology in Agriculture and Forestry, Vol. 50
L.E. Towill and Y.P.S. Bajaj (Eds.) Cryopreservation of Plant Germplasm II
© Springer-Verlag Berlin Heidelberg 2002

1.2 Conventional Storage Methods

Seed Banks. Seed banks represent a traditional means for establishing conservation collections of rare taxa in Australia. Most Western Australian species
possess orthodox seed, which can withstand drying to a low moisture content
and storage at –18 °C. Using these procedures, seeds of 286 endangered species
(and subspecies) from Western Australia have been stored at –18 °C in
regional facilities (Sweedman, Kings Park and Botanic Garden, pers. comm.).
However, in highly biodiverse floras, many species may not be amenable to
long-term storage using conventional seed-storage techniques. For example,
some endangered species may produce little or no viable seed or seed with
unusual dormancy mechanisms with high variability in viability and longevity
(Meney et al. 1994; Dixon et al. 1995; Roche et al. 1997). Such factors may not
have been considered when conventional seed-storage procedures were being
adopted for conserving taxa and where it is critical to protect remaining
genetic stocks of mother plants. For such species, conservation agencies should
supplement conventional germplasm collections by utilizing vegetative or
micropropagation techniques wherever possible.

Container Collections. Active growth collections and in situ garden collections
still account for significant numbers of rare and threatened plant taxa, predominantly housed in botanic gardens. These collections also represent
educational opportunities in promoting the conservation of rare plants, particularly through plantings in gardens. However, in regions with extreme
diversity, collections of endangered species particularly where genetic representation is necessary may result in an excessively large array of plants. For
example, in Western Australia, to maintain five containers of each of ten
genetic types for each of the 327 endangered species would require on-going
care and management of 16,350 potted plants. The space and curatorial costs
for such a sizable collection are prohibitive.

1.3 In Vitro Collections

In vitro germplasm collections are now utilized by many botanic gardens to
maintain rare and threatened plants. Micropropagation research at Kings Park
and Botanic Garden has resulted in in vitro maintenance protocols for over
100 rare and horticulturally significant native plant species. However, even
with the use of slow growth techniques (i.e., cool storage, growth retardant
media, etc.), space and subculturing routines limit the number of species that
can be actively maintained at any one time. In vitro collections are also prone
to losses through pathogenic contamination (Leifert et al. 1994). In addition,
somaclonal variation may occur where cultures are maintained for long
periods of continuous in vitro propagation (George and Sherrington 1993).
These limitations plus the increasing number of threatened taxa require more
efficient means for conserving key clonal material without compromising the
genetic integrity of the plant material.

1.4 Cryostorage

Cryogenic methods (cryostorage) provide a potentially useful means for effective and efficient conservation of rare and endangered species. Cryostorage is one of the few means for securing plant genetic resources while ensuring the maintenance of genetic fidelity on a long-term basis. Therefore, it is possible to establish long-term germplasm collections for endangered species that require little or no maintenance (Fay 1992).

2 A Role for Cryostorage in the Conservation of Australian Plants

2.1 Cryopreservation of Shoot Apices of Australian Species

Cryopreservation of shoot apices of Australian species has been limited to a small number of studies. Touchell et al. (1992), Touchell (1995), and Touchell and Dixon (1995, 1999) have demonstrated successful cryostorage of somatic tissues of 13 species from the southwest of Australia, while Monod et al. (1992) successfully cryostored shoot apices of a common eucalypt (*Eucalyptus gunnii*).

The first documented cryostorage of organized somatic tissues of a rare species was achieved for *Grevillea scapigera*, a critically endangered plant from the south west of Western Australia. These studies investigated the use of slow cooling techniques (Touchell et al. 1992) where shoot apices were pretreated with 5% dimethyl sulfoxide (DMSO) (v/v) for 48h followed by cooling at 0.5 °C/min, in a medium containing 10% DMSO, to a terminal temperature of –40 °C followed by plunging into liquid nitrogen. Although post-thaw survival of 25% was achieved using this method for *G. scapigera* (Fig. 1), the high level of callus formation in recovering tissues indicated that these methods were possibly sub-optimal. Alternative cryostorage procedures for *G. scapigera* included the use of encapsulation/dehydration and vitrification with tissue survival increasing to 33 and 79%, respectively (Fig. 1). The following methodologies describe procedures based on vitrification.

2.2 Methodology

Plant Material. In vitro grown plants of all species (Table 1) were maintained at Kings Park and Botanic Garden. Basal medium (BM) consisted of half-strength Murashige and Skoog (1962) salts supplemented with $3\,\mu$M thiamine hydrochloride, $2.5\,\mu$M pyridoxine hydrochloride, $4\,\mu$M niacin, $60\,$mM sucrose and 0.1% w/v agar. Stock shoot cultures were maintained on BM supplemented with $0.5\,\mu$M 6-benzylaminopurine (BAP) under standard culture conditions (i.e., 22–25 °C, irradiated with 16h light with PPFD of $40\,\mu$M m–2 s–1)

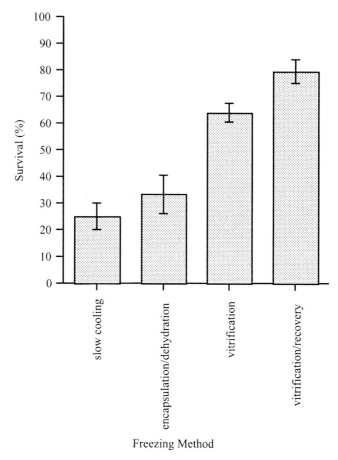

Fig. 1. Survival of frozen shoot apices following exposure to slow cooling, encapsulation/dehydration, vitrification and vitrification/optimized recovery medium in *Grevillea scapigera*

and 0.5-mm-long shoot apices were harvested at 3 weeks after each subculture.

2.2.1 Grevillea scapigera

Shoot apices were precultured for 48 h on BM supplemented with sorbitol (0, 0.2, 0.4, 0.6, 0.8 or 1.0 M) under standard culture conditions. Ten precultured shoot apices were placed in a 1.2-ml cryovial (Nunc) containing 1 ml of modified PVS2 (Yamada et al. 1991) cryoprotective media. Shoot apices were then incubated at either 23 or 0 °C for 0, 5, 10, 15, 20, 25, 30, 45 or 60 min before cooling by direct immersion in LN. Shoot apices were held in LN for 24 h and warmed rapidly in a water bath at 40 °C. The cryoprotective PVS2 media was drained from these shoot apices and replaced three times in 12 min with liquid

Table 1. Percentage survival of excised shoot tips for Australian species cryopreserved using vitrification procedures

Family	Species	Protocol	Survival	Reference
	Dicotyledon			
Chloanthaceae	*Pityrodia scabra*	1.2 M glycerol for 2 days, 30 mins PVS2	78.0%	Touchell (unpublished)
Goodeniaceae	*Lechenaultia formosa*	0.6 M sorbitol for 2 days, 30 mins PVS2	25.0%	Touchell (1996)
	Lechenaultia laricina	7 days ABA (culture media)	47.0%	Touchell (unpublished)
		1.2 M glycerol for 2 days, 30 mins PVS2		
Lamiaceae	*Hemiandra gardneri*	0.6 M sorbitol for 2 days, 30 mins PVS2	8.3%	Touchell (1995)
Myoporaceae	*Eremophila caerulea* spp *marellii*	0.6 M sorbitol for 2 days, 30 mins PVS2	27.0%	Touchell (1995)
	Eremophila resinosa	0.6 M sorbitol for 2 days, 30 mins PVS2	7.2%	Touchell (1995)
Myrtaceae	*Eucalyptus granticola*	1.2 M glycerol + 2 mM choline chloride, 30 mins PVS2	60.0%	Crowe (1998)
Papillionaceae	*Ptychosema pusillum*	0.6 M sorbitol for 2 days, 30 mins PVS2	12.1%	Touchell (1995)
Proteaceae	*Conospermum stoechadis*	0.6 M sorbitol for 2 days, 30 mins PVS2	5.4%	Touchell (1995)
	Grevillea cirsiifolia	0.6 M sorbitol for 2 days, 30 mins PVS2	46%	Tan (1998)
	Grevillea dryandroides spp *dryandroides*	0.6 M sorbitol for 2 days, 30 mins PVS2	74.0%	Bunn (unpublished)
	Grevillea dryandroides spp *hirsutus*	1.2 M glycerol for 2 days, 30 mins PVS2	75%	Tan (1998)
	Grevillea flexuosa	0.6 M sorbitol for 2 days, 30 mins PVS2	68.0%	Tan (1998)
	Grevillea maccutcheonii	0.6 M sorbitol for 2 days, 30 mins PVS2	50.0%	Bunn (unpublished)
	Grevillea scapigera	0.6 M sorbitol for 2 days, 30 mins PVS2	64.1%	Touchell (1995)
	Hakea aculeata	0.6 M sorbitol for 2 days, 30 mins PVS2	62.0%	Bunn (unpublished)
	Lambertia orbifolia	0.6 M sorbitol for 2 days, 30 mins PVS2	62.0%	Bunn (unpublished)
	Leucopogon obtectus	0.6 M sorbitol for 2 days, 30 mins PVS2	17.4%	Touchell (1995)
Rutaceae	*Eriostemon wonganensis*	0.6 M sorbitol for 2 days, 30 mins PVS2	37.0%	Touchell (1995)
Sterculiaceae	*Rulingia* sp. Trigwell Bridge	0.6 M sorbitol for 2 days, 30 mins PVS2	17.9%	Touchell (1995)
Tremandraceae	*Tetratheca deltoidea*	0.6 M sorbitol for 2 days, 30 mins PVS2	9.6%	Touchell (1995)
Monocotyledon				
Haemodoraceae	*Anigozanthos kalbarriensis*	0.4 M sorbitol for 2 days, 25 mins PVS2	6.3%	Turner (unpublished)
	Anigozanthos humilis spp *chrysanthus*	0.4 M sorbitol for 2 days, 25 mins PVS2	24.4%	Turner (unpublished)
	Anigozanthos manglesii	0.4 M sorbitol for 2 days, 25 mins PVS2	31.9%	Turner (1997)
	Anigozanthos pulcherrimus	0.4 M sorbitol for 2 days, 25 mins PVS2	17.1%	Turner (unpublished)
	Anigozanthos rufus	0.4 M sorbitol for 2 days, 25 mins PVS2	2.2%	Turner (1997)
	Anigozanthos viridis	0.4 M sorbitol for 2 days, 25 mins PVS2	57.0%	Turner (unpublished)
	Conostylis dielsii spp *teres*	0.4 M sorbitol for 2 days, 25 mins PVS2	20.8%	Turner (unpublished)
	Conostylis micrantha	0.4 M sorbitol for 2 days, 25 mins PVS2	14.6%	Turner (unpublished)
	Conostylis wonganensis	0.4 M sorbitol for 2 days, 25 mins PVS2	17.6%	Turner (unpublished)
	Macropidia fuliginosa	0.4 M sorbitol for 2 days, 25 mins PVS2	4.2%	Turner (unpublished)
Liliaceae	*Sowerbaea multicaulis*	0.6 M sorbitol for 2 days, 30 mins PVS2	10.9%	Touchell (1995)
Restionaceae	*Hopkinsia anoectocolea*	0.2 M sorbitol for 3 days, 25 mins PVS2	46.4%	Touchell (unpublished)

BM supplemented with 1.0 M sucrose (washing media). Shoot apices were then recovered on solidified BM under standard culture conditions and survival was determined by the percentage of treated shoot apices showing signs of growth within 2 weeks. Control shoot apices were treated in the same manner, but were not cooled. Three replicates of 20 shoot tips were used for each trial.

Selection of a Recovery Medium. After thawing, shoot tips were washed and recovered on one of the following media:

1. Basal medium (BM)
2. Basal medium +1 mM choline chloride (CC)
3. Basal medium +1 μM zeatin and 3 μM gibberellic acid (Z)
4. Basal medium +10 μM BA (BA10)
5. Basal medium +5 μM BA (BA5)

Recovery of shoot apices was then monitored over a 6-week period. Three replicates of 20 shoot tips were used for each trial.

2.2.2 *Anigozanthos viridis*

Optimum cryostorage procedures were developed for *Anigozanthos viridis* using the cryostorage method employed for *Grevillea scapigera*. Shoot apices were precultured for 48 h on basal media supplemented with 0–0.6 M sorbitol before being incubated at 0 °C for 0 (controls only) 15, 20, 25, and 30 min in PVS2 and then immersed in LN. Shoot apices were recovered on BM supplemented with 1 μM zeatin under standard culture conditions.

Post-Thaw Recovery of Anigozanthos viridis. A further trial was initiated to optimize post-thaw recovery conditions. The two factors investigated were absence/presence of light and the absence/presence of choline chloride in the recovery phase. Shoot apices were either exposed to light directly (8 h light) or maintained in complete darkness for 7 days following thawing. Additionally, 2 mM of choline chloride was added to the BM in two treatments (either 8 h of light or complete darkness). Three replicates of ten shoot apices were used for each combination.

2.2.3 *Application to Diverse Plant Taxa*

Modifications to standard protocols included the use of abscisic acid (ABA) in the subculture medium for *Lechenaultia*, the replacement of 0.6 M sorbitol with 1.2 M glycerol in the preculture media (see Table 1) and the addition of choline chloride to the preculture medium for *Eucalyptus graniticola*.

2.3 Results

2.3.1 Grevillea scapigera

Survival of shoot apices for all species was dependent on sorbitol concentration and exposure duration to PVS2 solution. For *Grevillea scapigera*, the highest survival (64.1%) was obtained by preculturing shoot apices on a medium supplemented with 0.6 M sorbitol for 48 h followed by exposure to PVS2 for 30 min at 0 °C before direct immersion in liquid nitrogen (Fig. 2). Different sorbitol concentrations in the preculture medium and exposure times to PVS2 gave significantly lower survival.

Recovery of Shoot Apices. The post-thaw survival of shoot apices depended on recovery media and regime. For *Grevillea scapigera*, recovery on media supplemented with 1 mM choline chloride significantly increased post-thaw survival to 79.6 ± 4.6% after 1 week. The addition of plant growth regulators to the recovery media had no effect on initial post-thaw survival. Although initial survival of shoot apices remained high, all recovery media, apart from media supplemented with zeatin, failed to promote shoot growth and regeneration 4 weeks after liquid nitrogen treatment (Table 2). After 4 weeks, all surviving shoot apices recovered on BA10, BM and 0.3 M sucrose medium had turned brown and died and only 13.75 ± 4% on BA5 media remained viable. Shoot apices on media supplemented with zeatin showed no decline in survival after 4 weeks; however, shoot apices that recovered on this medium had increased callus proliferation. Shoot regeneration increased to 85.8 ± 5.8% when treated shoot apices were cultured under reduced light conditions on basal medium with choline chloride for 7 days before being transferred to medium containing zeatin.

2.3.2 Anigozanthos viridis

For *Anigozanthos viridis* (a monocotyledon), a lower concentration of sorbitol in the preculture media was necessary for optimal post-thaw survival. Highest

Table 2. Post warming survival of cryostored shoot tips from *Grevillea scapigera* following exposure to various recovery media supplements (± standard error)

Treatment	Survival	4 week survival	Shoot regeneration	Callus
BA 5	62.5% (±1.4%)	13.8% (±4.0%)	0.0% (±0.0%)	13.8% (±4.0%)
BA 10	67.9% (±9.1%)	0.0% (±0.0%)	0.0% (±0.0%)	0.0% (±0.0%)
Zeatin	70% (±0.0%)	70.0% (±0.0%)	0.0% (±0.0%)	70.0% (±0.0%)
BM	64.1% (±3.5%)	0.0% (±0.0%)	0.0% (±0.0%)	0.0% (±0.0%)
CC	79.6% (±4.6%)	0.0% (±0.0%)	0.0% (±0.0%)	0.0% (±0.0%)
BM for 7 days followed by Z	64.1% (±3.5%)	64.1% (±3.5%)	85.8% (±5.8%)	0.0% (±0.0%)
0.3 M sucrose	41.7% (±6.1%)	0.0% (±0.0%)	0.0% (±0.0%)	0.0% (±0.0%)

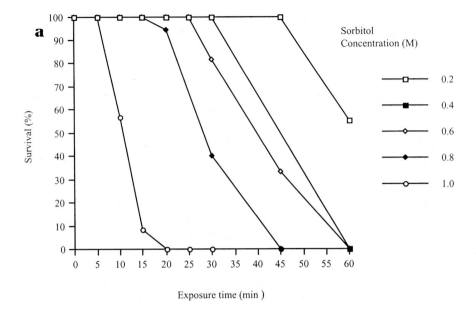

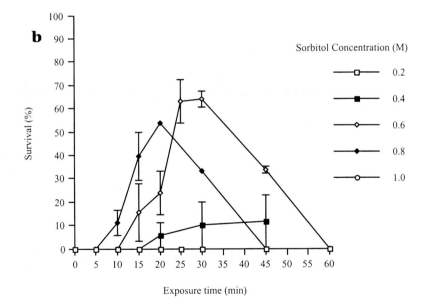

Fig. 2. Survival of *G. scapigera* shoot tips after exposure to the cryoprotective media PVS2 at
0 °C **a** without cooling and **b** after a cooling/warming cycle in LN

survival (41.4 ± 6.1%) was obtained when shoot apices were precultured for 48h on BM supplemented with 0.4M sorbitol followed by treatment with PVS2 at 0°C for 25min (Fig. 3). At this level of sorbitol and PVS2 exposure there was no significant difference between the treatment control and cooled treatment (*p* > 0.05). Other combinations of sorbitol concentration and PVS2 exposure time resulted in a significant decline in post-thaw survival (Fig. 3).

Optimization of Recovery Conditions for Anigozanthos viridis. Following optimization of sorbitol concentration and PVS2 exposure time, different recovery conditions were investigated. Highest survival (57.0 ± 9.9%) was obtained when shoot apices were cultured on BM supplemented with 2mM choline chloride and incubated in darkness for 14 days (Table 3), while the lowest survival rate occurred for shoot tips incubated in light on BM only following thawing (24.9 ± 5.5%). While the addition of 2mM choline chloride significantly (*p* < 0.05) improved survival, the impact of the photoperiod was not significantly different. The sorbitol concentration, PVS2 exposure time and recovery conditions that gave the highest survival for *A. viridis* were then applied to shoot apices from other monocotyledon species propagated in vitro (Table 1).

2.3.3 Post-Recovery Regeneration of Grevillea scapigera and Anigozanthos viridis

Shoot apices of *G. scapigera* regenerated to form shoots (20mm in length) within 7 weeks of thawing, and were then transferred to multiplication medium containing 0.5μM BA. After multiplication, shoots on root induction medium (BM supplemented with 0.25μM each of IBA and NAA) formed roots within 3 weeks and were subsequently transferred to soil. At this stage no morphological abnormalities were observed. Furthermore, recovered shoots subjected to genetic screening using random amplified polymorphic DNA techniques showed no detectable variations between treated and cooled shoots (Touchell 1995).

Emergence of green shoots was observed for *A.viridis* between 1 and 25 days after warming. Following a further 28 days, surviving shoots were approximately 20mm long and morphologically identical to shoot development from non-cooled apices. At this stage, developing plantlets were transferred to

Table 3. Percentage post warming survival for *Anigozanthos viridis* spp *terraspectans* incubated under different culture conditions (± standard error)

Recovery Media	Light Exposure	
	8h light/16h dark photoperiod	Darkness for 14 days
BM	24.9% (±5.5%)	27.7% (±6.2%)
BM + Choline Chloride	42.9% (±5.0%)	57.0% (±9.9%)

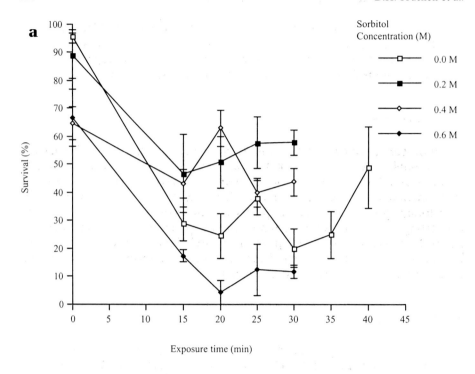

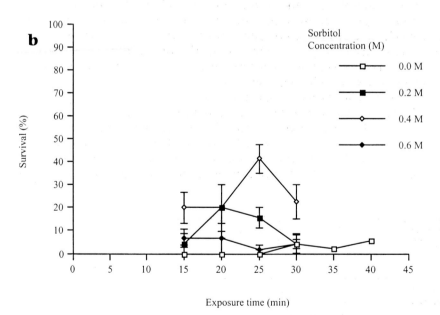

Fig. 3. Survival of *Anigozanthos viridis* shoot tips after exposure to the cryoprotective PVS2 medium at 0 °C **a** without cooling and **b** after a cooling/warming cycle in LN

hormone-free BM where roots spontaneously developed. Plantlets were transferred to soil 3 months after the first signs of growth and were grown under controlled conditions. There were no observable phenotypic differences between plantlets from cryostored apices and parent material.

2.3.4 Other Species

The protocol developed for cryostorage of shoot apices from *Grevillea scapigera* was applicable to other dicotyledonous species (Table 1) and with modification formed the basis for development of cryostorage procedures for monocotyledon species. For some species, only minor modifications to the cryostorage procedures were required to maximize survival. For example, converting from sorbitol to glycerol in the preculture medium increased survival of cryostored shoot apices in dicotyledonous species, such as the endangered *Eucalyptus graniticola*, from 27 to 60%. Species in which similar results were obtained are listed in Table 4.

2.4 Discussion

A variety of cryostorage procedures are applicable for a wide range of plant taxa (Kartha 1985; Bajaj 1995; Day and McLellan 1995). The development of a generally applicable cryostorage procedure would facilitate the widespread use of cryostorage as a conservation tool. In our current studies we have found that specificity of cryostorage procedures for particular species may be a result of the type of cryostorage procedure used. The results presented (Table 1) suggest that cryostorage utilizing a basic vitrification process can be achieved for diverse taxa with only minor modifications to protocols to achieve substantial post-thaw survival of somatic tissues.

In this study, shoot apices of *G. scapigera* produced the highest post-thaw survival after 48 h preculture on basal media supplemented with 0.6 M sorbitol (64.1 ± 3.5%) whereas, for *Anigozanthos viridis* shoot apices, post-thaw survival was optimized by reducing sorbitol in the preculture medium to 0.4 M. These protocols or slight modifications were then successfully applied to a

Table 4. Highest post-warming survival achieved for shoot apices of study species precultured on either 0.6 M sorbitol or 1.2 M glycerol basal media

Family	Species	Post-thaw survival	
		Sorbitol	Glycerol
Chloanthaceae	*Pityrodia scabra*	15.0%	78.0%
Goodeniaceae	*Lechenaultia. formosa*	25.0%	35.0%
Myrtaceae	*Eucalyptus graniticola*	27.0%	60.0%
Proteaceae	*Grevillea cirsiifolia*	46.0%	22.0%
	Grevillea dryandroides spp *dryandroides*	74.0%	63.0%
	Grevillea dryandroides spp *hirsutus*	41.0%	75.0%

further 33 taxa (21 dicotyledons and 12 monocotyledons) from 12 taxonomically diverse families and 17 genera. For all dicotyledons, optimum survival
was obtained using a procedure based on that developed for *Grevillea scapigera*, whereas all monocots responded best to the procedure developed for
Anigozanthos viridis. Furthermore, there are strong con-generic relationships
in cryostorage with five species of *Grevillea* and six species of *Anigozanthos*
being successfully revived after cooling utilizing genera specific methods.

If the con-generic responses observed in this study hold true then cryostorage procedures may be applicable to broader taxonomic units containing
endangered species, with minimal developmental costs. For example, the genus
Eucalyptus is a dominant tree genus in the Western Australian flora containing over 300 species (Hopper et al. 1990) of which 34 are listed as threatened
(Brown et al. 1998), with others being the mainstay of the forestry industry.
The cost in developing and maintaining essential germplasm for this genus
could be significantly reduced if cryostorage procedures developed for *Eucalyptus graniticola* (Table 1) can be adapted and used for other eucalypt species,
cultivars and provenance lines of commercial value. Although further studies
are required to investigate which cryostorage procedures can be applied based
on phylogenetic relationships, successful cryostorage of shoot apices appears
to be achievable for a large number of Western Australian species, using a
limited suite of in vitro procedures.

The key to successful vitrification as a cryostorage tool, as demonstrated
in this study, is the precise application of highly concentrated cryoprotectant
solutions so that tissues are less prone to freeze damage. The exact mechanisms by which vitrification solutions exert their protective qualities remain
somewhat speculative. However, the colligative action of cryoprotectants promoting cell desiccation and the penetration of cryoprotectants into cells leads
to viscosity of the cytosol, which, in turn, promotes intracellular glasses (vitrified solutions) rather than lethal intracellular ice crystal formation (MacFarlane et al. 1992). The application of PVS2 was necessary for survival of plants
during cooling and warming; however, PVS2 appeared to significantly inter-
act with the composition of the preculture medium.

For most species investigated, sorbitol was the predominant sugar
required in the preculture media to ensure post-thaw survival. Sorbitol concentration was also influential, with a concentration of 0.6 M proving most successful for dicots and 0.4 M for monocots. The replacement of sorbitol with
glycerol also improved survival in some species (Table 4). Similar findings have
been reported elsewhere (Phunchindawan et al. 1996). These studies were
reported for horseradish hairy root cultures, using in the preculture medium
1.0 M glycerol in combination with sucrose, and for embryonic axes of jackfruit (at 0.5 M), also in combination with sucrose (Thammasiri 1999). Preculture sugar composition was also critical for other species. In studies on
Trifolium repens (Yamada et al. 1991), meristems had to be cultured for 48 h
on media supplemented with 1.2 M sorbitol to maximize survival following
cryostorage. Other studies on banana (Panis et al. 1996), taro (Takagi et al.
1997) and sweet cherry (Niino et al. 1997) showed that a high concentration
of sucrose in the preculture medium was necessary to adequately desiccate

shoot apices for cryostorage. Glycerol is also a key component in most cry-oprotectant solutions. It is a key component in PVS2, and authors such as Ishikawa et al. (1997) and Charoensub et al. (1999) have shown that loading tissues with 2 M glycerol (plus 0.4 M sucrose) prior to treatment with PVS2 significantly increased post-thaw survival.

The action of sugars in the preculture media is also speculative. Sugars may act as an osmoticum and desiccate tissues. However, the response of shoot apices of different species to the different sugar types and concentrations used in this study suggests that sugars may have another mode of action. Yoshida (1984) and Tan et al. (1995) suggest that sorbitol and other sugars may act to prevent gross freeze injury to cell membranes by preventing cold-induced changes to lipid and protein membrane interactions during freezing.

Post-Thaw Recovery. In the early stages of post-thaw recovery, shoots are sus-ceptible to abnormal growth responses that can enhance the chances of genetic damage. Therefore, it is essential that recovery media enhance the rapid repair of damaged tissues and promote shoot production directly from the cryopre-served shoot apex (Harding and Benson 1994).

A careful balance of plant growth regulators in the recovery medium was necessary to ensure prolonged post-thaw survival of *Grevillea scapigera* shoot apices and promote shoot regeneration. Harding and Benson (1994) also observed that unfrozen shoot cultures of *Solanum tuberosum* can be routinely cultured on hormone-free media; however, the addition of hormones to media in the early post-thaw recovery period was essential for shoot apex survival and recovery.

For *Grevillea scapigera* shoot apices, as with *Solanum* spp. (Towill 1983; Harding and Benson 1994), the addition of zeatin to the post-thaw recovery media appeared essential for initiating morphogenesis. Towill (1983) specu-lates that this may be necessary because of the disruptions to the interactions between meristematic cells resulting from freeze damage. Zeatin may initiate cell division and repair of meristematic regions leading to de novo shoot growth.

The addition of choline chloride to the post-thaw recovery medium also significantly improves survival of shoot apices of both *Grevillea scapigera* and *Anigozanthos viridis*. Choline chloride esterifies with the phosphate group attached to a phospholipid chain to form phosphophatidyl choline, which is the most common phospholipid found in plant cell membranes (Leshem 1992). The addition of choline chloride to the post-thaw recovery medium may there-fore facilitate cell membrane repair mechanisms.

Light also affected post-thaw recovery of plant tissues. While the result was not statistically significant, shoot apices for *A. viridis* cultured in the dark did have improved survival as compared with apices cultured in light imme-diately after warming. Light can have an oxidative effect, which may induce tissue browning or necroses leading to cell death (Benson et al. 1989). Engel-mann et al. (1995) reported increased survival for oil palm apices when cul-tured under dark conditions for 10 days following thawing; however, this increased survival was only marginally better than the illuminated treatment.

Bajaj (1995) also reported increased survival and growth rates for sugarcane callus incubated in darkness following thawing, but the differences in this study were much more pronounced.

3 Conclusions

The plant genetic resources of the world are being swiftly eroded through human negligence and mismanagement of the natural environment. Many taxa have already been lost and many more will disappear before their economic and social values are realized. In order to conserve genetic resources, conservation efforts must nominate and integrate the most appropriate available technologies into conservation programs. Cryostorage, as featured in this study, is one technology that can be applied effectively and economically to diverse phylogenetic plant groups for conservation of germplasm stocks.

References

Bajaj YPS (ed) (1995) Biotechnology in agriculture and forestry, vol 32. Cryopreservation of plant germplasm I. Springer, Berlin Heidelberg New York, 512 pp

Bajaj YPS, Jian IC (1995) Cryopreservation of germplasm of sugarcane (*Saccharum* species). In: Bajaj YPS (ed) Biotechnology in agriculture and forestry, vol 32. Cryopreservation of plant germplasm I. Springer, Berlin Heidelberg New York, pp 256–265

Benson EE, Harding K, Smith H (1989) Variation in recovery of cryopreserved shoot tips of *Solanum tuberosum* exposed to different pre-and post-freeze light regimes. Cryo Lett 10:323–344

Brown A, Thomson-Dans C, Marchant N (eds) (1998) Western Australia's threatened flora. Department of Conservation and Land Management, Perth, Western Australia, 220 pp

Charoensub R, Phansiri S, Sakai A, Yongmenitchai W (1999) Cryopreservation of Cassava in vitro-grown shoot tips cooled to −196 °C by vitrification. Cryo Lett 20:83–88

Crowe M (1998) Survival of *Eucalyptus* species in cool storage and cryostorage: efficiency of choline chloride in cool storage media and glycerol and choline chloride in cryogenic preculture media. Honours Thesis, Murdoch University, Western Australia

Day JG, McLellan MR (eds) (1995) Cryopreservation and freeze drying protocols. Humana Press, New Jersey, USA, 254 pp

Dixon KW, Roche S, Pate JSP (1995) The promotive effect of smoke derived from burnt native vegetation on seed germination of Western Australian plants. Oecologia 101:185–192

Engelmann F, Assy-Bah B, Bagniol S, Dumet D, Michaux-Ferriere N (1995) Cryopreservation of date palm, oil palm and coconut. In: Bajaj YPS (ed) Biotechnology in agriculture and forestry, vol 32. Cryopreservation of plant germplasm I. Springer, Berlin Heidelberg New York, pp 148–169

Fay MF (1992) Conservation of rare and endangered plants using in vitro methods. In Vitro Cell Dev Biol 28:1–4

George EF, Sherrington PD (1993) Plant propagation by tissue culture, part 1. The technology, vol 1, 2nd edn. Exegentics, Edington, UK, 567 pp

Harding K, Benson EE (1994) A study of growth, flowering, and tuberisation in plants derived from cryopreserved potato shoot-tips: implications for in vitro germplasm collections. Cryobiology 15:59–66

Hopper SD, Van Leeuwen S, Brown A, Patrick S (eds) (1990) Western Australia's threatened flora. Department of Conservation and Land Management, Perth, Western Australia, 140 pp

Hopper SD, Harvey MS, Chappill JA, Main AR, York Main B (1996) The Western Australian biota as Gondwanan heritage – a review. In: Hopper SD (ed) Gondwanan heritage: past present and future of the Western Australian biota. Surrey Beatty, Chipping Norton, pp 1–16

Ishikawa K, Harata K, Mii M, Sakai A, Yoshimatsu K, Shimomura K (1997) Cryopreservation of zygotic embryos of a Japanese terrestrial orchid (*Bletilla striatta*) by vitrification. Plant Cell Rep 16:754–757

Kartha KK (ed) (1985) Cryopreservation of plant cell and organs. CRC Press, Boca Raton, 276 pp

Leifert C, Morris CE, Waites WM (1994) Ecology of microbial saprophytes and pathogens in tissue culture and field-grown plants: reasons for contamination problems in vitro. Crit Rev Plant Sci 13:139–183

Leshem YY (1992) Plant membranes: a biophysical approach to structure, development and senescence. Kluwer, Dordrecht, 266 pp

MacFarlane DR, Forsyth M, Barton CA (1992) Vitrification and devitrification in cryopreservation. In: Steponkus P (ed) Advances in low-temperature biology. JAI Press, London, pp 221–278

Meney KA, Nielssen GM, Dixon KW (1994) Seed bank patterns in restionaceae and epacridaceae after wildfire in Kwongan in southwestern Australian. J Veg Sci 5:5–12

Monod V, Poissonner M, Dereuddre J, Paques M (1992) Successful cryopreservation of *Eucalyptus gunnii* shoot tips in liquid nitrogen. In: Mass production technology for genetically improved fast growing forest tree species. Acetes Proceedings, Bordeaux, pp 133–145

Murashige T, Skoog F (1962) A revised medium for rapid growth and bio-assays with tobacco tissue cultures. Physiol Plant 15:473–497

Niino T, Tashiro K, Suzuki M, Ohuchi S, Magoshi J, Akihama T (1997) Cryopreservation of in vitro grown shoot tips of cherry and sweet cherry by one-step vitrification. Sci Hort 70:155–163

Panis B, Totte N, Van Nimmen K, Withers LA, Swennen R (1996) Cryopreservation of banana (*Musa* spp.) meristem cultures after preculture on sucrose. Plant Sci 121:95–106

Phunchindawan M, Hirata K, Sakai A, Miyaamota K (1996) Cryopreservation of encapsulated shoot primordia induced in horseradish (*Armoracia rusticana*) hairy root cultures. Plant Cell Rep 16:469–473

Roche S, Dixon KW, Pate JSP (1997) Seed ageing and smoke: partner cues in the amelioration of seed dormancy in selected Australian native species. Aust J Bot 45:783–815

Takagi H, Tien Thinh N, Islam OM, Senboko T, Sakai A (1997) Cryopreservation of in vitro grown shoot tips of taro (*Colocasia esculenta* (L.) Schott) by vitrification. 1. Investigation of basic conditions of the vitrification procedure. Plant Cell Rep 16:594–599

Tan S (1998) Cryopreservation for conservation of *Grevillea* species. Honours thesis, Murdoch University, Western Australia

Tan CS, Van Ingen CW, Talsma H, Van Miltenburg JC, Steffensen CL, Vlug IJA, Stalpers JA (1995) Freeze-drying of fungi: influence of composition and glass transition temperature of the protectant. Cryobiology 32:60–67

Thamasiri K (1999) Cryopreservation of embryonic axes of jackfruit. Cryo Lett 20:21–28

Touchell DH (1995) Principles of cryobiology for conservation of threatened species. PhD Thesis, University of Western Australia, Western Australia

Touchell DH, Dixon KW (1995) Cryopreservation for seedbanking of Australian species. Ann Bot 74:541–546

Touchell DH, Dixon KW (1996) Cryopreservation for the maintenance of commercial collections of Australian plants. In: Taji A, Williams R (eds) Tissue culture: towards the next century. Proceedings of 5th International Association for Plant Tissue Culture (Australian branch) Conference, University of New England, Armidale, New South Wales

Touchell DH, Dixon KW (1999) In vitro preservation. In: Bowes BG (ed) A colour atlas of plant conservation and propagation. Manson Publishing, London, pp 108–118

Touchell DH, Dixon KW, Tan B (1992) Cryopreservation of shoot tips of *Grevillea scapigera* (Proteaceae), a rare and endangered plant from Western Australia. Aust J Bot 40:305–310

Towill LE (1983) Improved survival after cryogenic exposure of shoot tips derived from in vitro plantlet cultures of potato. Cryobiology 20:567–573

Turner SR (1997) Cryopreservation of *Anigozanthos viridis* and related species. Honours Thesis, Curtin University of Technology, Western Australia

Yamada T, Sakai A, Matsumura T, Higgucho S (1991) Cryopreservation of apical meristems of white clover (*Trifolium repens* L.) by vitrification. Plant Sci 78:81–87

Yoshida S (1984) Studies on freezing injury of plants. Plant Physiol 75:38–42

IV.2 Cryopreservation of Australian Species – The Role of Plant Growth Regulators

Darren Touchell[1], S.R. Turner[1,2], T. Senaratna[1], E. Bunn[1], and K.W. Dixon[1]

1 Introduction

1.1 Background

Cryostorage of shoot apices has become an important tool for the preservation of plant tissues that cannot be maintained using conventional technologies. In recent years, the development and modification of procedures has led to the successful cryostorage of a large number of diverse agricultural, horticultural and endangered taxa (see Reinhoud et al. 2000; Sakai 2000). In particular, the vitrification procedure has been developed and employed prominently over the last decade, with over 140 species and cultivars being successfully cryostored (Sakai 2000).

Vitrification protocols aim to avoid freezing injury from ice formation by using highly concentrated cryoprotective solutions that act to penetrate and desiccate cells to form a highly viscous intracellular solution that vitrifies at low temperatures. Although tissues exposed to vitrification solutions are more amenable to cooling, they are more prone to injury through desiccation, including the loss of hydrophilic interactions (see Pammenter and Berjak 1999) and increases in free radical production (Senaratna and McKersie 1986). To reduce desiccation injury resulting from the exposure to vitrification solutions, tissues are treated to induce desiccation tolerance and provide cryoprotection. Most commonly used treatments involve osmoconditioning with sugars and sugar alcohols (Turner et al. 2001a).

The ability of shoot apices to withstand osmoconditioning, cryoprotection, desiccation and liquid nitrogen (LN) exposure may rest largely upon their physiological status prior to treatments. For most species, culture conditions developed for optimal plant growth will produce shoot apices amenable to cryostorage. For other species, manipulation of the growth conditions such as altering light regimes (Benson et al. 1988), cold conditioning (Reed 1990; Niino and Sakai 1992) and the use of plant growth regulators (Reed 1993) are required to physiologically condition shoot apices for cryostorage.

[1] Kings Park and Botanic Garden, West Perth, WA 6005, Australia
[2] Curtin University of Technology, Bentley, WA 6845, Australia
Current address: D. Touchell, School of Forestry and Wood Products, Michigan Technological University, Houghton, Michigan 49931, USA

Biotechnology in Agriculture and Forestry, Vol. 50
L.E. Towill and Y.P.S. Bajaj (Eds.) Cryopreservation of Plant Germplasm II
© Springer-Verlag Berlin Heidelberg 2002

Plant growth regulators (PGRs) are often incorporated in the pretreatment medium at similar levels to normal growth conditions. Commonly, cytokinins and auxins are used at various combinations and concentrations, depending upon species, to promote healthy, vigorous plant growth. However, there have been limited studies that have investigated the effects of PGRs in culture media prior to cryostorage. The use of 6-benzylaminopurine (BAP) in the preculture media did not influence the post-LN survival of blackberry shoot apices (Reed 1993). In contrast, kinetin had to be removed from the growth medium in order for alfalfa cells suspensions to tolerate low temperatures (Orr et al. 1985).

Abscisic acid (ABA) is perhaps the most widely investigated plant growth regulator in terms of cryostorage of shoot apices. In studies by Reed (1993) the application of ABA in the pretreatment medium increased post-LN survival of some *Rubus* sp. genotypes. For *Rubus* hybrid cv. Hillemeyer and *Rubus cissoids*, ABA in combination with cold acclimation promoted high survival after exposure to LN, whereas, for other *Rubus* species, there was a decrease in survival when plants were treated with ABA. In further studies ABA-responsive proteins improved survival and recovery of shoot apices of *Ribes* sp. exposed to liquid nitrogen (Luo and Reed 1997).

Recovery. One of the greatest challenges after cryostorage is the promotion of shoot proliferation and growth without inducing callus. Shoots regenerated from callus have been reported to have a high incidence of somaclonal variation (Scowcroft 1985; Potter and Jones 1991). Developing and optimizing recovery methods that promote rapid tissue recovery and growth can ensure the maintenance of the genetic stability of cryostored tissues.

Most post-warming recovery media are species-specific and often relate to the tissue culture conditions prior to cryostorage. In general, recovery media should optimize shoot apex survival and cell recovery so that the disturbances to the intercellular interactions within the integral structure of the shoot apex are minimized. This may be achieved through the addition of an optimal balance of PGR to the recovery medium.

The plant growth regulator composition of the recovery medium may influence both the survival and the shoot regeneration capacity of shoot apices. Cytokinins are the most common PGRs added to recovery medium and aid in shoot production. For *Solanum tuberosum* cv. Golden Wonder, a recovery medium supplemented with $2.3\,\mu M$ zeatin was optimal to produce a high number of shoots (Harding and Benson 1994). Towill (1983) also demonstrated the requirement of zeatin in the recovery media in recovering shoot apices of *S. tuberosum*. Other cytokinins, such as BAP and kinetin, have also been used to promote shoot production from cryostored shoot apices. For example, recovery media containing $0.44\,\mu M$ BAP promoted shoot production in *Wasabia japonica* (Matsumoto et al. 1994). A combination of $0.88\,\mu M$ BAP and $0.46\,\mu M$ kinetin has also been used to induce shoots from shoot apices of *Saccharum* sp. (Paulet et al. 1993). Alternatively, a combination of BAP with other PGRs such as auxins or gibberellic acid has been used to promote shoot production from cryopreserved shoot apices of *Ipomoea batatas* (Towill and Jarret 1992) and *Trifolium repens* (Yamada et al. 1991). Invariably, callus formation is significant when auxins are used in the recovery media (Harding and Benson 1994; Chang and Reed 1999).

2 Australian Species

2.1 Introduction

Shoot apices of a diverse range of Australian species have been successfully exposed to liquid nitrogen using vitrification procedures (see Touchell and Dixon 1999; Turner et al. 2001b; Touchell et al., this Vol.). The success of these procedures has been attributed to the precise application of the highly concentrated PVS2 cryoprotective solution (Yamada et al. 1991) in combination with appropriate preconditioning with sugars and sugar alcohols. However, it was observed that, although many species survived exposure to LN, there was often a low level of post-warming shoot formation (Touchell et al., this Vol.). Further, survival and initial regrowth of *Lambertia orbifolia* and *Conostylis wonganensis* is high during the first 7 day following warming before showing a rapid decline in survival (unpubl. observ.). Thus, improvements to cryostorage protocols are required to maximize survival and plant regeneration.

In earlier studies, cytokinin levels in growth media prior to the excision of shoot apices were shown to influence the survival of shoot apices of *A. viridis* (Turner et al. 2000). The application of cytokinins to the post-warming recovery has also been shown to influence the survival and shoot regeneration of *Grevillea scapigera* shoot apices. Based on these observations, this study explored the role of selected plant growth regulators in the cryostorage of the Australian species *Lechenaultia formosa*, *Grevillea scapigera* and *Anigozanthos viridis*.

2.2 Methodology

Plant Material. In vitro grown plants of all species (Table 1) were maintained at Kings Park and Botanic Garden. Basal medium (BM) consisted of half strength Murashige and Skoog (1962) salts supplemented with $3 \mu M$ thiamine hydrochloride, $2.5 \mu M$ pyridoxine hydrochloride, $4 \mu M$ niacin, 60 mM sucrose and 0.1% w/v agar. Stock shoot cultures were maintained on BM supplemented with $0.5 \mu M$ 6-benzylaminopurine (BAP) under standard culture conditions (i.e., 22–25 °C, irradiated with 16 h light with PPFD of $40 \mu M m^{-2} s^{-1}$) and 0.5-mm-long shoot apices were harvested at 3 weeks after each subculture.

2.2.1 Influence of ABA on Lechenaultia

Touchell and Dixon (1996) demonstrated cryostorage of a *Lecheanaultia* hybrid, although survival remained low. To increase post-LN survival of shoot

Table 1. Post LN survival after 28 days recovery for *A. viridis* after culturing for 21 days on BM supplemented with different concentrations of cytokinins (± standard error)

Concentration	Cytokinin Type			
	BAP	Kinetin	Zeatin	2 iP
0.0	83.3 (±4.2)	75.9 (±5.3)	86.5 (±2.1)	81.0 (±1.6)
0.25	96.5 (±3.5)	83.8 (±0.4)	83.6 (±5.0)	90.7 (±6.7)
0.5	93.1 (±1.5)	93.3 (±4.3)	88.0 (±2.4)	92.1 (±3.1)
1.0	90.9 (±1.5)	90.0 (±5.8)	85.8 (±4.3)	84.6 (±5.1)
2.0	90.0 (±0.0)	90.0 (±4.6)	90.7 (±4.3)	84.6 (±4.6)
4.0	93.3 (±4.1)	94.0 (±3.6)	83.2 (±11.0)	90.0 (±4.6)
8.0	90.2 (±7.1)	96.4 (±1.8)	89.2 (±3.0)	86.7 (±4.8)

apices of *Lechenaultia* the use of ABA in the shoot maintenance media was investigated. Plantlets were maintained on BM media for 14 day and then subcultured onto BM supplemented with 0, 2, 5, 10 and 50 µM ABA for 7 days. Shoot apices were excised and incubated on media supplemented with 1.2 M glycerol for 48 h under standard culture conditions. Ten precultured shoot apices were placed in a 1.2-ml cryovial (Nunc) containing 1 ml of modified PVS2 (Yamada et al. 1991) cryoprotective media. Shoot apices were then incubated at 0 °C for 30 min before cooling by direct immersion in LN. Shoot apices were held in LN for 24 h and warmed rapidly in a water bath at 40 °C. The cryoprotective PVS2 media was drained from these shoot apices and replaced three times in 12 min with liquid BM supplemented with 1.0 M sucrose (washing media). Shoot apices were then recovered on solidified BM under standard culture conditions and survival was determined by the percentage of treated shoot apices showing signs of growth within 2 weeks. Control shoot apices were treated in the same manner, but were not cooled. Three replicates of 20 shoot apices were used for each trial.

Chlorophyll Analysis. Chlorophyll content of shoot apices of plantlets treated with different concentrations of ABA was determined by the procedure of Moran and Porath (1980). Twenty shoot apices were weighed and placed in 1 ml *N,N*-dimethylformamide and incubated at 4 °C for 24 h to extract chlorophyll from the tissues. The sample was centrifuged for 5 min at 12,000 rpm and the supernatant and the absorbance read at 664 and 647 nm for chlorophyll *a* and *b*, respectively. Total chlorophyll content was calculated using the following equation:

$$\text{Total chlorophyll} + 7.04\, A_{664} + 20.27\, A_{647}$$

Antioxidant Determination. The relative activity of antioxidants in shoot apices treated with different concentrations of ABA was determined according to the procedures of Senaratna et al. (1985). Briefly, total lipid was extracted from approximately 40 shoot apices and quantity was determined gravimetrically. All samples were adjusted to the same lipid concentration (w/v) with ethanol. The relative quantity of antioxidants in lipid extracts was determined by monitoring the inhibition of linoleic acid oxidation (Senaratna et al. 1985).

2.2.2 Influence of Cytokinins on the Cryostorage of Grevillea scapigera

Preculture. Plantlets of *Grevillea scapigera* were subcultured on BM supplemented with either BAP, kinetin, 2-iP or zeatin at concentrations of 0, 0.5, 1.0, 2.5 and 8μM. Shoot apices were excised and incubated on media supplemented with 0.6M sorbitol for 48h under standard culture conditions. Shoot apices were cooled in LN as described earlier. Ten precultured shoot apices were placed in a 1.2-ml cryovial (Nunc) containing 1ml of modified PVS2 (Yamada et al. 1991) cryoprotective medium. Shoot apices were then incubated at 0°C for 25min before cooling by direct immersion in LN.

Recovery. In earlier studies, the application of zeatin to the post-warming recovery media was shown to be essential for the prolonged survival and shoot regeneration in *Grevillea scapigera* shoot apices. In this study, the application of zeatin to the recovery medium was further examined.

Shoot apices were exposed to LN as described above. After warming, shoot apices were recovered using one of the following regimes:

1. 1 week BM and transferred to medium containing 1 mM zeatin
2. 2 weeks BM and transferred to medium containing 1 mM zeatin
3. 3 week BM and transferred to medium containing 1 mM zeatin
4. 1 week CC medium and transferred to medium containing 1 mM zeatin

To test the effect of light on shoot apex recovery, cryostored shoot apices were warmed and cultured on BM for 1 week before being transferred to zeatin medium. Cultures were maintained under either standard light conditions (40 mmol m^{-2}s^{-1}) or a reduced PPFD of 4.3 mmol m^{-2}s^{-1}. Three replicates of 20 shoot apices were used for each trial.

Analysis of trans-Zeatin Riboside. Shoot apices, cryostored using procedures developed previously, were warmed and cultured on BM. Excised control shoot apices were cultured directly on BM. Samples of shoot apices for analysis of trans-zeatin riboside were taken at 0, 1, 2, 3 and 4 weeks after excision or LN exposure. Forty shoot apices were weighed and homogenized in 1.5 ml of 80% methanol. Samples were centrifuged at 12,000 rpm for 5 min and the supernatant removed. The residual was dissolved in 0.9% tris buffered saline (pH 7.5). Detection of *trans*-zeatin riboside was performed using a PGR immunoassay kit (Sigma).

2.2.3 Effects of Plant Growth Regulators on A. viridis

Modifications to A. viridis Culture Conditions. The influence of cytokinin and auxin in the preculture media on the survival of shoot apices of A. viridis subjected to LN was investigated. The cytokinins BAP (6-benzylaminopurine), zeatin (4-hydroxy-3-methyl-*trans*-2-butenylaminopurine), kinetin (6-furfurylaminopurine) and 2-iP [N^6-(2-isopentyl)adenine] were used at 0.0, 0.25, 0.5, 1.0, 2.0, 4.0 and 8.0 μM concentrations. Similarly, the auxins IAA (3-indolylacetic acid), IBA (3-indolebutyric acid) and NAA (1-naphthaleneacetic acid) at concentrations 0.0, 0.1, 0.25 and 0.5 µM, in combination with 0.5 µM BAP, were investigated. All PGRs were added to cooled BM following autoclaving. Plantlets were then grown on these media for 21 days prior to shoot apex extraction.

Liquid Nitrogen Treatment. Cryostorage procedures were modified from those described by Turner et al. (2001b) for *A. viridis* subsp. *terraspectans.* Shoot apices 1–1.5 mm long were excised from plantlets after 21 days of culture and then precultured on BM supplemented with 0.8 M glycerol for 72 h under standard culture conditions. Shoot apices were then incubated at room temperature for 20 min in 2 ml of loading solution consisting of BM supplemented with 0.4 M sucrose and 2 M glycerol (Charoensub et al. 1999; Turner et al. 2001b). Following pretreatment, shoot apices were exposed to modified PVS2 at 0 °C for 25 min (Sakai et al. 1990; Turner et al. 2001b), plunged into LN, warmed and rinsed and placed onto standard BM recovery medium as described above (unless otherwise stated). In recovery media experiments, shoot apices were removed 7 days later and placed onto alternative media (see below) for a further 3 weeks and incubated under standard culture conditions. Three replicates consisting of 15 shoot apices each were used for each treatment. Survival was determined as detectable growth of all or part of apical tissues 1–28 days after treatment.

Comparison of Five Different Recovery Media. Following 1-week incubation on recovery medium [BM + 2 mM choline chloride (CC)] shoot apices were randomly assigned to one of five different recovery media:

1. BM
2. BM supplemented with 1 µM zeatin
3. BM supplemented with 0.5 µM kinetin
4. BM supplemented with 0.5 µM kinetin +0.5 µM GA3 (gibberellic acid)
5. BM supplemented with 0.5 µM kinetin +0.5 µM GA$_3$ +0.5 µM IAA (indoleacetic acid)

Addition of PGRs at Day 0 and Day 7 to Recovering Shoot Apices. Warmed shoot apices were randomly placed onto two different recovery media at day 0 (BM + CC and BM + CC +0.5 µM kinetin/0.5 µM GA3), then incubated in darkness for 7 days. On day 7 both treatments were then transferred onto BM +0.5 µM kinetin/0.5 µM GA3).

2.3 Results

2.3.1 Influence of ABA on Lechenaultia formosa

Post-LN survival of shoot apices was dependent on ABA concentration in the preculture medium. Low levels of survival were obtained when shoot apices were from plantlets not treated with ABA or treated with 2µM ABA for 7 days. The highest survival was obtained when plantlets were maintained on BM medium for 14 days and then subcultured to media containing 5µM ABA for 7 days (Fig. 1). There was no survival if plantlets were treated with ABA for longer than 7 days.

In general, chlorophyll content of shoot apices declined with increasing concentrations of ABA. There was no significant difference between chlorophyll content of shoot apices from plantlets treated with 5, 10 or 50µM ABA (Fig. 1). Furthermore, ABA increased the relative quantity of lipid soluble antioxidants (Fig. 2).

2.3.2 Influence of Cytokinins on the Cryostorage of Grevillea scapigera

Preculture. Post-LN survival of shoot apices of *Grevillea scapigera* was dependent on the type and concentration of cytokinin in the culture media prior to shoot apex excision. The highest survival was obtained when plantlets were grown on media containing low levels of BAP for 21 days. Increased levels of BAP in the culture medium resulted in a decline in survival (Fig. 3). Replacing BAP with 2-iP or zeatin also resulted in lower survival, while plantlets cultured on kinetin produced significantly lower survival than any other treatment regardless of concentration (Fig. 3).

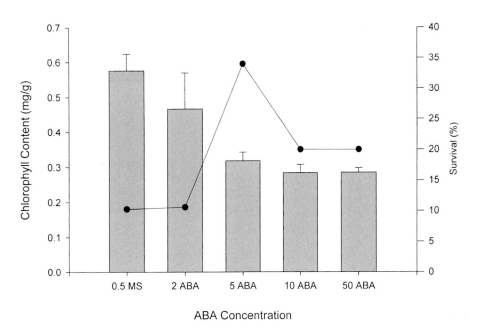

Fig.1. Survival of *Lechenaultia formosa* shoot apices after treatment with various concentrations of ABA. *Bars* represent the corresponding chlorophyll content

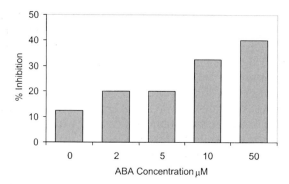

Fig. 2. Total lipid-soluble antioxidant potential, as estimated by the inhibition of linoleic acid oxidation, for shoot apices of *Lechenaultia formosa* from plantlets treated with different concentration of ABA

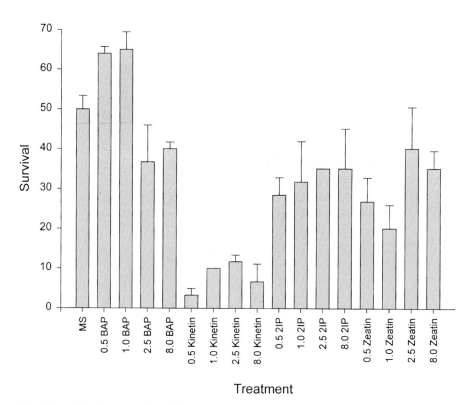

Fig. 3. Survival of shoot apices of *Grevillea scapigera* excised from plantlets treated with different types and concentrations of cytokinins for 21 days

Recovery. Transfer of shoot apices from BM to medium supplemented with 1 μM zeatin and 3 μM GA3 in the early stages of post-LN recovery was essential for prolonged survival of cryostored shoot apices (Fig. 4). However, shoot apices recovered on BM lost all ability to revive within 3 weeks of warming with transfer to zeatin medium having no effect on prolonged survival or promoting shoot regeneration. Transferring surviving cryostored shoot apices to zeatin medium at appropriate time intervals after warming also promoted direct shoot regeneration. Transferring shoot apices to zeatin medium 1 week after warming gave the highest direct shoot regeneration (28.9 ± 2.4%). Shoot regeneration was further increased to 85.8 ± 5.8% when shoot apex recovery occurred under reduced light.

The amount of endogenous *trans*-zeatin riboside present in both control and LN-treated shoot apices at different time intervals after excision or warming is shown in Fig. 5. The amount of *trans*-zeatin riboside present in control and cryostored shoot apices remained constant and low for 3 weeks after treatments. After 4 weeks there was a significant increase in tra*ns*-zeatin riboside in both control and LN-treated shoot apices. However, this increase was approximately three times greater in control shoot apices.

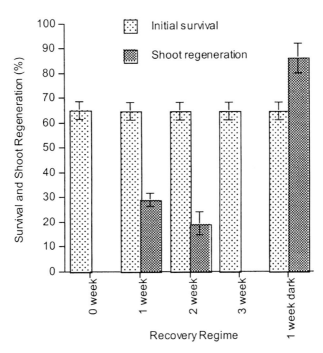

Fig. 4. Effect on shoot tip survival and shoot regeneration when *Grevillea scapigera* shoot apices were transferred to medium supplemented with zeatin and GA₃ after different post-LN recovery times on BM

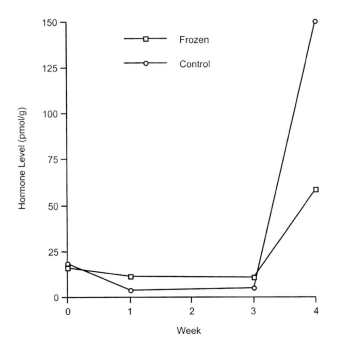

Fig. 5. Indicative amounts of *trans*-zeatin riboside produced by treatment controls or LN-cooled *Grevillea scapigera* shoot apices at different time intervals after excision or warming

2.3.3 Effects of PGRs in Culture Media on A. viridis

Influence of Different Types and Concentrations of Cytokinins. Post-LN survival varied slightly when shoot apices were harvested from plantlets cultured on different media. Highest survival was from shoot apices harvested from plantlets grown on media supplemented with 8 µM kinetin or 0.25 µM BAP treatment (96%; Table 1). In general, the lowest survival occurred from shoot apices that had been harvested from plantlets grown on media with no growth regulator supplements (76–86%). However, there were no significant differences in survival between the different treatments (Table 1). There were some differences in the appearance of plantlets after 21 days of culture on the different treatments. Plantlets grown on media supplemented with any cytokinins had high numbers of shoot apices and were stunted in growth compared with plantlets grown on BM. The higher concentrations (2, 4 and 8 µM) also exhibited some signs of hyperhydricity.

Influence of Different Types and Concentrations of Auxins. High LN survival was achieved regardless of the type and concentration of auxin in the pretreatment media. Survival varied (Table 2) from 79% (0 µM NAA) to 95% (0 µM IAA). Plantlets were also similar in appearance and produced similar numbers of shoot apices.

Table 2. Post LN survival after 28 days recovery for *A. viridis* after culturing for 21 days on BM supplemented with 0.5 μM BAP and different concentrations of auxins (± standard error)

Concentraion	Auxin Type		
	IAA	IBA	NAA
0.0	94.6 (±3.0)	83.3 (±3.9)	78.8 (±1.3)
0.1	85.8 (±4.8)	89.0 (±2.4)	91.3 (±5.1)
0.25	87.5 (±5.5)	83.8 (±5.6)	81.3 (±2.4)
0.5	87.5 (±5.1)	91.2 (±2.0)	88.2 (±5.9)

2.3.4 Effects of PGRs in Recovery Media on A. viridis

Comparison of Five Different Recovery Media. Survival of shoot apices was high 28 days after warming and ranged from 87.1–89.9% (Fig. 6). There were no significant differences between the survival of shoot apices 28 days after warming that were recovered on different medium (p > 0.05). Although the observed survival after 7 days was lower than that observed after 28 days, there were no significant differences within the same treatment after 7 and 28 days of recovery (*p* > 0.05).

Shoot apices recovered on the different media showed different growth responses during recovery. Shoot apices recovered on media supplemented with GA$_3$ showed greater elongation, while those recovered on media supplemented with zeatin were generally large and had increased shoot production. Differences between treatments were detected 3 days after transfer and became more pronounced over the following 18 days.

Addition of PGRs at Day 0 and Day 7 to Recovery Media. Survival 7 days after warming was similar for shoot apices recovered on BM or media supplemented with 0.5 μM kinetin and 0.5 μM GA3. Transferring shoot apices after 7 days after warming to media containing 0.5 μM kinetin and 0.5 μM GA$_3$ did not significantly increase survival (Fig. 7). For both recovery regimes direct shoot formation was obtained from shoot apices without the formation of callus. Shoots recovered directly onto media containing 0.5 μM kinetin and 0.5 μM GA$_3$ showed greater elongation after 28 days (data not shown).

2.4 Discussion

The key to successful cryostorage using vitrification procedures is the prevention of lethal ice crystal formation through application of concentrated cryoprotective solutions, which act to desiccate tissue. Thus, to prevent desiccation injury, tissues require conditioning with high levels of sugars and sugar alcohols to induce or provide desiccation tolerance. However, preculture alone may not be sufficient to maximize plant growth and recovery after exposure to vitrification solutions and cryostorage.

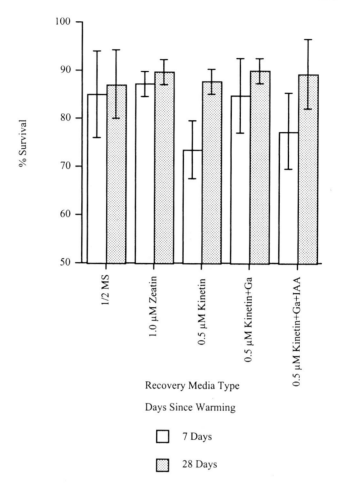

Fig. 6. Post-LN survival after 7- and 28-day recovery on five different recovery media for *A. viridis* (± SE)

For *Grevillea scapigera*, post-LN survival of shoot apices was significantly influenced by the type and concentration of cytokinin in the growth media prior to shoot apex excision. In comparison, shoot apices from *Anigozanthos viridis* cultured on media with similar types and concentrations of cytokinins showed no significant differences in post-LN survival. Furthermore, the presence of auxin in the pregrowth media also had no effect on survival of shoot apices exposed to LN. These results for *A. viridis* were unexpected, as previous studies by Turner et al. (2000) have indicated that high levels of cytokinins in culture prior to cryostorage produced a significant reduction in post-LN survival. The increase in survival of shoot apices from plantlets cultured on high levels of cytokinins observed in the present study can be attributed to improvements in the vitrification protocol.

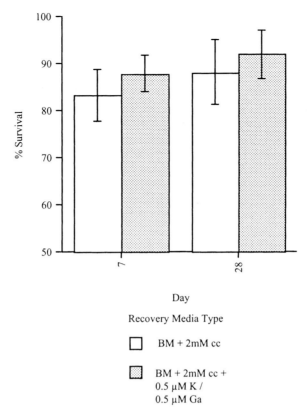

Fig. 7. Post-LN survival after 7- and 28-day recovery following culture on two different recovery media from days 0–7

For both *G. scapigera* and *A. viridis*, highest post-warming survival (79 and 94.4%, respectively) was obtained when shoot apices were excised from plantlets grown on low levels of BAP. However, survival of *G. scapigera* shoot apices significantly decreased when plantlets were grown either at high levels of BAP or on an alternative cytokinin. In particular, shoot apices from plantlets cultured on medium containing kinetin, regardless of concentration, had significantly low survival. Similarly, the use of kinetin in culture media reduced the freeze tolerance of alfalfa cell suspensions (Orr et al. 1985), whereas BAP had no effect on the survival of shoot apices of blackberry (Reed 1993).

The role of cytokinins and auxins in growth media prior to cryostorage remains unclear and may vary greatly between species. However, it has been noted than in order to achieve successful cryopreservation plant tissues need to be in optimal physiological condition. The growth parameters to obtain optimal physiological conditioning for the cryostorage of shoot apices may vary between species. For example, culture media for almond contained $1.33\,\mu M$ BAP and $0.28\,\mu M$ IAA prior to cryostorage (Shatnawi et al. 1999).

Pretreatment culture media for *Zoysia* and *Lolium* species contained 2.33 and 0.46 µM kinetin, respectively (Chang et al. 2000), while Bajaj (1995) used a combination of 0.44 µM BAP, 1.07 µM NAA and 0.14 µM GA$_3$ in the preculture media for cassava. In these studies, the relationship of plant growth regulators to shoot apices survival was not explored. Thus, careful manipulation of growth conditions and use of auxins and cytokinins is required in order to condition shoot apices for cryostorage. In contrast, suboptimal growth conditions and plant growth regulator supplements can lead to reduced post-LN survival, as observed for *G. scapigera* in this study. For example, high levels of cytokinins can produce increased hyperhydricity and tissue water content, which may result in disorganized tissues or increased water content (Rossetto et al. 1992) and, thus, tissues are likely to be in suboptimal condition for cryostorage.

Abscisic acid (ABA) is also readily used in culture medium to condition shoot apices for cryostorage. ABA is believed to promote a number of stress-related responses in plant tissues including the synthesis of proteins (Taiz and Zeiger 1991). Reed (1993) and Luo and Reed (1997) have shown increase in survival of shoot apices following liquid nitrogen treatments of *Rubus* and *Ribes* species treated with ABA and ABA-responsive proteins, respectively. In this study it was demonstrated that pretreatment with ABA for 7 days prior to harvesting shoot apices increased survival of shoot apices of *Lechenaultia formosa*. Studies by Tanino et al. (1990) have indicated that maximum effects of ABA may occur within 7 days of treatment. The precise effects of ABA were not examined in this study; however, *L. formosa* shoot apices from plantlets treated with ABA for 7 days had lower chlorophyll contents and increased levels of lipid-soluble antioxidants. These two physiological changes may provide increased protection from oxidative damage in shoot apices recovering from liquid nitrogen exposure. Increased in levels of oxidative damage after cryostorage have been observed for *Brassica napus* shoot apices (Benson and Noronha-Dutra 1988) and *Zizania palustris* embryos (Touchell and Walters 2000). Increases in free radicals are attributed to disruptions of the photosynthetic electron transport systems (Wise 1995). Thus, by reducing the light harvesting capabilities, a reduction in free radicals may be possible. Furthermore, increased levels of lipid-soluble antioxidants may protect cell membranes from free radical injury (Senaratna et al. 1985).

PGRs in Recovery Medium. Recovery of tissues after exposure to liquid nitrogen involves a series of events (Benson and Noronha-Dutra 1988). Severely damaged tissues are subject to degradative processes, which ultimately lead to tissue death. On the other hand, partially damaged or sublethally damaged tissues may recover and return to normal growth if provided with appropriate recovery conditions. Thus, it is of paramount importance that recovery media are supplemented with appropriate PGR compositions that promote recovery and differentiated growth.

In previous studies (Touchell et al., this Vol.), initial, high post-LN survival was obtained for shoot apices of *Grevillea scapigera*; however, over a 4-week period following warming, over 90% of the initial surviving shoot apices senesced. Similar results have been observed for other Australian species such as *Lambertia orbifolia* and *Conostylis wonganensis*. However, for *G. scapigera*, the addition of zeatin to recovery media ensured prolonged shoot apex survival and growth. For *Populus alba*, the use of zeatin in the recovery media also markedly improved survival (Lambardi et al. 2000).

In this study, the application of zeatin to the recovery media of *G. scapigera* shoot apices was further investigated. Endogenous synthesis of *trans*-zeatin riboside 4 weeks after excision is substantially lower in cryostored shoot apices than in control shoot apices. This correlates strongly with the senescence of cryostored shoot apices as well as shoot development in control shoot apices (unpubl. observ.). This further highlights the importance of zeatin in shoot development and alludes to the necessity of the exogenous application of zeatin for recovering cryostored shoot apices. However, the time interval after warming in which zeatin was exogenously applied to cryostored shoot apices was also critical. Transferring cryostored shoot apices to medium supplemented with zeatin 3 weeks after warming did not reduce senescence of shoot apices; however, placing warmed shoot apices directly on medium supplemented with zeatin promoted unwanted callus formation. Optimal results were obtained when warmed shoot apices were placed on BM for 1 week before being transferred to medium containing zeatin. Demeulemeester et al. (1993) showed related results for *Cichorium intybus* shoot apices. However, shoot apices of this species were cultured on media supplemented with hormones for 2 weeks before being transferred to hormone-free media.

For *A. viridis*, results presented here suggest that early introduction of PGRs to the recovery medium is essential to obtain healthy, vigorous plants from cryopreserved shoot apices, and, furthermore, the first 7 days of recovery appears to be the critical time for the introduction of appropriate plant growth regulation. However, presence or absence of a PGR does not appear to influence post-LN survival.

From the five different recovery media initially evaluated for *A. viridis* it was found that a combination of cytokinin and GA$_3$ was superior in producing large actively growing plantlets during the first month of recovery. GA$_3$ is commonly used in recovery media. Normah and Chin (1995) and Grospietsch et al. (1999) have reported the use of GA$_3$ post-LN recovery medium for rubber (*Hevea brasiliensis*) and potato (*Solanum tuberosum*) shoot apices. Both also reported using it in conjunction with IAA (or NAA) and kinetin.

GA$_3$ stimulates both seed germination and internode elongation (Pierik 1987). It may also be used for "breaking" stunting in various species when grown in vitro (Torres and Carlisi 1986). Increased growth rates observed from the GA$_3$ treatments may be the result of cell elongation. However, following cryopreservation, it is important to encourage active cell growth into organized tissues rather than undifferentiated callus. GA$_3$ may also suppress callus formation (George 1993) that, if allowed to form, may increase the probability of somaclonal variation.

In comparison, cytokinins stimulate cell division and also control morphogenesis (George 1993), and are therefore vitally important to the newly developing plantlet. However, for *A. viridis*, the use of the auxin IAA at

0.5 µM (in conjunction with kinetin/GA$_3$) appears to have had no effect and its addition was probably unnecessary. Chang and Reed (1999) found that 0.49 µM IBA when added to recovery medium for *Rubus* sp. resulted in callus formation of shoot apices post-LN immersion and its use was therefore considered unwarranted.

As with *G. scapigera*, the time at which PGRs were introduced to recovery medium was also found to be important to obtaining large plantlets. There was no significant difference in survival after warming when shoot apices were placed directly onto kinetin/GA$_3$ recovery medium rather than 1 week later. However, shoot apices incubated on BM + CC + kinetin/GA$_3$ during the first week were larger than the controls, with no visible signs of callusing. These differences in size were still visible after a further 3 weeks on recovery medium indicating the beneficial effect for this species of PGR introduction from day 1.

3 Conclusions

The results of this study show that there are many factors that influence the survival and growth of shoot apices after exposure to LN. The use of vitrification protocols involving conditioning with sugar alcohols and cryoprotection with highly concentrated solutions has been successful for a growing number of Australian species. However, based on this study, plant growth regulators also have a significant and important impact on the survival and growth of shoot apices exposed to LN. The use of PGRs in growth media prior to cryostorage appears to aid in conditioning tissues to withstand desiccation and freeze stress inherent to cryostorage. Further, the use of an appropriate PGR composition in the recovery medium will promote tissue repair and growth.

References

Bajaj YPS (1995) Cryopreservation of germplasm of potato (Sola*num tuberosum* L.) and Cassava (*Manihot esculenta* Crantz). In: Bajaj YPS (ed) Biotechnology in agriculture and forestry, vol 32. Cryopreservation of plant germplasm I. Springer, Berlin Heidelberg New York, pp 398–416

Benson EE, Noronha-Dutra AA (1988) Chemiluminescence in cryopreserved plant tissues. The possible role of singlet oxygen in cryoinjury. Cryo Lett 18:65–76

Benson EE, Harding K, Smith H (1988) Variation in recovery of cryopreserved shoot tips of *Solanum tuberosum* exposed to different pre-and post-freeze light regimes. Cryo Lett 10: 323–344

Chang Y, Barker RE, Reed B (2000) Cold acclimation improves recovery of cryopreserved grass (*Zoysia* and *Lolium* spp.). Cryo Lett 2:107–116

Chang Y, Reed BM (1999) Extended cold acclimation and recovery medium alteration improve regrowth of rubus shoot tips following cryopreservation. Cryo Lett 20:371–376

Charoensub R, Phansiri S, Sakai A, Yongmenitchai W (1999) Cryopreservation of Cassava in vitro-grown shoot tips cooled to −196 °C by vitrification. Cryo Lett 20:89–94

Demeulemeester MAC, Vandenbussche B, Proft MPD (1993) Regeneration of chicory plants from cryopreserved in vitro shoot tips. Cryo Lett 14:57–64

George EF (1993) Plant propagation by tissue culture, part 1. The technology, 2nd edn. Exegetics, London, 574 pp

Grospietsch M, Stodulkova E, Zamecnik J (1999) Effect of osmotic stress on the dehydration tolerance and cryopreservation of *Solanum tuberosum* shoot tips. Cryo Lett 20:339–346

Harding K, Benson EE (1994) A study of growth, flowering, and tuberisation in plants derived from cryopreserved potato shoot-tips: implications for in vitro germplasm collections. Cryobiology 15:59–66

Lambardi M, Fabbri A, Caccavale A (2000) Cryopreservation of white poplar (*Populus alba* L.) by vitrification of in vitro-grown shoot tips. Plant Cell Rep 19:213–218

Li CJ, Bangerth F (1992) The possible role of cytokinins, ethylene and indoleacetic acid in apical dominance. In: Karssen CML, Vanluon C, Vreugdenhill D (eds) Current plant science and biotechnology in agriculture, vol 13. Progress in plant growth regulators. Kluwer, Amsterdam

Luo J, Reed BM (1997) Abscisic acid-responsive protein, bovine serum albumin, and proline pretreatments improve recovery of in vitro currant shoot-tip meristems and callus cryopreserved by vitrification. Cryobiology 34:240–250

Matsumoto T, Sakai A, Yamada K (1994) Cryopreservation of in vitro-grown apical meristems of wasabi (*Wasabia japonica*) by vitrification and subsequent high plant regeneration. Plant Cell Rep 13:442–446

McComb JA, Bennett IJ, Tonkin C (1996) In vitro propagation of *Eucalyptus* species. In: Taji A, Williams R (eds) Tissue culture of Australian plants. University of New England, Armidale, pp 112–156

Moran R, Porath D (1980) Chlorophyll determination in intact tissues using N,N-dimethylformamide. Plant Physiol 65:478–479

Murashige T, Skoog F (1962) A revised medium for rapid growth and bioassays with tobacco tissue cultures. Physiol Plant 15:473–497

Niino T, Sakai A (1992) Cryopreservation of alginate-coated in vitro grown shoot-tips of apple, pear and mulberry. Plant Sci 87:199–206

Normah MN, Chin HF (1995) Cryopreservation of germplasm of rubber (*Hevea brasiliensis*). In: Bajaj YPS (ed) Biotechnology in agriculture and forestry, vol 32. Cryopreservation of plant germplasm I. Springer, Berlin Heidelberg New York, pp 180–190

Orr W, Singh J, Brown DCW (1985) Induction of freezing tolerance in alfalfa cell suspension cultures. Plant Cell Rep 4:15–18

Paulet F, Engelmann F, Glaszmann JC (1993) Cryopreservation of apices of in vitro plantlets of sugarcane (*Saccharum* sp. hybrid) using encapsulation/dehydration. Plant Cell Rep 12:525–529

Pammener NW, Berjak P (1999) A review of recalcitrant seed physiology in relation to desiccation-tolerance mechanisms. Seed Sci Res 9:13–37

Pierik RLM (1987) In vitro culture of higher plants. Nijhoff, Dordrecht, pp 45–82

Potter R, Jones MGK (1991) An assessment of genetic stability of potato in vitro by molecular and phenotypic analysis. Plant Sci 76:239–248

Reed BM (1990) Survival of in vitro grown apical meristems of pyrus following cryopreservation. Hortscience 25:111–113

Reed BM (1993) Responses to ABA and cold acclimation are genotype dependent for cryopreserved blackberry and raspberry meristems. Cryobiology 30:179–184

Reinhoud PJ, Van Iren F, Kijne JW (2000) Cryopreservation of differentiated plant cells. In: Engelmann F, Takagi H (eds) Cryopreservation of tropical plant germplasm. Current research progress and application. Japan International Research Centre for Agricultural Sciences, Tsukuba, Japan/International Plant Genetic Resources Institute, Rome, Italy, pp 91–102

Rosetto M, Dixon KW, Bunn E (1992) Aeration: a simple method to control vitrification and improve in vitro culture of rare Australian plants. In Vitro Cell Dev Biol 28:65–67

Sakai A (2000) Development of cryopreservation techniques. In: Engelmann F, Takagi H (eds) Cryopreservation of tropical plant germplasm. Current research progress and application. Japan International Research Centre for Agricultural Sciences, Tsukuba, Japan/International Plant Genetic Resources Institute, Rome, Italy, pp 1–20

Sakai A, Kobayashi S, Oiyama I (1990) Cryopreservation of nucellar cells of navel orange (*Citrus sinesis* Osb. var *brasiliensis* Tanaka) by vitrification. Plant Cell Rep 9:30–33

Scowcroft WR (1985) Somaclonal variation: the myth of clonal uniformity. In: Hohn B, Dennis ES (eds) Genetic flux in plants. Springer, Vienna New York, pp 217–245

Senaratna S, McKersie BD (1986) Lossod desiccation tolerance during seed germination: a free radical mechanism of injury. In: Leopold AC (ed) Membranes, metabolism and dry organisms. Cornell Univ Press, Ithaca, pp 85–101

Senaratna T, McKersie BD, Stinson RH (1985) Antioxidant levels in germinating soybean seed axes in relation to free radical and dehydration tolerance. Plant Physiol 78:168–171

Shatnawi MA, Engelmann F, Frattarelli A, Damiano C (1999) Cryopreservation of apices of in vitro plantlets of almond (*Prunus dulcis* Mill). Cryo Lett 20:13–20

Taiz L, Zeiger E (1991) Plant physiology. Benjamin/Cummings Publishing, Redwood City, pp 473–489

Tanino KK, Chen THH, Fuchigami LH, Weiser CJ (1990) Metabolic alterations associated with abscisic acid-induced frost hardiness in bromegrass suspension culture cells. Plant Cell Physiol 31:505–511

Torres KC, Carlisi JA (1986) Enhanced shoot multiplication and rooting *of Camellia sasanqua*. Plant Cell Rep 5:381–384

Touchell DH, Dixon KW (1996) Cryopreservation for the maintenance of commercial collections of Australian plants. In: Taji A, Williams R (eds) Tissue culture: towards the next century. Proceedings of 5th International Association for Plant Tissue Culture (Australian branch) Conference, University of New England, Armidale, New South Wales, pp 169–172

Touchell DH, Dixon KW (1999) In vitro preservation. In: Bowes BG (ed) A colour atlas of plant propagation and conservation. Manson Publishing, London, pp 108–118

Touchell DH, Walters C (2000) Recovery of embryos of *Zizania palustris* following exposure to liquid nitrogen. Cryo Lett 21:261–270

Towill LE (1983) Improved survival after cryogenic exposure of shoot tips derived from in vitro plantlet cultures of potato. Cryobiology 20:567–573

Towill LE, Jarret RL (1992) Cryopreservation of sweet potato (*Ipomoea batatas* [L.] Lam.) shoot tips by vitrification. Plant Cell Rep 11:175–178

Turner SR, Touchell DH, Dixon K, Tan B (2000) Cryopreservation of *Anigozanthos viridis* subsp. *viridis* and related taxa from the south west of Western Australia. Aust J Bot 48:739–744

Turner SR, Senaratna T, Touchell DH, Bunn E, Dixon KW, Tan B (2001a) Stereochemical arrangement of hydroxyl groups in sugar and polyalcohol molecules as an important factor in effective cryopreservation. Plant Sci 160:489–497

Turner SR, Senaratna T, Bunn E, Tan B, Dixon KW, Touchell DH (2001b) Cryopreservation of shoot tips from six endangered Australian species using a modified vitrification protocol. Ann Bot 87:371–378

Wise RR (1995) Chilling-enhanced photooxidation – the production, action and study of reactive oxygen species produced during chilling in the light. Photosynth Res 45:79–97

Yamada T, Sakai A, Matsumura T, Higgucho S (1991) Cryopreservation of apical meristems of white clover (*Trifolium repens* L.) by vitrification. Plant Sci 78:81–87

Subject Index

Printing (Computer to Film): Saladruck Berlin
Binding: Stürtz AG, Würzburg